MULTIVARIATE DATA ANALYSIS

Using SPSS and AMOS

MULTIVARIATE DATA ANALYSIS

Using SPSS and AMOS

R. Shanthi

Assistant Professor, Department of Commerce,
University of Madras, Chennai.

Chennai **New Delhi** **Tirunelveli**

ISBN 978-81-8094-412-3 **MJP Publishers**

No. 44, Nallathambi Street,
Printed and bound in India Triplicane, Chennai 600 005

MJP 376

Publisher : C. Janarthanan

Project Editor : C. Ambica

Preface

Multivariate Data Analysis is the leading textbook as it provides a holistic approach involved in both theory and practice. This particular book covers the fundamental topics such as uses, techniques and applicability of multivariate analysis. It familiarizes the SPSS and AMOS Application software to the learners and introduces the concept of Outliers, Normality, Linearity test, Data Transformation and Bootstrapping. The book also focuses on Multivariate Analysis of Variance (MANOVA). The ground-breaking research to understand the need of the learners paved way to include further topics like Multivariate Regression Analysis and concept of Binary Logistic Regression.

The textbook reflects the effort of the author with a goal of creating the best edition to the learners. An extensive focus group was conducted to fully understand the course and classroom needs as being a responsible instructor having an experience of more than 20 years. Based on this understanding input was designed to preserve the strengths of the material and further enhance learning. Statistics is of interest to most of them and must be kept up-to-date and contemporary. Both students and instructors should feel that the book is talking directly to them in terms of both content and delivery.

This edition of Multivariate Data Analysis has been both streamlined and expanded to bring essentials and classic examples into sharper focus, while covering new ideas in depth. The book brings key material upfront, where the instructor and learner can find key points and assignments that can be incorporated into lecture and classroom practice.

Acknowledgement

The completion of this undertaking could not have been possible without the participation and assistance of so many people. Their contributions are sincerely appreciated and gratefully acknowledged.

I acknowledge with thanks the kind of patronage, inspiration and guidance which I have received from the Vice Chancellor and Registrar of University of Madras.

I express my deep sense of gratitude to our Head of the Department of Commerce, my colleagues, and Research Scholars for their constant support and encouragement.

I thank my family and friends for their endless support and understanding spirit during the course of work.

I thank MJP publishers for their meticulous planning and delivery as promised.

Above all, I thank the Great Almighty, for his countless love and blessing.

Contents

Chapter - 1

Multivariate Data Analysis

Learning Objectives

This chapter helps to understand the following

- Meaning of Multivariate Analysis
- Types of Measurement Scale – Metric and Non-Metric Scale
- Method of Dependence and Interdependence Technique
- Structural Approach to Multivariate Model Building
- Multivariate Statistical Analysis

Key Terms

1. **Neural Network** - A neural network can approximate a wide range of predictive models with minimal demands on model structure and assumption
2. **Structural Equation Modeling (SEM)** - Structural equation modeling (SEM) examines multiple relationships between sets of variables simultaneously.
3. **Canonical Correlation** - Canonical correlation analysis is a method for exploring the relationships between two multivariate sets of variables (vectors), all measured on the same individual.
4. **Conjoint Analysis** - Conjoint analysis is often referred to as "trade-off analysis," since it allows for the evaluation of objects and the various levels of the attributes to be examined.
5. **Correspondence Analysis** - Correspondence analysis is a statistical visualization method for picturing the associations between the levels of a two-way contingency table.

6. **Multivariate Analysis** - Multivariate analysis refers to the entire statistical techniques that simultaneously analyze multiple measurements on individuals or objects under investigation
7. **Non-Metric Data** - Non metric data describe differences in type by indicating the presence or absence of a characteristic
8. **Nominal Scale** - A nominal scale assigns numbers as a way to label or identify subject.
9. **Ordinal Scales** - Ordinal scales are the subsequently "higher" level of measurement precision.
10. **Metric Data** - Metric data are used when subjects differ in amount or degree on a particular attribute.
11. **Interval Scales** - Interval scales and ratio scales (both metric) provide the highest level of measurement precision, permitting nearly any mathematical operation to be performed.
12. **Ratio Scales** - Ratio scales represent the highest form of measurement precision because they possess the advantages of all lower scales plus an absolute zero point.
13. **Dependence Technique** - A dependence technique is one where a single variable or a set of variables is identified as the dependent variable to be explained or predicted by other variables known as explanatory or independent variables
14. **Interdependence Technique** - An interdependence technique involves the simultaneous analysis of all variables in the data set; variables are not classified as dependent or independent
15. **Multiple Regression** - Multiple regression is a forecasting tool which is commonly utilized multivariate technique to examine the relationship between a metric dependent variable and two or more metric independent variables
16. **Logistic Regression** - Logistic Regression also referred to as "choice models," this technique is a variation of multiple regression that allows for the prediction of an event.
17. **Discriminant Analysis** - Discriminant analysis is to correctly classify observations or community into homogeneous groups.
18. **MANOVA** - MANOVA examines the dependence relationship between a set of dependent measures across a set of groups.

19. **Factor Analysis** - Factor analysis is used when there are too large variables in a research design, which is often helpful to reduce the variables to a smaller set of factors.
20. **Cluster Analysis** - Cluster analysis is to reduce a large data set to meaningful subgroups of individuals or objects.
21. **Multidimensional Scaling** - MDS is to transform consumer judgments of similarity into distances represented in multidimensional space.

Introduction

Multivariate analysis refers to the entire statistical techniques that simultaneously analyze multiple measurements on individuals or objects under investigation. Thus, any simultaneous analysis of more than two variables can be loosely considered multivariate analysis.

Multivariate analysis methods typically used for

- Consumer and market research
- Quality control and quality assurance across a range of industries such as food and beverage, paint, pharmaceuticals, chemicals, energy, telecommunications, etc
- Process optimization and process control
- Research and development

Most of the multivariate techniques are the extensions of univariate analysis and bivariate analysis and this extension to the multivariate domain introduces additional concepts and issues of particular significance. Each concept plays a significant role in the successful application of any multivariate technique.

These concepts range from the need for a conceptual understanding of the basic building block of multivariate analysis - the variate -to specific issues dealing with the types of measurement scales used and the statistical issues of significance testing and confidence levels. Data analysis involves the identification and measurement of variation in a set of variables, either among themselves or between a dependent variable and one or more independent variables.

Scales of Measurement

Measurement is important in accurately representing the concept of interest and is instrumental in the selection of the appropriate multivariate method of analysis. Data can be classified into one of two categories based on the type of attributes or characteristics it represents. They are

1. Non-Metric Data (Qualitative)
2. Metric Data (Quantitative)

Non-Metric Measurement Scales

Non-metric data describe differences in type by indicating the presence or absence of a characteristic. These characteristics are discrete in nature; for example, if the answer is true, it cannot be false or the gender of person should be recognized as male or female. Non-metric measurements can be made with either a nominal or an ordinal scale.

Nominal Scales

A nominal scale assigns numbers as a way to label or identify subject. The numbers assigned to the objects have no quantitative meaning beyond indicating the presence or absence of the attribute or characteristic under investigation. Therefore, nominal scales, also known as categorical scales provide the number of occurrences in each class or category of the variable being studied. The nominal data represent categories; they may be coded as numbers but the numbers has no real meaning, it's just a label they have no default or natural order, e.g. Gender, Education, Occupation, and Religion etc.

Ordinal Scales

Ordinal scales are the subsequently "higher" level of measurement precision. In this case, the variables can be ordered or ranked in relation to the amount of the attributes it possesses. For example, every subject can be compared in the term of a greater than or lesser than relationship. The Numbers used in ordinal scale is non-quantitative.

These are data that can be put in an order, but don't have a numerical meaning beyond the order. So for instance, the difference between 2 and 4 in the example of a Likert scale below might not be the same as the difference between 2 and 5. Examples: Questionnaire responses coded: 1 = strongly disagree, 2 = disagree, 3 = indifferent, 4 = agree, 5 = strongly agree.

Metric Measurement Scales

Metric data are used when subjects differ in amount or degree on a particular attribute. Metrically measured variables reflect relative quantity or degree and are appropriate for attributes involving amount or magnitude, such as the level of customer satisfaction or level of job commitment. The two different metric measurement scales are interval and ratio scales. The only real difference between interval and ratio scales is that interval scales use an arbitrary zero point, whereas ratio scales include an absolute zero point.

Interval Scales

Interval scales and ratio scales (both metric) provide the highest level of measurement precision, permitting nearly any mathematical operation to be performed. These two scales have constant units of measurement, so differences between any two adjacent points on any part of the scale are equal. These are numerical data where the distances between numbers have meaning, but the zero has no real meaning. With interval data it is not meaningful to say that one measurement is twice another, and might not still be true if the units were changed. For example, Level of Job Satisfaction, Level of Customer Satisfaction, Temperature etc.

Ratio Scales

Ratio scales represent the highest form of measurement precision because they possess the advantages of all lower scales plus an absolute zero point. All mathematical operations are permissible with ratio-scale measurements. These are numerical data where the distances between data and the zero point have real meaning. With such data it is meaningful to say that one value is twice as much as another, and this would still be true if the units were changed.

Objectives of Multivariate Analysis

- Classification - dividing variables or samples into groups with shared properties.
- Identifying gradient or trends in multivariate data.
- Identifying which environmental variables are most influential in determining community structure.
- Finally and usually most importantly – aims to extract from a set of data derived from an almost infinitely complex world the most important features so these can be presented clearly to others

Methods of Multivariate Data Analysis

Multivariate analysis methods are defined as statistical procedures that involve the analysis of more than one variable at a time, with an emphasis on modeling the relationships between such variables.

Multivariate techniques are classified into dependence or interdependence techniques.

Dependence Techniques

A dependence technique is one where a single variable or a set of variables is identified as the dependent variable to be explained or predicted by other variables known as explanatory or independent variables. Dependence techniques are statistical procedure which requires you to identify variables as either 'Independent Variables' or 'Dependent Variables'. There are four dependence techniques - multiple regression, discriminant analysis, Multivariate analysis of variance and conjoint analysis. It is used to assess the degree of relationship between dependent variable and independent variables, It differs in the kind and character of the relationship as reflected in the measurement properties of the variables. The dependence techniques are

- Multiple regression
- Discriminant analysis
- Logistic regression – logit analysis
- Multivariate analysis of Variance
- Conjoint analysis
- Canonical correlation
- Structural Equation Modeling
- Path analysis

Interdependence Techniques

An interdependence technique involves the simultaneous analysis of all variables in the data set; variables are not classified as dependent or independent. Interdependence techniques are statistical procedures where there are no dependent or independent variables and where all variables are considered simultaneously and each are related to each other. Underlying these interrelated items or variables are a "latent set of factors" or dimensions that are themselves made up of all other variables. The interdependence techniques cover factor analysis, cluster analysis and perceptual mapping. These techniques are suited for assessing structure by focusing on the portrayal of the relationships among and between characters, whether they are respondents or objects. The Interdependence techniques are

- Factor analysis
- Cluster analysis

- Multidimensional scaling
- Correspondence analysis

Types of Multivariate Data Analysis

The type of Multivariate Data Analysis is discussed briefly.

Multiple Regression Analysis

Multiple regression is a forecasting tool which is commonly utilized multivariate technique to examine the relationship between a metric dependent variable and two or more metric independent variables. The technique also relies upon evaluating the linear relationship with the lowest sum of squared variances; as a result, assumptions of normality, linearity, and equal variance are carefully observed. The beta coefficients (weights) are the marginal impacts of each variable, and the size of the weight can be interpreted directly. Multiple regression is often used as a forecasting tool.

Logistic Regression Analysis

Logistic Regression also referred to as "choice models," this technique is a variation of multiple regression that allows for the prediction of an event. It is allowable to utilize non-metric (typically binary) dependent variables, as the objective is to arrive at a probabilistic assessment of a binary choice. The independent variables can be either discrete or continuous. A contingency table is produced, which shows the classification of observations as to whether the observed and predicted events match. The sum of events that were predicted to arise which actually did occur and the events that were predicted not to occur which actually did not occur, divided by the total number of events, is a measure of the effectiveness of the model.

Discriminant Analysis

The purpose of discriminant analysis is to correctly classify observations or community into homogeneous groups. The independent variables must be metric and must have a high degree of normality. It also builds a linear discriminant function, which can be used to classify the observations. The overall fit is assessed by looking at the degree to which the group means differ (Wilkes Lambda or D2) and how well the model classifies. To determine which variables have the most impact on the discriminant function, it is possible to look at partial F values. The higher the partial F, the more impact that variable has on the discriminant function.

Multivariate Analysis of Variance (MANOVA)

This technique examines the relationship between several categorical independent variables and two or more metric dependent variables. Whereas analysis of variance (ANOVA) assesses the differences between groups (by using T tests for two means and F tests between three or more means), MANOVA examines the dependence relationship between a set of dependent measures across a set of groups. Typically this analysis is used in experimental design, and usually a hypothesized relationship between dependent measures is used. This technique is slightly different in that the independent variables are categorical and the dependent variable is metric. Sample size is an issue, with 15-20 observations needed per cell. However, too many observations per cell (over 30) and the technique loses its practical significance. Cell sizes should be roughly equal, with the largest cell having less than 1.5 times the observations of the smallest cell. That is because, in this technique, normality of the dependent variables is important. The model fit is determined by examining mean vector equivalents across groups. If there is a significant difference in the means, the null hypothesis can be rejected and treatment differences can be determined.

Factor Analysis

Factor analysis is used when there are too large variables in a research design, which is often helpful to reduce the variables to a smaller set of factors. This is an independence technique, in which there is no dependent variable. Rather, the researcher is looking for the underlying structure of the data matrix. Ideally, the independent variables are normal and continuous, with at least three to five variables loading onto a factor. The sample size should be over 50 observations, with over five observations per variable. Multicollinearity is generally preferred between the variables, as the correlations are key to data reduction. Kaiser's Measure of Statistical Adequacy is a measure of the degree to which every variable can be predicted by all other variables. An overall MSA of .80 or higher is very good, with a measure of under .50 deemed poor.

There are two main factor analysis methods: common factor analysis, which extracts factors based on the variance shared by the factors, and principal component analysis, which extracts factors based on the total variance of the factors. Common factor analysis is used to look for the latent (underlying) factors, whereas principal component analysis is used to find the fewest number of variables that explain the most variance. The first factor extracted explains the most variance. Typically, factors are extracted as long as the eigen values are greater than 1.0 or the Screen test visually indicates how many factors to extract. The factor loadings are the correlations between the factor and the variables. Typically a factor loading of .4 or higher is required to attribute a specific variable to a factor. An orthogonal rotation assumes no correlation between the factors, whereas an oblique rotation is used when some relationship is believed to exist.

Cluster Analysis

The purpose of cluster analysis is to reduce a large data set to meaningful subgroups of individuals or objects. The division is accomplished on the basis of similarity of the objects across a set of specified characteristics. Outliers are a problem with this technique, often caused by too many irrelevant variables. The sample should be representative of the population, and it is desirable to have uncorrelated factors. There are three main clustering methods: hierarchical, which is a treelike process appropriate for smaller data sets; nonhierarchical, which requires specification of the number of clusters a priori; and a combination of both. There are four main rules for developing clusters: the clusters should be different, they should be reachable, they should be measurable, and the clusters should be profitable (big enough to matter). This is a great tool for market segmentation.

Multidimensional Scaling (MDS)

The purpose of MDS is to transform consumer judgments of similarity into distances represented in multidimensional space. This is a decompositional approach that uses perceptual mapping to present the dimensions. As an exploratory technique, it is useful in examining unrecognized dimensions about products and in uncovering comparative evaluations of products when the basis for comparison is unknown. Typically there must be at least four times as many objects being evaluated as dimensions. It is possible to evaluate the objects with nonmetric preference rankings or metric similarities (paired comparison) ratings. Kruskal's Stress measure is a "badness of fit" measure; a stress percentage of 0 indicates a perfect fit, and over 20% is a poor fit. The dimensions can be interpreted either subjectively by letting the respondents identify the dimensions or objectively by the researcher.

Correspondence Analysis

Correspondence analysis is a statistical visualization method for picturing the associations between the levels of a two-way contingency table. This technique provides for dimensional reduction of object ratings on a set of attributes, resulting in a perceptual map of the ratings. However, unlike MDS, both independent variables and dependent variables are examined at the same time. This technique is more similar in nature to factor analysis. It is a compositional technique, and is useful when there are many attributes and many companies. It is most often used in assessing the effectiveness of advertising campaigns. It is also used when the attributes are too similar for factor analysis to be meaningful. The main structural approach is the development of a contingency (crosstab) table. This means that the form of the variables should be non-metric. The model can be assessed by examining the Chi-square value for the model. Correspondence analysis is difficult to interpret, as the dimensions are a combination of independent and dependent variables.

Conjoint Analysis

Conjoint analysis is often referred to as "trade-off analysis," since it allows for the evaluation of objects and the various levels of the attributes to be examined. It is both a compositional technique and a dependence technique, in that a level of preference for a combination of attributes and levels is developed. A part-worth, or utility, is calculated for each level of each attribute, and combinations of attributes at specific levels are summed to develop the overall preference for the attribute at each level. Models can be built that identify the ideal levels and combinations of attributes for products and services.

Canonical Correlation

Canonical correlation analysis is a method for exploring the relationships between two multivariate sets of variables (vectors), all measured on the same individual. The most flexible of the multivariate techniques, canonical correlation simultaneously correlates several independent variables and several dependent variables. This powerful technique utilizes metric independent variables, unlike MANOVA, such as sales, satisfaction levels, and usage levels. It can also utilize non-metric categorical variables. This technique has the fewest restrictions of any of the multivariate techniques, so the results should be interpreted with caution due to the relaxed assumptions. Often, the dependent variables are related, and the independent variables are related, so finding a relationship is difficult without a technique like canonical correlation.

Structural Equation Modeling

Unlike the other multivariate techniques discussed, structural equation modeling (SEM) examines multiple relationships between sets of variables simultaneously. This represents a family of techniques, including LISREL, latent variable analysis, and confirmatory factor analysis. SEM can incorporate latent variables, which either are not or cannot be measured directly into the analysis. For example, intelligence levels can only be inferred, with direct measurement of variables like test scores, level of education, grade point average, and other related measures. These tools are often used to evaluate many scaled attributes or to build summated scales.

Neural Network

A neural network can approximate a wide range of predictive models with minimal demands on model structure and assumption. The form of the relationships is determined during the learning process. If a linear relationship between the target and predictors is appropriate, the results of the neural network should closely approximate those of a traditional linear model. If a nonlinear relationship is more appropriate, the neural network will automatically approximate the "correct" model structure. The trade-off for this flexibility is that the neural network is not easily interpretable. If you are trying to explain an underlying process that produces the relationships between the target and predictors, it would be better to use a more traditional statistical model. However, if model interpretability is not important, you can obtain good predictions using a neural network. Neural networks can be considered as nonlinear function approximating tools (i.e., linear combinations of nonlinear basis functions), where the parameters of the networks should be found by applying optimization methods.

Neural networks are used to recognize:-

- Patterns by repeated exposure to many different examples.
- Patterns or salient characteristics whether they are hand-written characters, profitable loans or good trading decisions.
- Patterns in data that are inexact and incomplete.

Neural networks find this relationship through a learning cycle where many hundreds of samples are presented repeatedly to the network. Neural network cannot guarantee an optimal solution to a problem. However, properly configured and trained neural networks can often make consistently good classifications, generalizations or decisions in a statistical sense.

A Structured Approach to Multivariate Model Building

The stages of structured approach to multivariate model building is as follows

Define Research Problem, Objectives, and Multivariate Techniques to be Used

In Multivariate analysis, defining the research problem and analyzing objectives in conceptual terms is important before specifying any variables or measures. Define the concepts and identify the fundamental relationships to be considered. It need not be complex and detailed but a sample representation of the relationships is important to be studied; if a dependence relationship is proposed as the research objective, the researcher needs to specify the dependent and independent concepts. For an application of an interdependence technique, the dimension of structure or similarity should be specified. In both the dependence and interdependence situations, the researcher first identifies the ideas or topics of interest rather than focusing on the specific measures to be used.

Develop the Analysis Plan

In multivariate analysis, it is important to develop a specific analysis plan to address the set of issues particular to its purpose and design. The issues range from the general consideration of minimum or desired sample sizes, to suitable or required types of variables.

Evaluate the Assumptions Underlying the Multivariate Technique

With the collected data, evaluating the underlying assumptions is considered to be important in multivariate statistical analysis. For the techniques based on statistical inference, the assumptions of multivariate normality, linearity, independence of the error terms, and equality of variances in a dependence relationship must all be met.

Estimate the Multivariate Model and Assess Model Fit

In multivariate data analysis, with the satisfied assumptions, the analysis proceeds for the actual estimation of the multivariate model and an assessment of the overall model fit.

Interpret the Variates

In multivariate analysis, interpreting the variates reveals the nature of multivariate relationship.

Validate the model

In multivariate analysis, it is important to validate the model by demonstrating the generalizability of the results to the total population.

Multivariate Statistical Analyses

Method	Purpose	Level of Measurement		
		Independent variable	Dependent Variable	Covariates
Multiple Regression	To test and predict the relationship between two or more dependent variable and one independent variable	Two or more independent variable can be either Nominal, Ordinal, Interval	One dependent variable	No covariate
Analysis of Covariance (ANCOVA)	To test the difference between the means of two or more groups while controlling for one or more covariates	One or more independent variable which is nominal	One dependent variable which is either interval or ratio	One or more covariates which can be either nominal, interval or ratio
Multivariate Analysis of Variance	To test the difference between the means of two or more groups for two or more dependent variables simultaneously	One or more independent variable which is nominal.	Two dependent variable which can be either interval or ratio	No covariates
Multivariate analysis of Covariance	To test the difference between the means of two or more groups for two or more dependent variable while controlling for one or more covariates	One or more independent variable which is nominal	Two or more dependent variable which can be either interval or ratio	One or more covariates which can be either nominal, ordinal or ratio

Method	Purpose	Level of Measurement		
		Independent variable	Dependent Variable	Covariates
Canonical correlation	To test the relationship between two sets of variables (variables on the right and on the left)	Two or more independent variable which can be either nominal, interval or ratio	Two or more dependent variable which can be either nominal, interval or ratio	No covariates
Factor analysis	To determine the dimensions/ structure on a set of variables	Nil	Nil	Nil
Discriminant analysis	To test the relationship between two or more independent variables and one dependent variable To predict group membership To classify cases into groups	Two or more independent variable which can be either nominal, interval or ratio	One dependent variable which is nominal	No covariates
Logistic Regression	To test the relationship between two or more independent variable and one dependent variable To predict the probability of an event To estimate the relative risk	Two or more independent variable which can be either nominal, interval or ratio	One dependent variable which is nominal	No covariates

Chapter - 2

Introduction to SPSS

Learning Objectives

This chapter helps to understand the following

- SPSS Data Editor Window
- SPSS Data View and Its Components
- SPSS Variable View and Its Components
- SPSS Output Window
- Modify and Save Files in SPSS

Introduction

SPSS Statistics is a computer application to assist statistical analysis of data. It permits for in-depth data access and preparation, analytical reporting, graphics and modeling. Long produced by SPSS Inc., it was acquired by IBM in 2009. The software name stands for Statistical Package for the Social Sciences (SPSS), reflecting the original market, although the software is now popular in other fields as well, including the health sciences and marketing.

SPSS is a widely used program for statistical analysis in social science. It is also used by market researchers, health researchers, survey companies, government, education researchers, marketing organizations, data miners and others.

Data Editor Window

In SPSS, Data Editor is the dynamic window which is used to records all the data to be evaluated / processed. It has hold two views such as variable view and data view. The variable view allows the user to name each column in the data table as well specify what sort of values the column contains. The data view contains a table with a large number of cells in rows and columns. The table can be very large with only a small part of it visible, in which case use the scroll bars on the edges of the windows to move around the table. Each row represents an individual respondents or samples and each column represents variables.

In data editor window, the accurate way of entering data depends on the details of concern study, the term 'variable' means a column in the data; as well it does not have the same meaning as in experimental design.

Starting SPSS

Go to the Applications folder, and select SPSS from the list of programs (or Start > Programs > SPSS, or click on the SPSS icon in Desktop). A window will appear as given in figure - 1. There are several options, often the data is import from Excel. In that case, select the option "Open another type of file", select "More files..." and select the excel file to work with or To just open it up for the first time, click "Type in data" and select "OK".

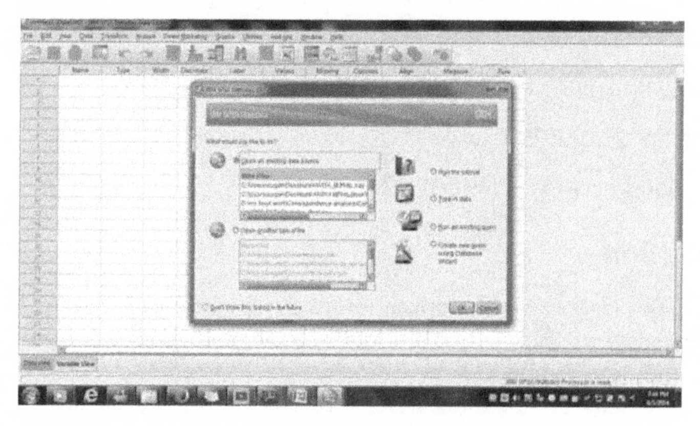

Figure – 1

Defining a variable in SPSS

In the data editor window, the bottom left hand corner contains two tabs i.e. data view and variable view. These two tabs contain two different pages of information.

Data View

The data view screen allows entering data into SPSS. Initially, the data view shows an empty data table in with each of the variables which is in column is labeled as 'var'. Before entering data into data table, it is must to set up the variables so it is ready to receive data. In SPSS, all the variables to be named will be inserted at the top of the column in data table.

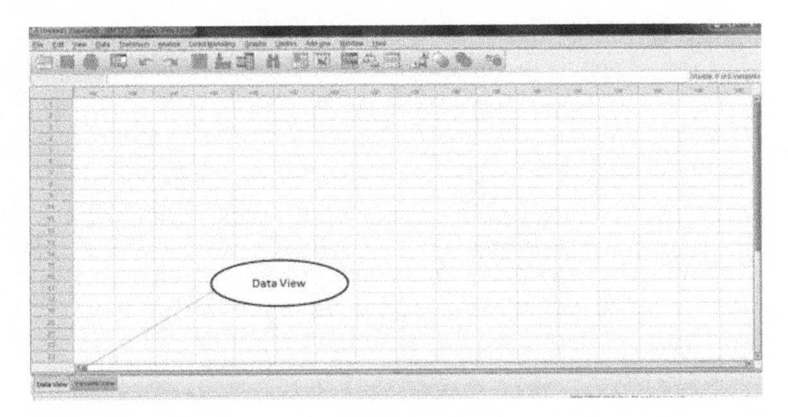

Figure – 2

Variable view

In SPSS, each of the variable information is to be given. The process of defining the variables is undertaken in the variable view. The variable view tab contains the following heads; they are named as Name, Type, Width, Decimals, label values etc. In this view of data table, the variables are arranged down the side of the table and each column gives information about a variable.

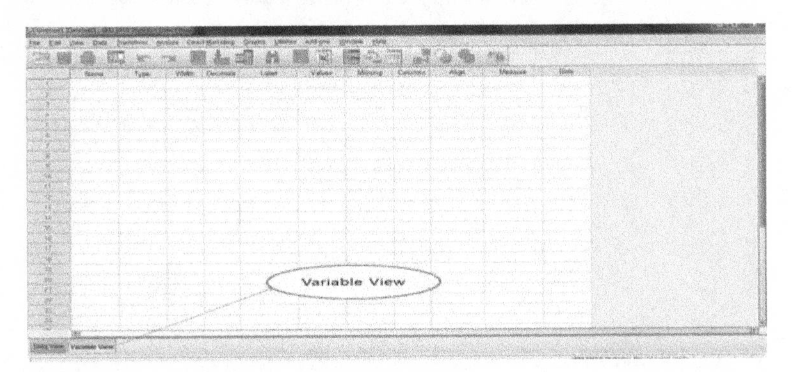

Figure – 3

Variable Name

Giving meaningful name for the variable is important. Type the name of first variable into the first row of the Name column, the variable name should make sense or easy to understand. The variable name can be of any length, but it is suggested to keep it short and simple, so it will be easy to read. The two important aspect to be noted in variable name is that

1. It should start in alphabet and not in numbers.
2. It should not contain spaces or some special characters such as colons, semicolons, hyphens, full stops, comma etc., but full stops, the @, #, $ and _ characters are allowed. In SPSS, the invalid variable names are not accepted, it will warn to move from the name column.

Once the variable name is entered, move to another column in the table. The other column will be filled with either words or numbers. For example, in the variable name – type Gender and leave the default settings as they are or change some or all of them before moving on to define next variable.

Variable Type

The second column in the variable view table is headed type. In SPSS, the variables are classified into different type, such as numeric, string, date etc. The type column is used to indicate type of variable in the data. The default type is numeric; these can be changed by clicking on the button that appears in the cells which is nearer to numeric; Variable type dialogue box will appear. It is strongly recommended to use numeric variables to represent categories i.e. for example, instead of using "M" and "F" for Gender. Use "1" and "2". The variable types are explained below in detail.

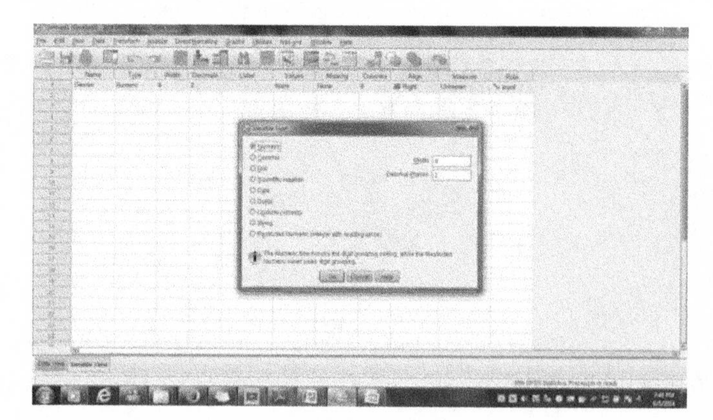

Figure – 4

Numeric

The variables are expressed in term of numbers such as standard numeric values or scientific notation which is accepted by data editor window.

Comma

Numeric variables are expressed with commas delimiting every three places and displayed with the period as a decimal delimiter which is acceptable by data editor window by with or without comma or in scientific notation. Values cannot contain commas to the right of the decimal indicator.

Dot

A numeric variable whose values are displayed with periods delimiting every three places and with the comma as a decimal delimiter. The Data Editor accepts numeric values for dot variables with or without periods or in scientific notation. Values cannot contain periods to the right of the decimal indicator.

Scientific notation

A numeric variable whose values are displayed with an embedded E and a signed power-of-10 exponent. The Data Editor accepts numeric values for such variables with or without an exponent. The exponent can be preceded by E or D with an optional sign or by the sign alone. for example, 183, 1.73E2, 1.23D2, 1.23E+2, and 1.23+2.

Date

A numeric variable whose values are displayed in one of several calendar-date or clock-time formats. Select a format from the list. i.e. the dates can be entered with slashes, hyphens, periods, commas, or blank spaces as delimiters. The century range for two-digit year values is determined by the Options settings (from the Edit menu, choose Options, and then click the Data tab).

Dollar

A numeric variable displayed with a leading dollar sign ($), commas delimiting every three places, and a period as the decimal delimiter. The data can be entered with or without the leading dollar sign.

Custom Currency

A numeric variable whose values are displayed in one of the custom currency formats that is to be defined on the Currency tab of the Options dialog box. Defined custom currency characters cannot be used in data entry but are displayed in the Data Editor.

String

A variable whose values are not numeric and therefore are not used in calculations. The values can contain any characters up to the defined length.

Uppercase and lowercase letters are considered distinct. This type is also known as an alphanumeric variable.

Restricted Numeric

A variable whose values are restricted to non-negative integers. Values are displayed with leading zeros padded to the maximum width of the variable. Values can be entered in scientific notation.

Width

In SPSS, Width column in variable view helps to alter the number of digits displayed in the column by clicking in the appropriate Width cell. The default value for width is 8.

Decimal

In SPSS, Decimal Column in variable view enables to alter the number of digits after the decimal place by clicking in the appropriate Width cell. The default decimal value is 2.

Variable Label

Variable label used to define a label for a variable makes output easier to read but does not have any effect on the actual analysis. For example, the label "Customer Satisfaction" is easier to understand than the name of the variable, Cus_Sat.

Value labels

The value label is similar to variable labels. Whereas "variable labels" define the label to use instead of the name of the variable in output, "value labels" enable the use of labels instead of values for specific values of a variable, thereby improving the quality of output. For example, for the variable gender, the labels "Male" and "Female" are easier to understand than "0" or "1". In effect, using value labels indicates to SPSS that: use the label "Male" instead of the value "1" and the label "Female" instead of the value "2"."

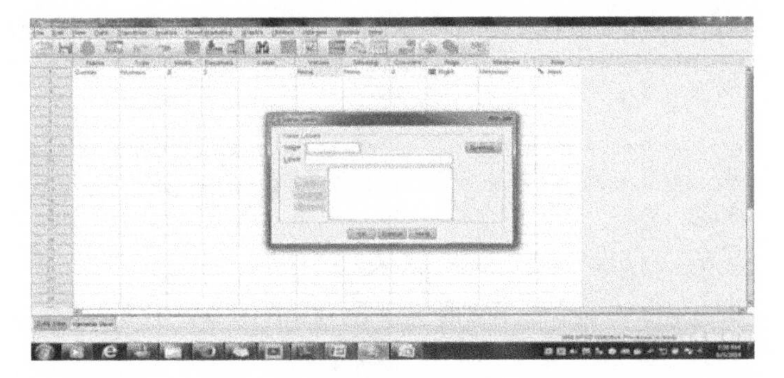

Figure – 5

Steps for Creating Value Label

- Enter a value and a label.
- Click on the Add button.
- Then, click on OK.
- It is also possible to return here in the future and change value labels or remove them.

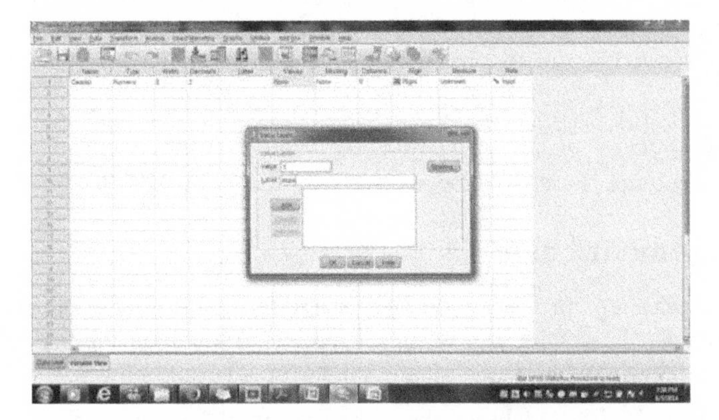

Figure – 6

Missing Value Declaration

Missing Values are used to define specified data values as user-missing. For example, to distinguish between data that are missing because a respondent refused to answer and data that are missing because the question didn't apply to that respondent. Data values that are specified as user-missing are flagged for special treatment and are excluded from most calculations.

- User-missing value specifications are saved with the data file. It is not necessary to redefine user-missing values each time you open the data file.
- In this option it is possible to enter upto three discrete (individual) missing values, a range of missing values, or a range plus one discrete value.
- Ranges can be specified only for numeric variables.
- All string values, including null or blank values, are considered to be valid unless you explicitly define them as missing.
- Missing values for string variables cannot exceed eight bytes. (There is no limit on the defined width of the string variable, but defined missing values cannot exceed eight bytes.)
- To define null or blank values as missing for a string variable, enter a single space in one of the fields under the Discrete missing values selection.

Column Format

Column format assist in improving the on-screen viewing of data by using appropriate column sizes (width) and displaying appropriate decimal places which does not affect or change the actual stored values.

Align

The Align column is used to choose alignment of data to the left, right or center.

Measurement Level

In SPSS, Measurement column enables to specify the level of measurement as scale (numeric data on an interval or ratio scale), ordinal, or nominal. Nominal and ordinal data can be either string (alphanumeric) or numeric.

- **Nominal**: A variable can be treated as nominal when its values represent categories with no intrinsic ranking (for example, the department of the company in which an employee works). Examples of nominal variables include region, zip code, and religious affiliation.
- **Ordinal**: A variable can be treated as ordinal when its values represent categories with some intrinsic ranking (for example, levels of service satisfaction from highly dissatisfied to highly satisfied). Examples of ordinal variables include attitude scores representing degree of satisfaction or confidence and preference rating scores.

- **Scale**: A variable can be treated as scale (continuous) when its values represent ordered categories with a meaningful metric, so that distance comparisons between values are appropriate. Examples of scale variables include age in years and income in thousands of dollars.

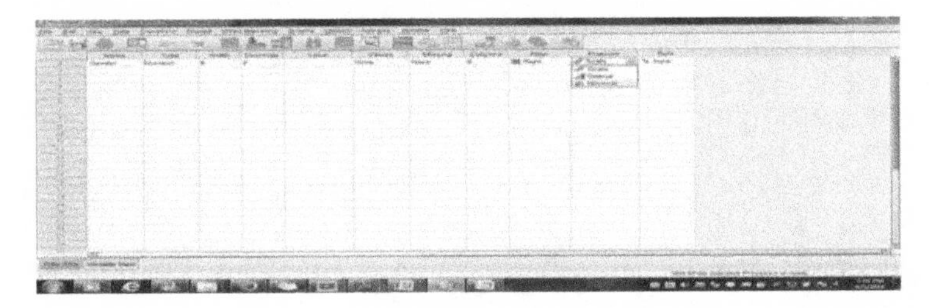

Figure – 7

Role

Role can also be used to specify the type of variable such as input, target, both, none, partition, split.

- **Input** is the variables that can be used as input (e.g., predictor, independent variable). By default all variables are assigned the Input role.
- **Target** are the variables used as output or target(e.g. dependent variable),
- **Both** are variables that can be used as input and output,
- **None** are variables that have no role assignment.
- **Partitions** are variables that can be used to partition the data into separate samples for training, testing and validation.
- **Split** is the variable type that is included for round-trip compatibility with SPSS modeler.

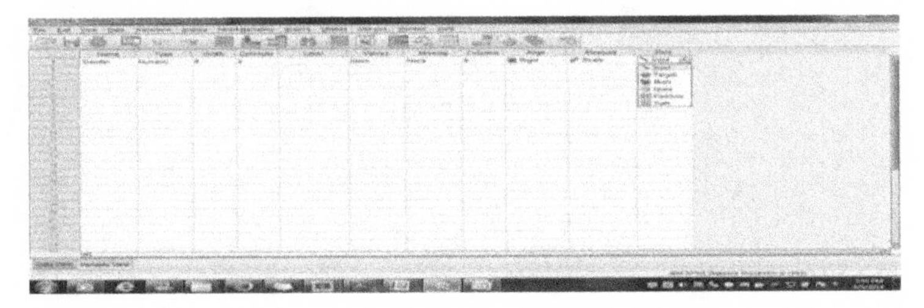

Figure – 8

The output view

The output window enables to see the results of various analyses such as frequency distributions, cross-tabs, statistical tests, and charts. In SPSS, each window handles a separate task.

Figure – 9

Modifying Data File

The created data file can be altered/modified at any time by way of add. Inserting, Deleting the variables in the data editor window.

Saving Data File and Output

As mentioned earlier, SPSS works with different windows for different tasks i.e. Data Editor Window to manage data, and the SPSS Viewer to examine the results of analyses. So the files need to save each window separately.

The SPSS working file is saved by File > Save in either window; for saving the file at first time the window will be asked to name the file and choose where to save it. The data editor file/working file is saved in .sav. and the output file is saved as .spo. The output file can be exported to MS Word or as MS Excel or also stored in PDF format.

Outliers

Learning Objectives

This chapter helps to understand the following

- Meaning of outliers
- Effects of Outliers
- Causes of Outliers
- Procedure to identify outliers using SPSS
- Results and Discussion of Outliers Output

Introduction

The occurrence of outliers can lead to magnified error rates and substantial misrepresentations of parameter and statistic estimates in both parametric and nonparametric tests. In general, an outlier is considered to be a data point that is far outside the norm for a variable or population. According to the Dictionary of Statistics, outlier is "an observation that appears to deviate markedly from the other observations of the sample in which it appears". Statistical measures that are not extremely affected by outliers are called robust.

Definition of Outliers

Hawkins defines an outlier is "an observation that deviates so much from other observations as to arouse suspicions that it was generated by a different mechanism". According to Dixon, Outliers are defined as the values that are "dubious in the eyes of the researcher". It is also considered as contaminants.

Effects of outliers

Mean:Even a single outlier can have a huge effect on the mean.

Median: The median, or the number that is higher than half the numbers and lower than half, is much less affected by outliers than the mean.

Mode: Unless two or more of the outliers have the exact same value, the outliers will have no effect at all on the mode, which is the most common value.

Multivariate: Multivariate outliers are traditionally analyzed when conducting correlation and regression analysis. The multivariate outlier analysis is somewhat complex to identify multivariate outliers to correlation and regression

Causes of Outliers

Outliers can arise from several different methods or causes. In 1960, Anscombe identified that outliers occurs from two categories i.e. those are arising from errors in the data, and those arising from the natural variability of the data.

Outliers from data errors

Outliers are often caused by human error, such as errors in data collection, recording, or entry. Data from an interview can be recorded incorrectly, or mistyped upon data entry. For example, in a data collection, weekly salary of the employee is requested, by mistake some employees give their monthly salary. In this case, the error can be rectified by checking; the original document/ instrument/questionnaire even if the respondent corrected wrongly, it is possible to overcome the error or by recalculating the salary by different cadres or else there is a possibility for occurring outliers. It is purely data collection error. But most of the researchers do anonymous survey, so it is not possible to identify the respondent. In certain cases, it is advisable to eliminate the particular data from the valid population.

Outliers causes Intentionally

Sometimes participants purposefully report incorrect data to the surveyors and researchers. This will have effects on inflation of all estimates or production of outliers. If all subjects respond the same way, the distribution will shift upward, not generally causing outliers. However, if only a small subsample of the group responds this way to the experimenter, or if multiple researchers conduct interviews, then outliers can be created.

Outliers from sampling error

Another cause of outliers is sampling. It is possible that a few members of a sample were unintentionally drawn from a different population than the rest of the sample. These cases should be removed as they do not reflect the target population.

Outliers from standardization failure

Outliers can be caused by research methodology, particularly if something anomalous happened during a particular subject's experience. These data can be legitimately discarded if the researchers are not interested in studying the particular phenomenon in question

Outliers from defective distributional assumptions

Incorrect assumptions about the distribution of the data can also lead to the presence of suspected outliers. Depending upon the goal of the research, these extreme values may or may not represent an aspect of the inherent variability of the data, and may have a legitimate place in the data set.

Outliers as legitimate cases sampled from the correct population

Finally, it is possible that an outlier can come from the population being sampled legitimately through random chance. It is important to note that sample size plays a role in the probability of outlying values. Within a normally distributed population, it is more probable that a given data point will be drawn from the most densely concentrated area of the distribution, rather than one of the tails. As a researcher casts a wider net and the data set becomes larger, the more the sample resembles the population from which it was drawn, and thus the likelihood of outlying values becomes greater.

In other words, there is only about a 1% chance you will get an outlying data point from a normally-distributed population; this means that, on average, about 1% of your subjects should be 3 standard deviations from the mean.

In the case that outliers occur as a function of the inherent variability of the data, opinions differ widely on what to do. Due to the deleterious effects on power, accuracy, and error rates that outliers and fringeliers can have, it might be desirable to use a transformation or recoding/truncation strategy to both keep the individual in the data set and at the same time minimize the harm to statistical inference.

Outliers as potential focus of inquiry

The remarkable research is often as much a matter of coincidence as planning and inspiration. Outliers can represent a nuisance, error, or legitimate data. They can also be inspiration for inquiry. In a study the first author was involved with, a teenager reported 100 close friends. Is it possible? Yes. Is it likely? Not generally, given any reasonable definition of "close friends." So this data point could represent either motivated mis-reporting, an error of data recording or entry (it wasn't), a protocol error reflecting a misunderstanding of the question, or something more interesting. This extreme score might shed light on an important principle or issue. Before discarding outliers, researchers need to consider whether those data contain valuable information that may not necessarily relate to the intended study, but has importance in a more global sense.

STEPS to explore Outliers using SPSS

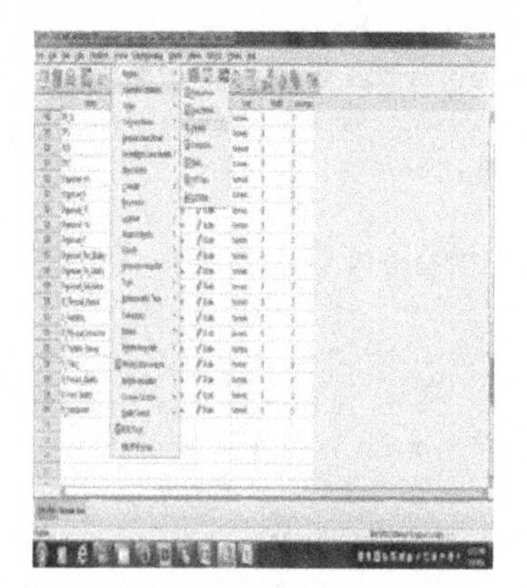

Figure – 1

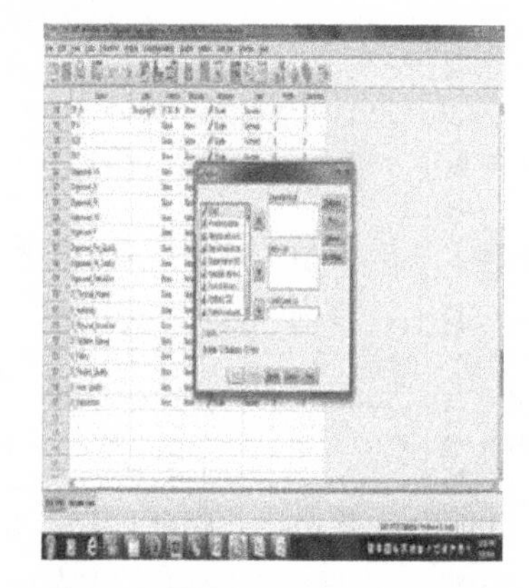

Figure – 2

Step – 1: Click Analyze > Descriptive Statistics > Explore as shown in Figure -1, the Explore Dialogue Box will appear as given in Figure -2.

Step – 2: In Explore Dialogue Box, transfer the variable, which has to be checked for outliers. In this example, outliers to be checked for satisfaction. So the variable is transferred to dependent list as given in figure -3.

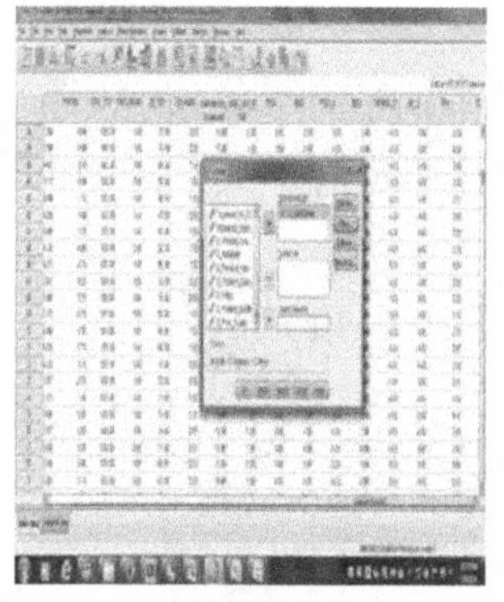

Figure – 3

Figure – 4

Step – 3: In Explore dialogue box, Click on Statistics tab. The Explore-Statistics dialogue box will appear, Descriptive option is selected in default as given in Figure -4

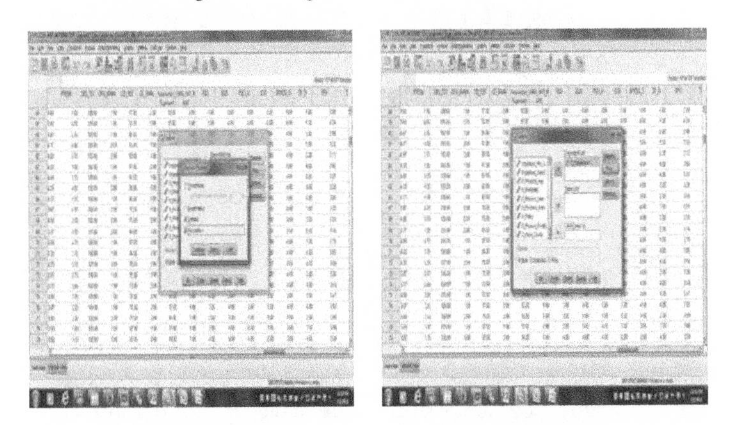

Figure – 5 **Figure – 6**

Step – 4: In Explore: Statistics tab. Deselect Descriptive option and select Outliers and Percentiles option. Click on continue tab as given in Figure -5 to return to Explore Dialogue box as given in Figure – 6.

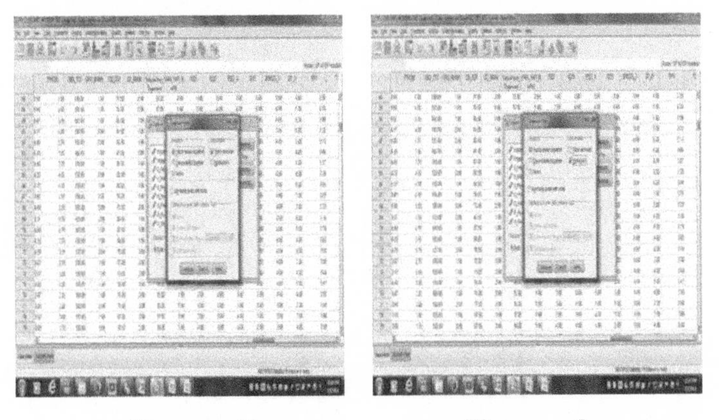

Figure – 7 **Figure – 8**

Step – 5: In the Explore Dialogue Box Click on Plots, the dialogue box will appear as given in Figure -7. In Default, Box plots Factor level together and in Descriptives, Stem and leaf was selected in default. Select Histogram and deselect Stem and Leaf option and click on continue to return to Explore dialogue box (Figure -9).

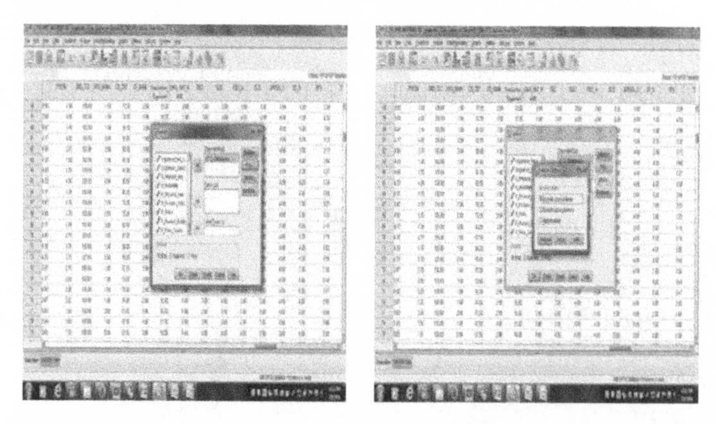

Figure – 9 **Figure – 10**

Step – 6: In Explore Dialogue Box, Click on Option. The Explore Option Dialogue box will appear, but leave it as a default and click on continue to return to Explore Dialogue box.

Step – 7: In Explore Dialogue box, Select Both in Display option and click on Ok tab as given in Figure -11. To generate the outliers output.

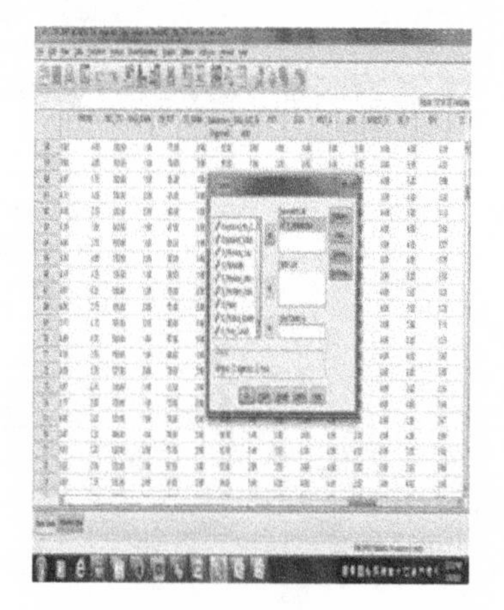

Figure – 11

SPSS Output of the Outliers Analysis

Case Processing Summary

	Cases					
	Valid		Missing		Total	
	N	Percent	N	Percent	N	Percent
O_Satisfaction	516	100.0%	0	0.0%	516	100.0%

Percentiles

		Percentiles						
		5	10	25	50	75	90	95
Weighted Average (Definition 1)	O_Satisfaction	2.6000	2.9400	3.2000	3.6000	3.8000	4.0000	4.4000
Tukey's Hinges	O_Satisfaction			3.2000	3.6000	3.8000		

Extreme Values

			Case Number	Value
O_Satisfaction	Highest	1	184	5.00
		2	192	5.00
		3	194	5.00
		4	300	5.00
		5	309	5.00[a]
	Lowest	1	435	1.40
		2	417	1.40
		3	411	2.20
		4	287	2.20
		5	432	2.40[b]

a. Only a partial list of cases with the value 5.00 are shown in the table of upper extremes.

b. Only a partial list of cases with the value 2.40 are shown in the table of lower extremes.

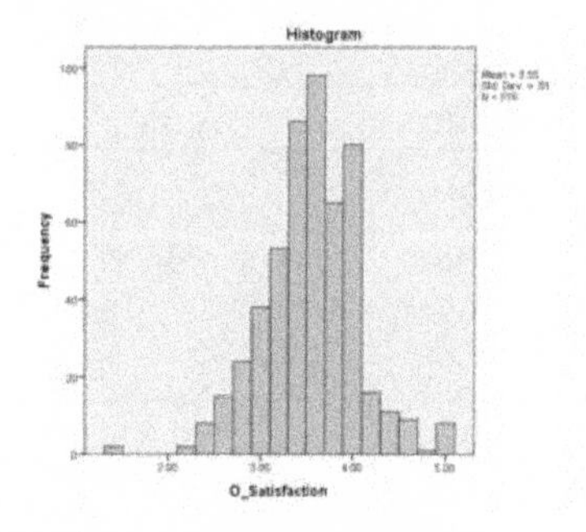

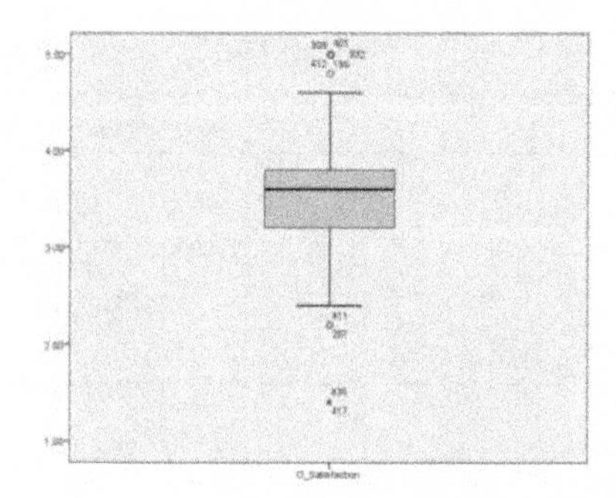

Results & Discussion

Case Processing Summary

In this example, 100% of the respondents answered the question. There is no missing number. So the total sample size is 516.

Percentiles

In the percentile table, Tukey Hinges box plots have to be considered. In this example, the 25th percentile is 3.2000 and the 75th percentile is 3.8000, the median is 3.6000. A circle represents each outliers- the number next to the outlier is the observation numbers.

Extreme Value

The Extreme Values table is very helpful and reports the highest and lowest five cases for the variable specified. The above table given only partial lists of cases with the value 5.00 are shown in upper extremes and a partial list of cases with the value 2.40 are shown in the table of lower extremes.

Histogram

Outliers are often easy to spot in histograms. In this study, the point on the far left in the above figure is an outlier. A convenient definition of a outlier is a point which falls more than 1.5 times the interquartile range above the third quartile or below the first quartile.

Box plots

A box plot is built by drawing a box between the upper and lower quartiles with a solid line drawn across the box to locate the median. A point away from an inner boundary marker on either side is considered a mild outlier. A point beyond an outer fence is considered an extreme outlier.

The extreme values in the dataset were removed and the outlier test was proceeded again. The results were given below

Case Processing Summary

	Cases					
	Valid		Missing		Total	
	N	Percent	N	Percent	N	Percent
O_ Satisfaction	506	100.0%	0	0.0%	506	100.0%

Percentiles

		Percentiles						
		5	10	25	50	75	90	95
Weighted Average (Definition 1)	O_ Satisfaction	2.8000	3.0000	3.2000	3.6000	3.8000	4.0000	4.2000
Tukey's Hinges	O_ Satisfaction			3.2000	3.6000	3.8000		

Extreme Values

			Case Number	Value
O_ Satisfaction	Highest	1	326	5.00
		2	395	5.00
		3	405	5.00
		4	196	4.80
		5	151	4.60[a]
	Lowest	1	371	2.40
		2	366	2.40
		3	364	2.40
		4	343	2.40
		5	333	2.40[b]

a. Only a partial list of cases with the value 4.60 are shown in the table of upper extremes.

b. Only a partial list of cases with the value 2.40 are shown in the table of lower extremes.

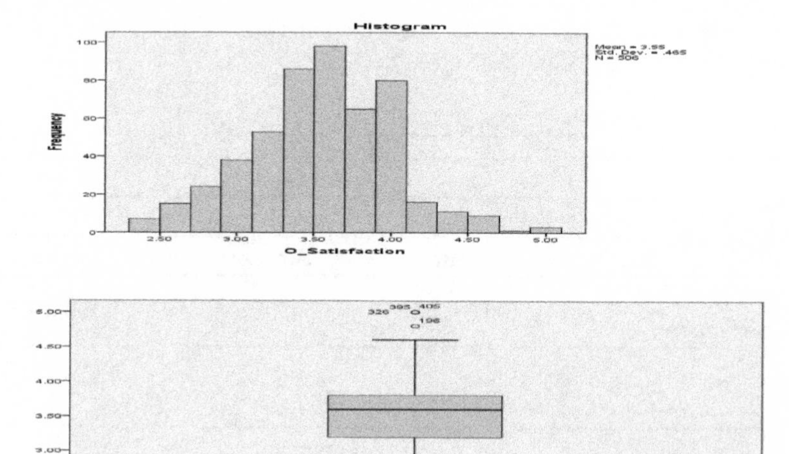

Results & Discussion

By checking the case processing summary, after removing the extreme values, the numbers of respondents came down to 506. There are no changes in percentiles. The extreme value table depicts the lowest boundary is static to 2.40, and only a partial list of cases with the value 4.60 were shown in upper extremes. The histogram shows the point at the right end is outliers, which can be seen in box plot, some of the cases were seen above the upper quartiles.

Again the outlier test is done for removing the extreme values. The results were given below

Case Processing Summary

	Cases					
	Valid		Missing		Total	
	N	Percent	N	Percent	N	Percent
O_Satisfaction	501	100.0%	0	0.0%	501	100.0%

Percentiles

		Percentiles						
		5	10	25	50	75	90	95
Weighted Average (Definition 1)	O_ Satisfaction	2.8000	3.0000	3.2000	3.6000	3.8000	4.0000	4.2000
Tukey's Hinges	O_ Satisfaction			3.2000	3.6000	3.8000		

Extreme Values

			Case Number	Value
O_Satisfaction	Highest	1	182	4.60
		2	183	4.60
		3	193	4.60
		4	310	4.60
		5	331	4.60[a]
	Lowest	1	368	2.40
		2	363	2.40
		3	361	2.40
		4	340	2.40
		5	330	2.40[b]

a. Only a partial list of cases with the value 4.60 are shown in the table of upper extremes.

b. Only a partial list of cases with the value 2.40 are shown in the table of lower extremes.

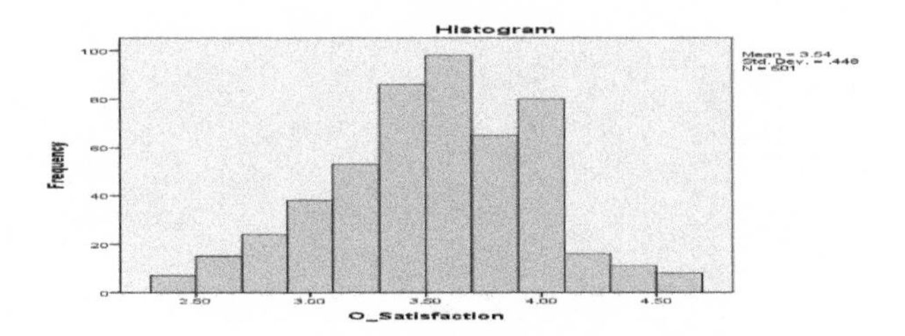

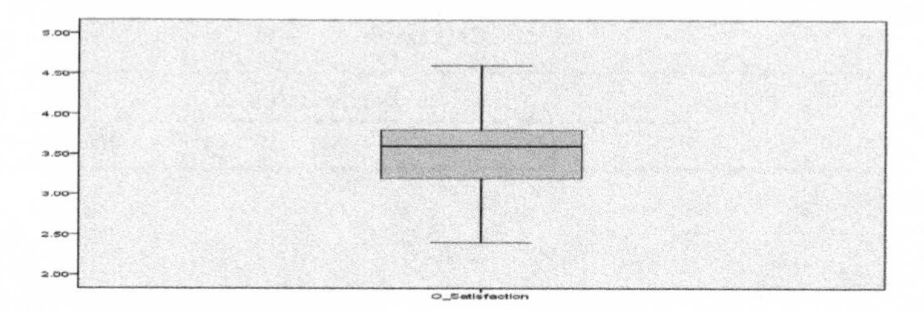

Results & Discussion

From the above results, it is evident that after removing the extreme values of the cases number of respondents came down to 501. There were no changes in percentiles. There values are static, no extreme value is shown. The histogram is normally distributed. By checking the box plots, there is no extreme case.

Normality

Learning Objectives

This chapter helps to understand the following

- Meaning of Normality
- Properties of Normality
- Running Normality Test Using SPSS
- Results and Discussion of Normality Output

Introduction

A normal distribution is a frequency distribution that represents the relative number of occurrences at each value of a variable. The shape of the distribution resembles a bell, and so gets its nickname as the bell-shaped curve. The statistical procedures assume that the errors associated with the scores on the dependent variable are normally distributed, when this assumption is met, the dispersion of the scores themselves also tends to be normally distributed.

Normal distributions have a variety of properties, including the following:

- The normal distribution is horizontally symmetric around the mean of distribution. The middle value represents the mean, median, and mode of the distribution.
- The standard deviation is the distance between the mean and the inflection point (change of the direction of slope) on the side of the curve.
- Standard deviation is an interval-level scale of measurement. Knowing the distance of one standard deviation allows us to fill in the rest of the X-axis. The count of standard deviation units in terms of distance from the mean is a z score scale.
- The mean has a z score of zero.
- Between ± 1.00 standard deviation units, there is approximately 68.26% of the area (or scores); this range corresponds to percentile scores of approximately 16–84.

- Between ±1.96 standard deviation units, there is approximately 95% of the area; this range corresponds to percentile scores of approximately 2.5–97.5.
- Between ±3.00 standard deviation units, there is approximately 99% of the area; this range corresponds to percentile scores of approximately 0.1–99.9.
- As the distance from the mean increases, the curve approaches but never reaches the X-axis

The shape of a distribution of continuous variable in multivariate analysis should correspond to a Univariate normal distribution i.e. the variable's frequency distribution of values should roughly approximate a bell shaped curve. The statistical approaches that assess Univariate normality often begin with the measures of skewness and kurtosis. A normally distributed variable will generate skewness and kurtosis values that hover around zero which is obtained through the explore procedure. The tests of Kolmogorov-Smirnov test and Shapiro-Wilk Test is additional statistical test to find normality in the data. Shapiro-Wilk test appears to be the most powerful in detecting departure from normality. If the normality assumption appears to be violated, thus a statistically significant result ($P<.05$) is an indicative of normality violation. As well, it is possible to repair the problem through transformation process.

Normality Test Using SPSS

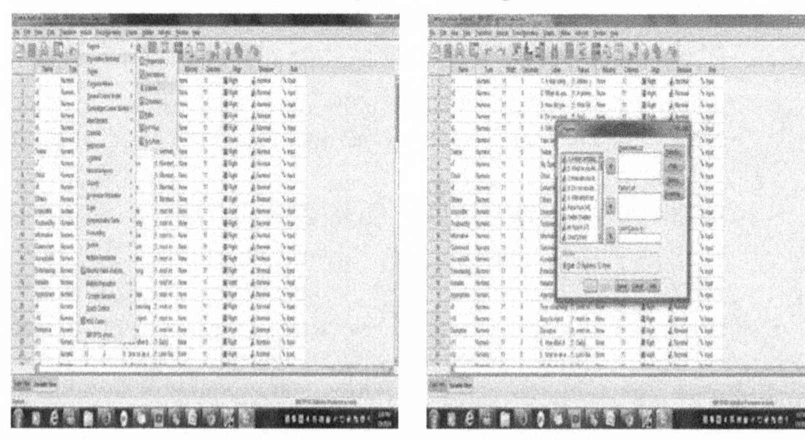

Figure – 1 **Figure – 2**

Step -1: Click Analyze > Descriptive Statistics > Explore as shown in Figure -1, the Explore Dialogue Box will appear as given in Figure -2.

Step -2: In Explore Dialogue Box, transfer the variable in dependent list for normality. In this example, statement total is transferred into dependent list as given in figure -3.

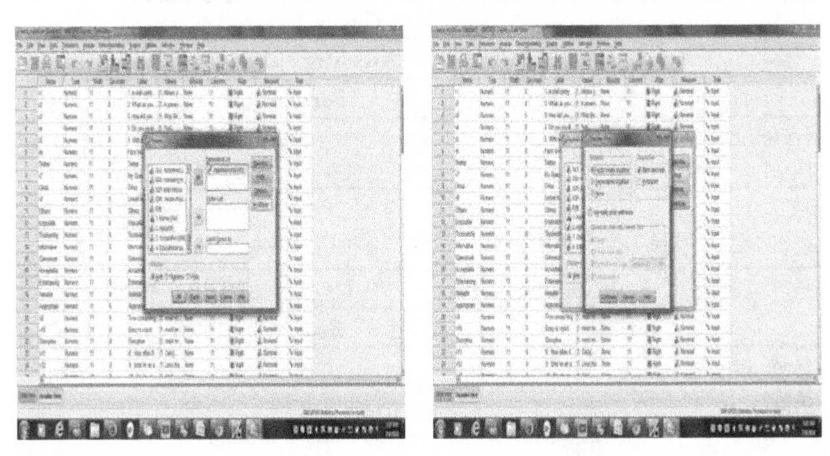

Figure – 3 **Figure – 4**

Step – 3: Click on plots tab in explore dialogue box, a dialogue box will appear as given in figure 4. In boxplots select factor level together, in descriptives deselect stem-and-leaf and select histogram option. Select Normality plots with tests as given in figure -5 and click on continue tab to return to explore dialogue box.

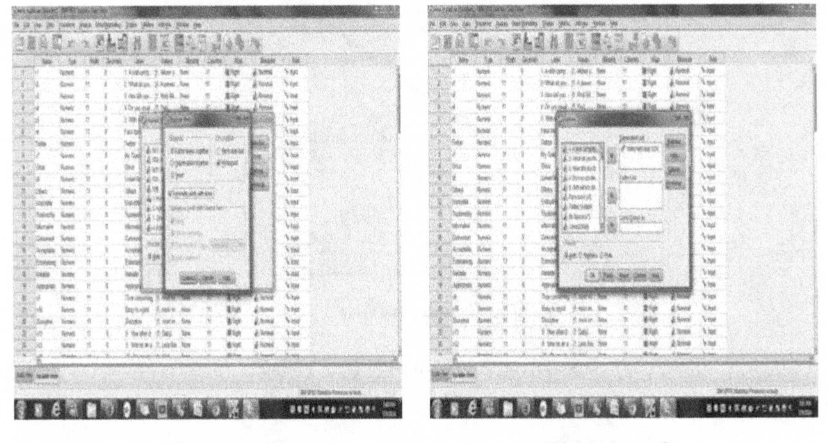

Figure – 5 **Figure – 6**

Step – 4 : In Explore dialogue box, click on both in the Display option. And click Ok to generate output.

SPSS OUTPUT FOR THE NORMALITY TEST

Table -1 Case Processing Summary

	Cases					
	Valid		Missing		Total	
	N	Percent	N	Percent	N	Percent
statement total	150	100.0%	0	0.0%	150	100.0%

Table -2 Tests of Normality

	Kolmogorov-Smirnov[a]			Shapiro-Wilk		
	Statistic	df	Sig.	Statistic	df	Sig.
statement total	.051	151	.200*	.992	151	.560

*. This is a lower bound of the true significance.
a. Lilliefors Significance Correction

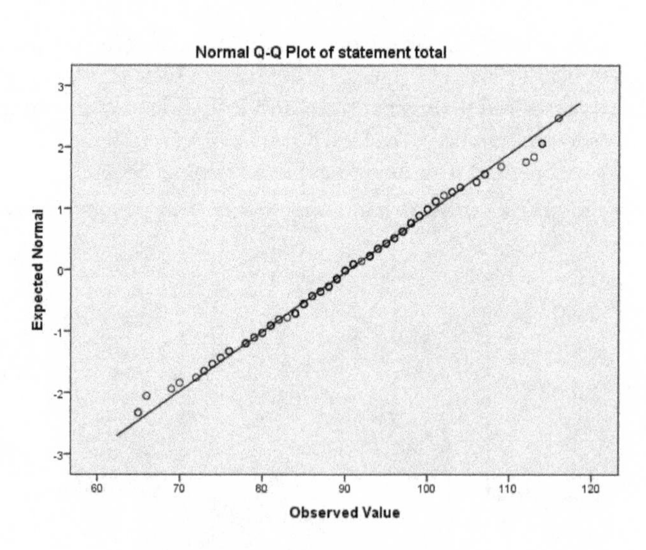

Figure 7 Normal Q-Q Plot

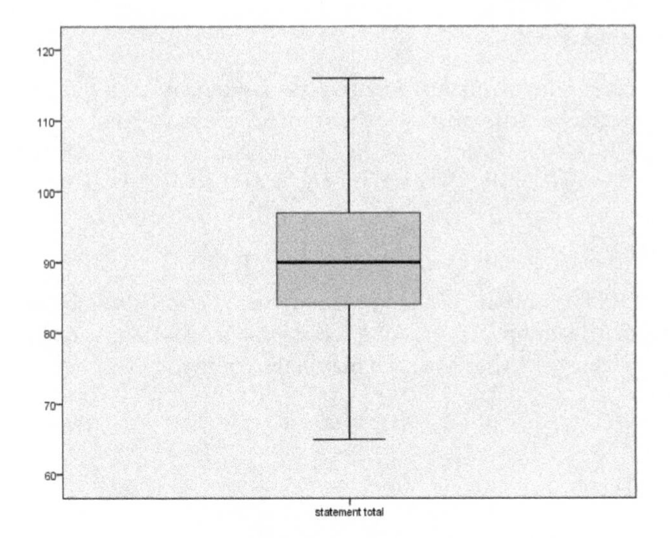

Figure 8 BoxPlot

Results and Discussion

In normality test, Normality table and the Normal Q-Q Plots are numerical and graphical method to test for the normality of data.

Case processing Summary

In this example, 100% of the respondents answered the question. There is no missing number. So the total sample size is 150.

Test of Normality

Test of Normality table presents the results from two well-known normality test namely Kolmogorov-Smirnov test and the Shapiro-Wilk Test.

The Shapiro -Wilk test is more appropriate for small sample size, it also handles larger sample as 2000. The Kolmogorov-Smirnov test is used when the sample is more than 2000. In this study, the Shapiro - Wilk test is used to assess normality, because the sample size is 151.

In this test the sig. value which is above or greater than 0.05 is considered to be Normal. If it is below 0.05, the data significantly deviate from normal distribution.

In this example, the significant value is .560. It is concluded that the data is normally distributed.

Normal Q-Q Plot

In order to determine normality graphically, the output of a normal Q-Q plot is used. When the data are normally distributed, the data points are close to the diagonal line. If the data points lose the way or wander away from the line in an obvious non-linear fashion, the data are not normally distributed.

Box Plot

A boxplot that is symmetric with the median line in approximately the center of the box and with symmetric whiskers somewhat longer than the subsections of the center box suggests that the data may have come for a normal distribution.

Test of Linearity

Learning Objectives

This chapter helps to understand the following

- Meaning of Linearity
- Ways to evaluate Linearity
- Procedure to run Test of Linearity using SPSS
- Results and Discussion of Linearity Output

Introduction

Linearity means that the amount of change, or rate of change, between scores on two variables are constant for the entire range of scores for the variables. Linearity can be evaluated by

- Both graphical and statistical methods
- Graphical methods include the examination of scatterplots, often overlaid with a trendline. While commonly recommended, this strategy is difficult to implement.
- Statistical methods include diagnostic hypothesis tests for linearity, a rule of thumb that says a relationship is linear if the difference between the linear correlation coefficient (r) and the nonlinear correlation coefficient (eta) is small, and examining patterns of correlation coefficients.

Multivariate techniques such as MANOVA, Multiple Regression assume that the variable in the analysis relates to each other in a linear manner i.e. they assume that the best fitting function representing the scatter plot is a straight line. These procedures often compute the Pearson correlation coefficient as part of the calculations required for multivariate statistical analysis. Because the Pearson r assesses the degree to which a pair of variables is linearly related, it is useful to make a rough determination of the degree to which the relation is linear before calculating the Pearson r. This can be accomplished by examining the scatterplot of the two variables. To the extent that such non linearity is present; the observed Pearson r would be a less contributing index to strengthen the association between the variables accounted for i.e. it gives less relationship strength than existed because it would capture only the linear component of the relationship.

The bivariate scatter plots are the most typical way of accessing Linearity between the variables. Assessing linearity through bivariate scatter plots is also possible if the possible pairs are examined. The cure for non linearity lies in data transformation.

Test of Linearity Using SPSS

Step - 1: Click on Graph > Legacy Dialogs > Scatter /Dot as given in figure -1, Scatter/Dot dialogue box will appear as given in figure -2.

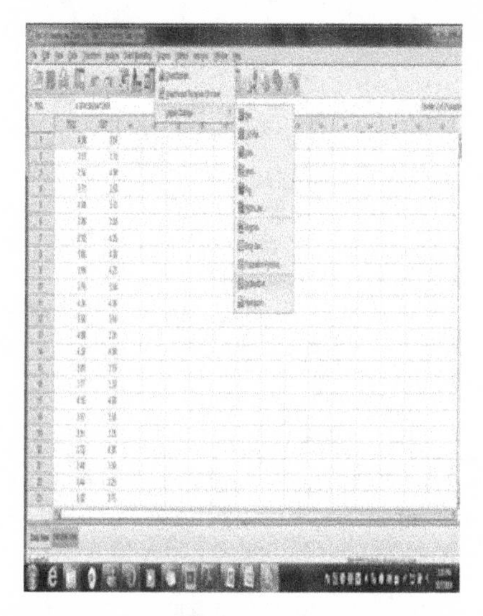

Figure – 1

Step - 2: Click Simple Scatter on Scatter /Dot Dialogue Box and click on Define Tab (figure-3), a Simple Scatterplot Dialogue box will appear as given in Figure-4.

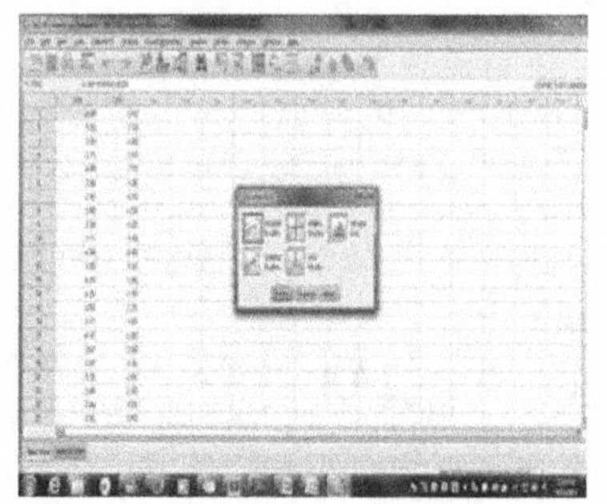

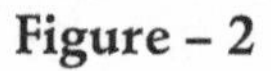

Figure – 2

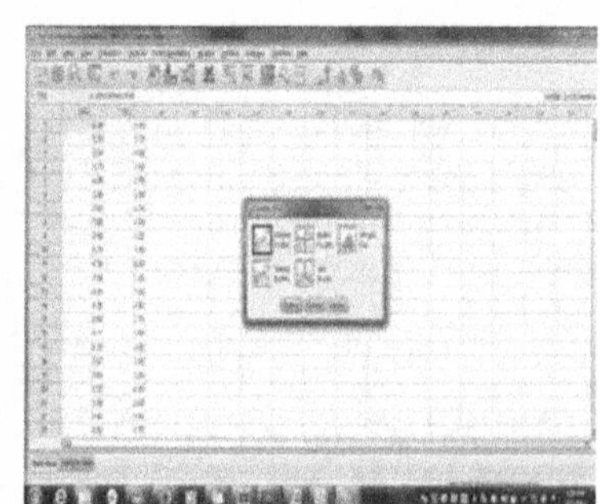

Figure – 3

Step - 3: In the simple scatterplot dialogue box, transfer the required variables in y axis and x axis column as given in figure -5. In this example, Service quality is transferred in y axis and customer satisfaction is transferred in x axis (see figure-5) and click OK to generate output.

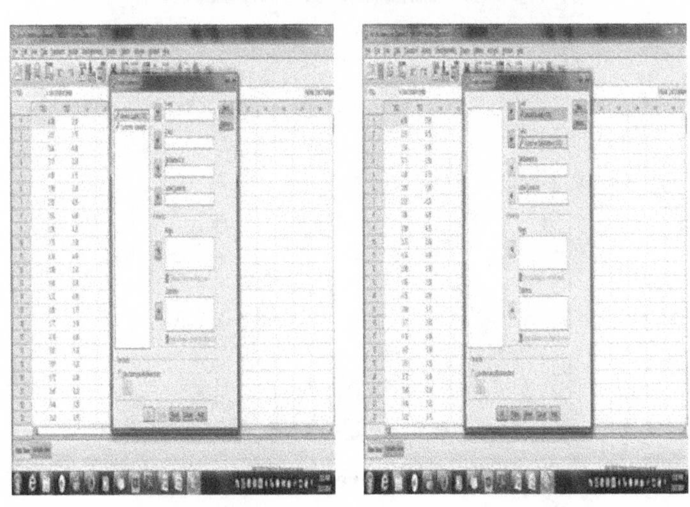

Figure – 4 **Figure – 5**

SPSS Output for Test of Linearity

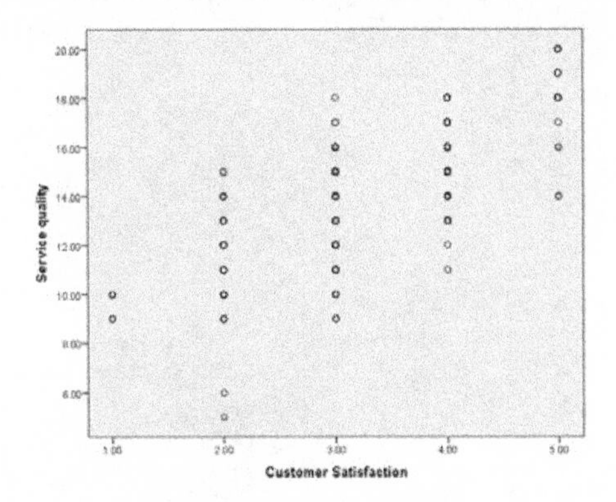

Figure 6 Scatter plot

RESULT & DISCUSSION

The above scatter plot is generated in test of linearity between two variables such as Service Quality and Customer Satisfaction. The scatter plot is produced in the SPSS output viewer given in figure-6. The pointers in a scatter plot are considered linear. The pointer in this scatter plot is not really clear. In this case double click on the image in the SPSS output viewer, Chart editor dialogue box will appear as given in Figure - 7.

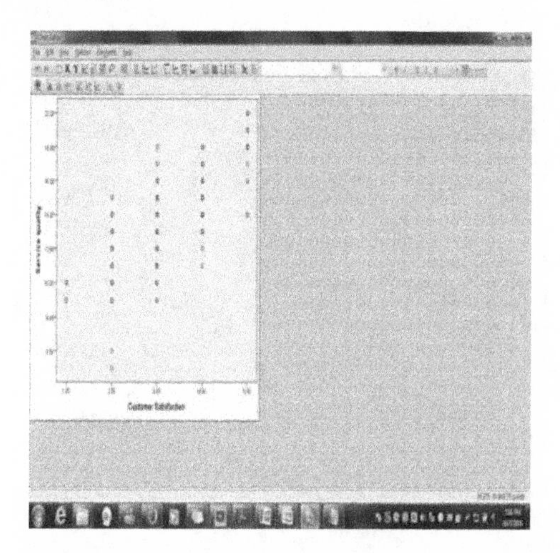

Figure – 7

In the Chart editor dialogue box click on Element > Fitline at Total as given in Figure -8

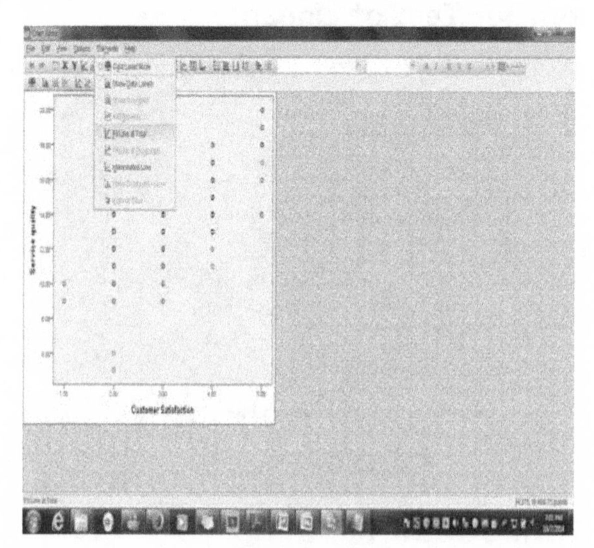

Figure – 8

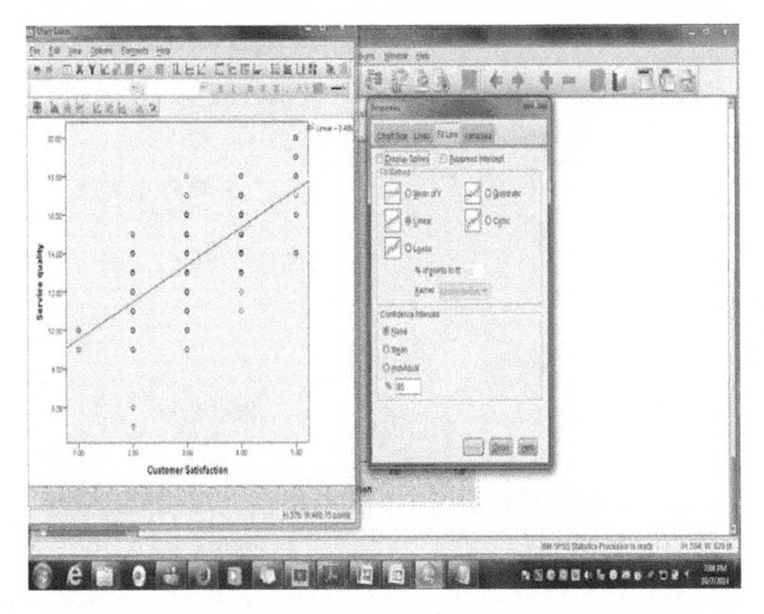

Figure – 9

By Clicking of Fit Line at Total, a properties dialogue box will appear as given in figure -9, as well if it is to be noted the scatter plot a line will appear. It is set as default in the properties; if it's needed it can be modified. The graph line in the scatter plot diagram shows that there is positive linear relationship between the variable service quality and customer satisfaction (Figure -10).

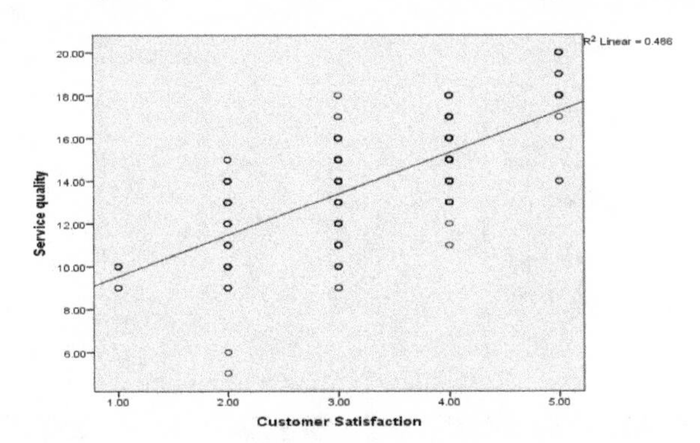

Figure – 10

If it is required or with the reference to linearity, the Pearson correlation between the variables can be confidently generated by Clicking on Analyse > Correlation > Bivariate as given in

Figure -11, Bivariate Dialogue Box will appear as given in figure -12.

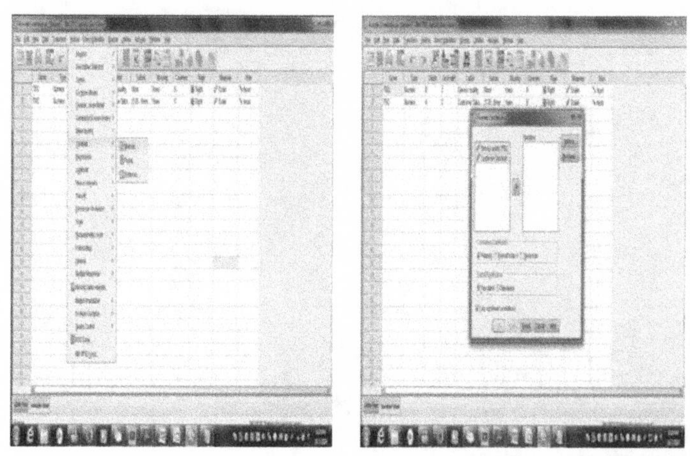

Figure – 11 **Figure – 12**

Transfer the required variables in variable box by selecting and clicking the arrow tab. Click on options tab. Bivariate correlation: option dialogue box will appear as given in figure-13, Select Mean and Standard Deviations and Click on Continue to return to Bivarate dialogue box; by default in correlation coefficient Pearson is selected, click on OK tab to generate output as given in figure-14.

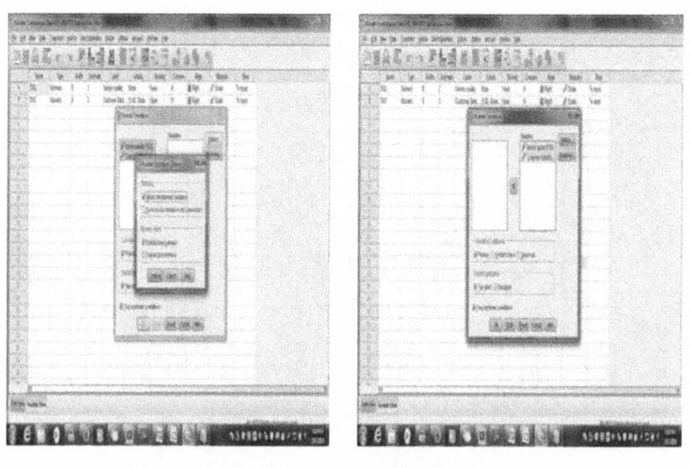

Figure – 13 **Figure – 14**

The Bivariate correlation analysis generates two output table: Descriptive Statistics Table and Correlation Table. In descriptive statistics (table -1), Mean and standard deviation value is ascertained. The correlation table (Table -2) gives

Pearson correlation coefficient value which is .697 and the sig. value is .000. The result reveals there is 69.7% positive relationship between customer satisfaction and service quality and it is statistically significant.

Table 1 Descriptive Statistics

	Mean	**Std. Deviation**	**N**
Service quality	14.2151	2.12663	516
Customer Satisfaction	3.4205	.76545	516

Table 2 Correlations

		Service quality	**Customer Satisfaction**
Service quality	Pearson Correlation	1	.697**
	Sig. (2-tailed)		.000
	N	516	516
Customer Satisfaction	Pearson Correlation	.697**	1
	Sig. (2-tailed)	.000	
	N	516	516

**. Correlation is significant at the 0.01 level (2-tailed).

Data Transformation

Learning Objectives

This chapter helps to understand the following

- Meaning of Data Transformation
- Ways to deal with Data Transformation

Introduction

Data transformation is a mathematical procedure which is used to modify variables that violates the statistical assumption of normality, linearity and Homoscedasticity or that have unusual outlier patterns. To proceed with data transformation, one has to determine, the extent to which the assumptions are violated; then to decide whether the data transformation is correct for the situation or suits for this matter or not. Once the data transformation is decided it is important to instruct SPSS to carry out the changes with every value of the variable or variables to be transformed. Most of the experts and statistician compared data transformation with double-edged sword. Because one has to significantly improve the precision of a multivariate analysis, at the same time, transformation can pose a difficult data interpretation problem.

IBM SPSS provide many ways for data transformation such as

1. Combine values of a variable into several categories,
2. Create new variables out of old variables,
3. Select particular cases and analyze only these cases,
4. Weight cases so that some cases count more heavily than others
5. Recoding of data set.
6. Computation of data etc

For data transformation, click on Transform in SPSS Menu. Wide options are there to proceed with. Transforming data is performed for a complete multitude of different reasons, but one of the most common is to apply a transformation to data that is not normally distributed so that the new, transformed data is normally distributed. Transforming a non-normal distribution into a normal distribution is performed in a number of different ways depending on the original distribution of data, but a common technique is to take the log of the data. In practice, there are infinite possible ways to transform data, although there are some approaches that are much more common than others. In SPSS Statistics transform data for "square", "square root", "reflect and square root", "reflect and log", "reciprocal", "reflect and inverse" and "log" transformations.

Bootstrapping

Learning Objectives

This chapter helps to understand the following

- Meaning of Bootstrapping
- Procedures of Bootstrapping

Introduction

In SPSS, Bootstrapping is an efficient way to ensure that analytical models are reliable and will produce accurate results. It can be used to test the stability of analytical models and procedures found throughout the practice and usage of SPSS Statistics including descriptive, means, crosstabs, correlations, regression and in other techniques. Bootstrapping also enables the following features, such as

1. Quickly and easily estimate the sampling distribution of an estimator by re-sampling with replacement from the original sample.
2. Create thousands of alternate versions of a data set for a more accurate view of what is likely to exist in the population.
3. Reduce the impact of outliers and anomalies, helping to ensure the stability and reliability of your models.
4. Estimate the standard errors and confidence intervals of a population parameter such as the mean, median, proportion, odds ratio, correlation coefficient, regression coefficient and more.

Procedures of Bootstrapping

The following procedures hold up bootstrapping i.e. Bootstrapping does not work with multiply imputed datasets. If there is an Imputation_ variable in the dataset, the Bootstrap dialog is disabled. Bootstrapping uses listwise deletion to determine the case basis; that is, cases with missing values on any of the analysis variables are deleted from the analysis, so when bootstrapping is in effect, listwise deletion is in effect even if the analysis procedure specifies another form of missing value handling. The statistic base options which help with booststrap estimated are explained below

- In Frequency and Descriptive Table, bootstrap estimates supports for percent, mean, standard deviation, variance, median, skewness, kurtosis, and percentiles in statistics table as same as in descriptive statistics.
- In Explores option, The Descriptives table supports bootstrap estimates for the mean, 5% Trimmed Mean, standard deviation, variance, median, skewness, kurtosis, and interquartile range. As well the M-Estimators table supports bootstrap estimates for Huber's M-Estimator, Tukey's Biweight, Hampel's M-Estimator, and Andrew's Wave. Also the Percentiles table supports bootstrap estimates for percentiles.
- In Crosstabs, the bootstrap estimates reflects in directional measures, symmetric measures, risk estimates, Mantel-Haenszel common odds ratio.
    1. The Directional Measures table supports bootstrap estimates for Lambda, Goodman and Kruskal Tau, Uncertainty Coefficient, and Somers'd.
    2. The Symmetric Measures table supports bootstrap estimates for Phi, Cramer's V, Contingency Coefficient, Kendall's tau-b, Kendall's tau-c, Gamma, Spearman Correlation, and Pearson's R.
    3. The Risk Estimate table supports bootstrap estimates for the odds ratio.
    4. The Mantel-Haenszel Common Odds Ratio table supports bootstrap estimates and significance tests for ln(Estimate).
- In Mean, The Report table supports bootstrap estimates for the mean, median, grouped median, standard deviation, variance, kurtosis, skewness, harmonic mean, and geometric mean.
- The T-Test table supports bootstrap estimates by testing the significant difference of the mean. In one-sample t test, independent sample t test and paired sample t test table bootstrap estimates supports by mean, standard deviation and correlation.
- In One Way ANOVA, descriptive statistics table, multiple comparisons table and contrast test table supports bootstrap estimates by mean, standard deviation, mean differences and significance tests for value of contrast.
- In GLM Univariate, Descriptive Statistics table, Parameter estimates table, contrast table, estimates marginal mean - estimate table and estimates marginal mean pairwise comparison table, post hoc test supports bootstrap estimates by the way of Mean, standard deviation,

significance test for the coefficient,B, Significant test differences, mean differences etc.,.

- In Bivariate and Partial Correlation, the Descriptive Statistics table and correlation table supports bootstrap estimates for the mean standard deviation and by significance tests for correlations.
- In Linear Regression, The Descriptive Statistics table, Correlations table, Model Summary table, Coefficients table, Correlation Coefficients table, Residuals Statistics table supports bootstrap estimates for the mean, standard deviation, correlations, Durbin-Watson, significance tests for the coefficient, B, correlations.
- In Ordinal Regression, Parameter Estimates table supports bootstrap estimates by significance tests for the coefficient, B.
- In Discriminant Analysis, Standardized Canonical Discriminant Function Coefficients table supports bootstrap estimates for standardized coefficients; Canonical Discriminant Function Coefficients table supports bootstrap estimates for unstandardized coefficients; Classification Function Coefficients table supports bootstrap estimates for coefficients.
- In GLM Multivariate, the Parameter Estimates table supports bootstrap estimates by significance tests for the coefficient, B.
- In Linear Mixed Models, The Estimates of Fixed Effects table and Estimates of Covariance Parameters table supports bootstrap estimates by significance tests for the estimate.
- In Generalized Linear Model, The Parameter Estimates table supports bootstrap estimates by significance tests for the coefficient, B.
- In Cox Regression, The Variables in the Equation table supports bootstrap estimates by significance tests for the coefficient, B.
- In Binary Logistic Regression, The Variables in the Equation table supports bootstrap estimates by significance tests for the coefficient, B.
- In Multinomial Logistic Regression, The Parameter Estimates table supports bootstrap estimates and significance tests for the coefficient, B.

Homoscedasticity

Learning Objectives

This chapter helps to understand the following

- Meaning of Homoscedasticity
- Evaluate Homoscedasticity

Introduction

Homoscedasticity refers to the assumption that that the dependent variable exhibits similar amounts of variance across the range of values for an independent variable. It is simply stated as as "a property of a set of random variables where each variable has the same finite variance". The Homoscedasticity is also referred as homogeneity of variance. The assumption of homoscedasticity simplifies mathematicalandcomputationaltreatment.Seriousviolationsinhomoscedasticity (assuming a distribution of data is homoscedastic when in actuality it is heteroscedastic may result in overestimating the goodness of fit as measured by the Pearson coefficient). It is central to linear regression models which describes the circumstances in which the error term is same across all variables i.e. random disturbance in the relationship between the independent variables and the dependent variable.

Heteroscedasticity which is simply described as the violation of homoscedasticity presents when the size of the error term differs across values of an independent variable. The impact of violating the assumption of homoscedasticity is a matter of degree, increasing as heteroscedasticity increases. While it applies to independent variables at all three measurement levels, i.e. the methods used for evaluation of homoscedasticity requires that the independent variable be non-metric (nominal or ordinal) and the dependent variable be metric (ordinal or interval). When both variables are metric, the assumption is evaluated as part of the residual analysis in multiple regression.

Evaluating Homoscedasticity

1. Homoscedasticity is evaluated for pairs of variables.
2. There are both graphical and statistical methods for evaluating homoscedasticity.
3. The graphical method is called a boxplot.
4. The statistical method is the Levene statistic which SPSS computes for the test of homogeneity of variances.
5. Neither of the methods is absolutely definitive.

Introduction to IBM SPSS – AMOS

Learning Objectives

This chapter helps to understand the following

- Meaning of AMOS
- Steps to Run AMOS Graphics
- Usage of Drawing Tools in AMOS Graphics

Introduction

IBM SPSS AMOS implements the general approach to data analysis known as **structural equation modeling** (SEM), also called as **analysis of covariance structures**, or **causal modeling**. This approach includes, as special cases, many well-known conventional techniques, including the general linear model and common factor analysis. IBM SPSS Amos (Analysis of Moment Structures) is an easy-to-use program for visual SEM. With Amos, the user can quickly specify, view, and modify model graphically using simple drawing tools. Then it helps to assess model's fit, make any modifications, and print out a publication-quality graphic of your final model. Simply specify the model graphically (left). Amos quickly performs the computations and displays the results (right).

IBM SPSS Amos was originally designed as a tool for teaching this powerful and fundamentally simple method. For this reason, every effort was made to see that it is easy to use. Amos integrates an easy-to-use graphical interface with an advanced computing engine for SEM. The publication-quality path diagrams of Amos provide a clear representation of models for students and fellow researchers. The numeric methods implemented in Amos are among the most effective and reliable available.

Running AMOS Graphics

To initiate AMOS Graphics, first follow the usual Windows procedure as given below **Start > Programs >AMOS (Version) > AMOS Graphics (or) In the SPSS Data Set > Select Analyse > Click on IBM SPSS AMOS**

A dialogue box will appear as given below in Figure -1, the large area on the right is to draw path diagrams. The toolbar on the left provides one-click access to the most frequently used buttons, either the toolbar or menu commands can be used for most operations.

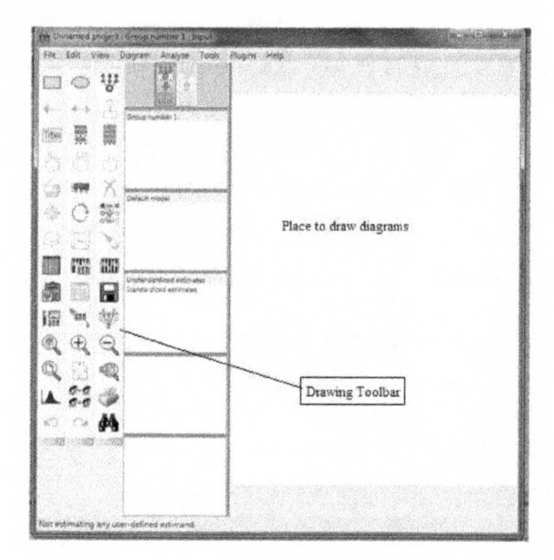

Figure – 1

Opening the data file

The next step is to open the file that contains the data. This package supports excel as well SPSS dataset. If the user launch Amos from the Analyze menu in SPSS Statistics, Amos automatically uses the file that is open in SPSS Statistics.

- From the menus, choose File → Data Files.
- In the Data Files dialog box, click File Name.
- In the Data Files dialog box, click OK.

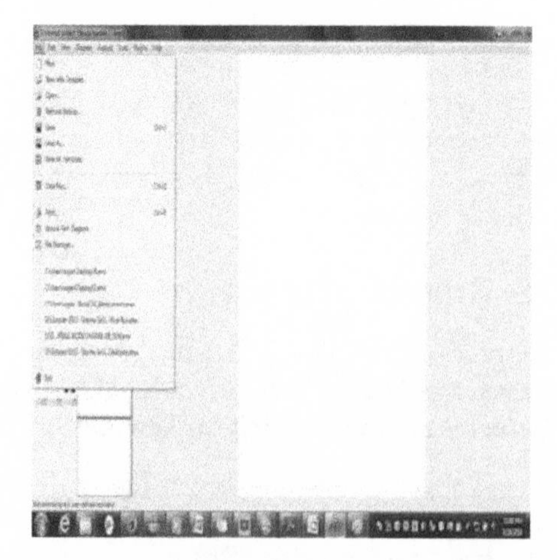

Figure – 2

Drawing Tool Icons and Its Purpose

Drawing Tool Icon	Purpose
	Rectangle icon used to draws observed (measured) variables
	Oval icon used to draws unobserved (latent, unmeasured) variables
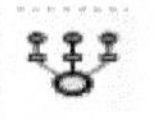	Indicator icon helps to draws a latent variable or adds an indicator variable
	Path icon used to draws a regression path
	Covariance icon helps to draws covariances
	Error icon helps to adds an error/uniqueness variable to an existing observed variable
 Title	Title icon helps to adds figure caption to path diagram
	Variable list (i) icon shows the lists of variables in the model
	Variable list (ii) icon shows the lists of variables in the data set
	Single selection icon helps to selects one object at a time
	Multiple selection icon helps to selects all objects
	Multiple deselection icon helps to deselects all objects
	Duplicate icon help to make or take multiple copies of selected object(s)
	Move icon enables to move selected object(s) to an alternate location
	Erase icon helps to delete selected object(s)
	Shape change icon helps to alter shape of selected object(s)

Drawing Tool Icon	Purpose
	Rotate icon helps to change orientation of indicator variables
	Reflect icon helps to reverse direction of indicator variables
	Move parameter icon enables to move parameter values to alternate location
	Scroll icon helps to reposition path diagram to another part of the screen
	Touch-up icon enables rearrangement of arrows in path diagram
	Data file icon helps to select and read data file(s)
	Analysis properties icon helps to do request for additional calculations
	Calculate estimates icon enables to calculate default and/ or requested Estimates
	Clipboard icon helps to copy path diagram to windows clipboard
	Text output icon helps to view output in textual format
	Save diagram icon helps to save the current path diagram
	Object properties icon defines properties of variables
	Drag properties icon helps to transfer selected properties of an object to one Or more target objects
	Preserve symmetry icon helps to maintain proper spacing among a selected Group of objects
	Zoom select icon magnifies selected portion of a path diagram
	Zoom-in icon helps to view smaller area of path diagram
	Zoom-out icon helps to view larger area of path diagram

Drawing Tool Icon	Purpose
	Zoom page icon shows entire page on the screen
	Fit-to-page icon helps to resize path diagram to fit within page boundary
	Loupe icon examines path diagram with a loupe (magnifying glass)
	Bayesian icon enables to analyse based on bayesian statistics
	Multiple group icon enables analyses of multiple groups
	Print icon prints selected path diagram
	Undo (i) icon helps to undo previous change
	Undo (ii) icon helps to undo previous undo
	Specification search enables modeling based on a specification Search
	Deselect the output in graphic
	To view graphic output

Specifying the Model and Drawing Variables

The next step is to draw the variables in model. First, draw three rectangles to represent the observed variables, and then draw an ellipse to represent the unobserved variable.

- From the menus, choose Diagram → Draw Observed .
- In the drawing area, move mouse pointer to draw rectangle to appear. Click and drag to draw the rectangle. Exact size or placement of the rectangle can change later.
- From the menus, choose Diagram → Draw Unobserved

Naming the Variables

- In the drawing area, right-click the top left rectangle  and choose Object Properties from the pop-up menu.

- Click the Text tab. In the Variable name text box, type Variable name, Example Customer Satisfaction.
- Use the same method to name the remaining variables. Then close the Object Properties dialog box.

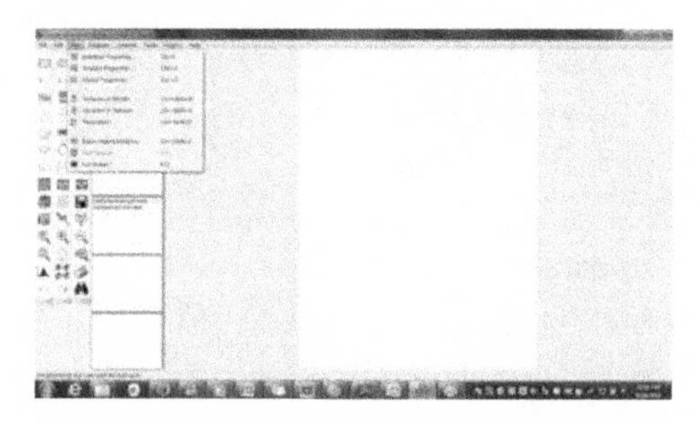

Figure – 3

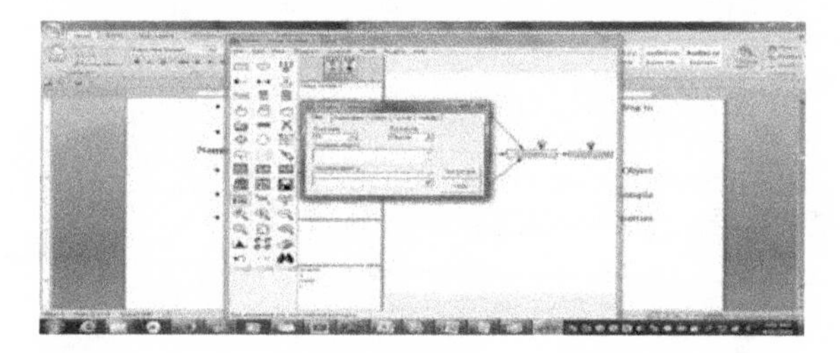

Figure – 4

Drawing latent factor with three indicator variables and their associated error terms

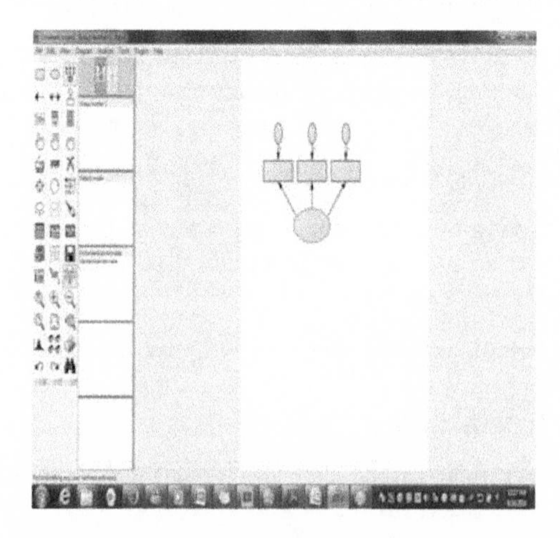

Rotating the Latent Factor Using Icon

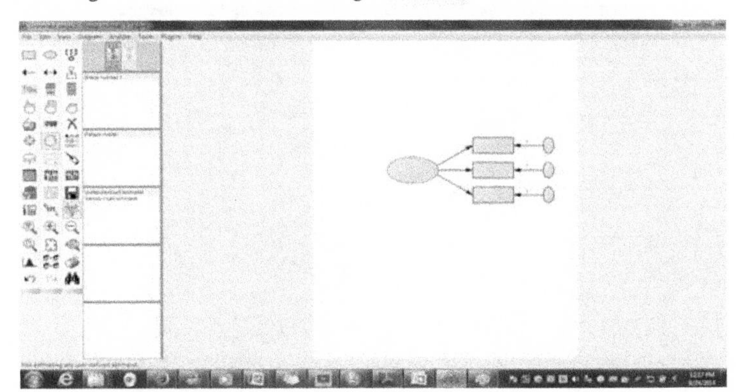

Moving the latent Factor using moving icon

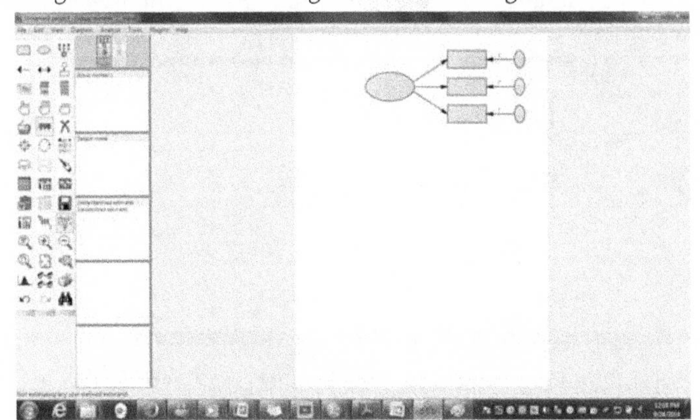

Duplicating the latent Factor using duplicate icon

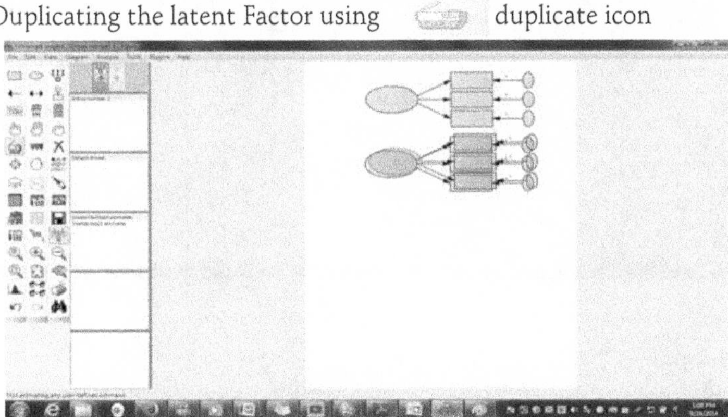

Five latent factors has been created using the Duplicate icon

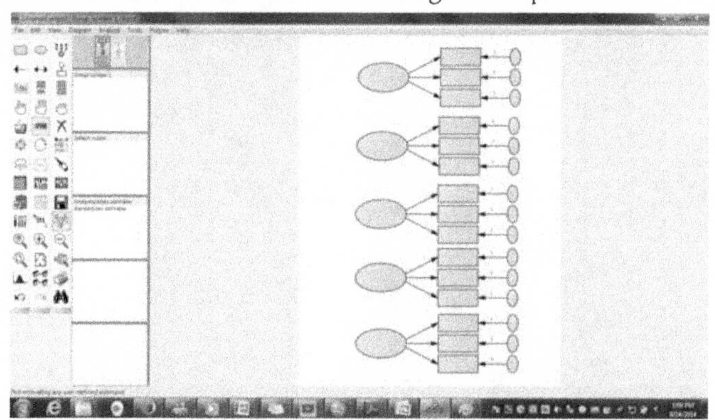

Drawing Covariance between the Latent Factor Model Using ⟷ covariance icon

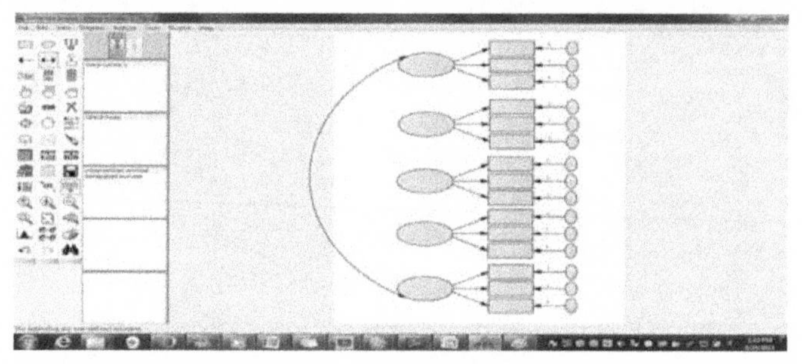

Likewise, Drawing Covariance for Connecting with all the latent factor using Covariance icon

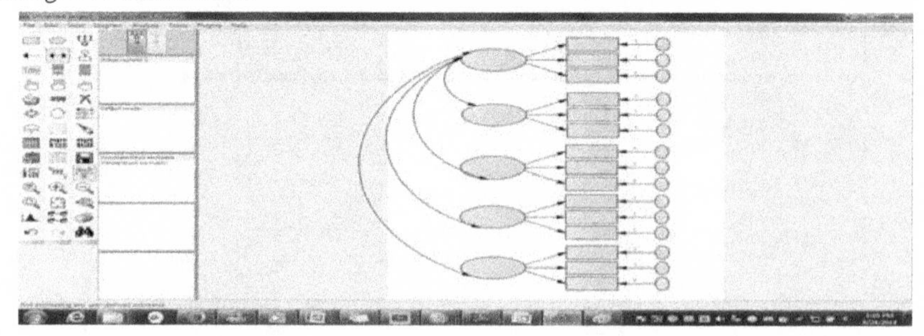

Using Touch-up icon, the latent factor model was arranged in a perfect manner

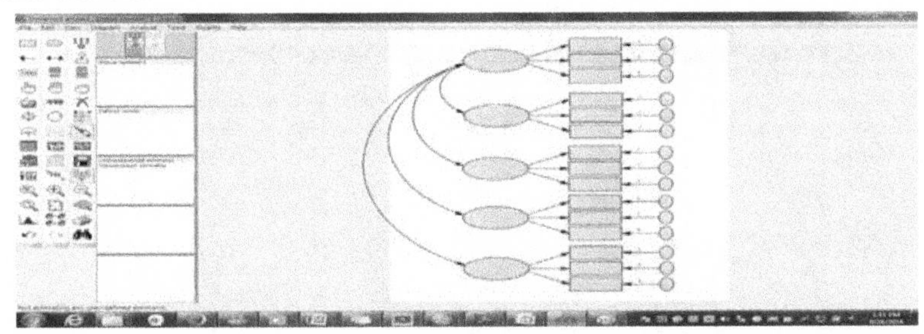

Using Single Selection Icon , Selecting the oval Shapes in the latent factor model.

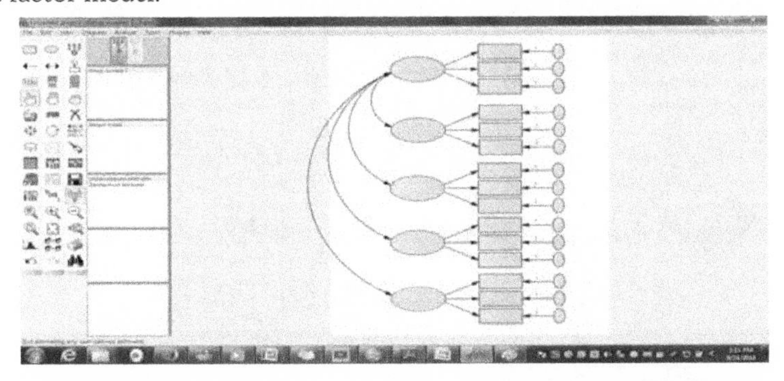

Goto Plugins in Amos Graphics toolbar > click on Drawing Covariance in a easy manner using

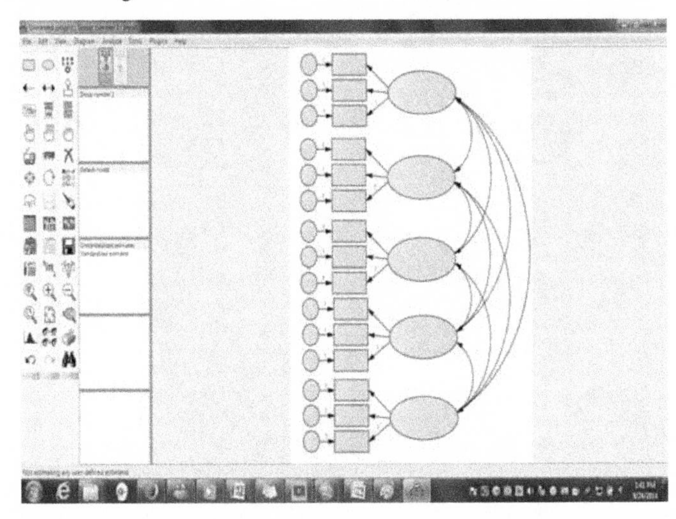

Second order CFA Specimem, Using error icon, Creating residual errors for each latent factors and connecting it with the unobserved variable using oval icon.

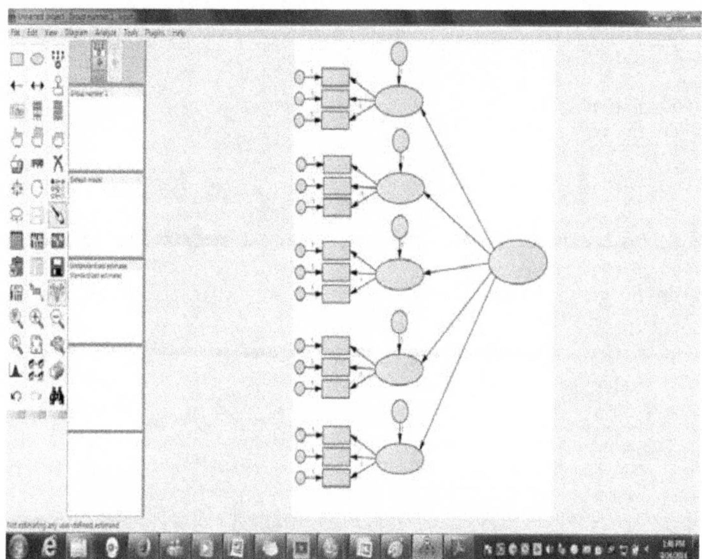

Specimen Path Diagram Model

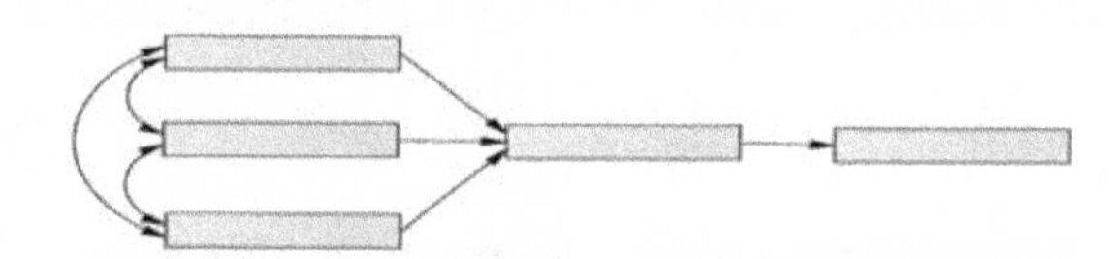

Specimen Structural Equation Modeling with AMOS

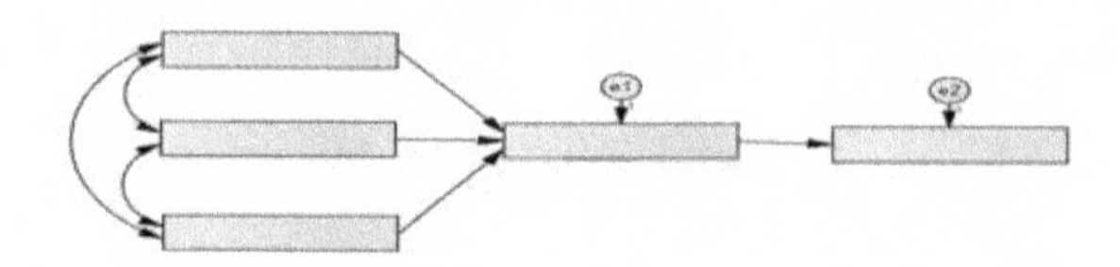

Chapter - 3

Multivariate Analysis of Variance (MANOVA)

Learning Objectives

This chapter helps to understand the following

- Introduction to MANOVA
- Needs, Advantage and Disadvantage of MANOVA
- Assumptions of MANOVA
- Procedure to Run MANOVA using SPSS
- Results and Discussion of MANOVA Output

Introduction

Multivariate analysis of variance (MANOVA) is simply an ANOVA with several dependent variables. That is to say, ANOVA tests for the difference in means between two or more groups, while MANOVA tests for the difference in two or more vectors of means. The general purpose of multivariate analysis of variance (MANOVA) is to determine whether multiple levels of independent variables on their own or in combination with one another have an effect on the dependent variables. MANOVA requires that the dependent variables meet parametric requirements.

Although we mention Wilks' λ here, there are also other statistics that may be used, including Hotelling's trace and Pillai's criterion. The "covariance" here is included because the two measures are probably correlated and we must take this correlation into account when performing the significance test. Testing the multiple dependent variables is accomplished by creating new dependent variables that maximize group differences. These artificial dependent variables are linear combinations of the measured dependent variables.

When do you need MANOVA?

MANOVA is used under the same circumstances as ANOVA but when there are multiple dependent variables as well as independent variables within the model which the researcher wishes to test. MANOVA is also considered a valid alternative to the repeated measures ANOVA when sphericity is violated.

Multivariate analyses of variance (MANOVA) differs from univariate analyses of variance (ANOVA) in the number of dependent variables utilized. The major purpose of univariate analyses of variance is to determine the effects of one or more independent variables upon ONE dependent variable. However, there arise situations in which measurements are available on MORE THAN ONE dependent variable.

For example, say you are interested in designing an experiment to test the effects of success of street kids exiting street life. "Success" can be measured in more than one way.

It can be measured by both 'the number of days a kid "sleeps" at home' or 'the number of days a kid returns to school'. In such a case, there are TWO dependent variables involved in the analysis.

In analyzing such data, one obvious (and popular) approach is to perform an ANOVA on each dependent variable separately. In our example above, two ANOVAs could be carried out (i.e., one ANOVA with "home" scores Dr. Robert Gebotys January 2000 and one ANOVA with "school" scores). However, it is quite apparent that both the home and school variables are highly correlated. Thus, the two analyses are not independent. One's 'return to school' is most likely positively correlated with 'one's return to home'. The independent ANOVA ignores the interrelation between variables. Consequently, substantial information may be lost (SPSS, Release 9.0), and the resultant p values for tests of hypothesis for the two independent ANOVA's are incorrect.

The extension of univariate analysis of variance to the case of multiple dependent variables is known as Multivariate Analysis of Variance (MANOVA). MANOVA allows for a direct test of the null hypothesis with respect to ALL the dependent variables in an experiment. In MANOVA, a linear function (y) of the dependent variables in the analysis is constructed, so that "inter-group differences" on y are maximized. The composite variable y is then treated in a manner somewhat similar to the dependent variable in a univariate ANOVA, with the null hypothesis being accepted or rejected accordingly.

Essentially, the multivariate technique known as Discriminant analysis is quite similar to MANOVA. In discriminant analysis we consider the problem of finding the best linear combination of variables for distinguishing several groups.

The coefficients (Betas) are chosen in order that the ratio of the 'between-groups sums of squares' to the 'total sums of squares' is as large as possible. MANOVA can be viewed as a problem of first finding linear combinations of the dependent variables that 'best separate the groups' and then testing 'whether these new variables are significantly different for the groups'. In the MANOVA situation, we already know which categories the cases belong to, and thus are not interested in classification BUT in identifying a composite variable (y) which illuminates the differences among the groups.

Comparison to the Univariate

- Analysis of Variance allows for the investigation of the effects of a categorical variable on a continuous outcome
- We can also look at multiple grouping variables, their interaction, and control for the effects of exogenous factors (Ancova)
- Just as Anova and Ancova are special cases of regression, Manova and Mancova are special cases of canonical correlation

Multivariate Analysis of Variance

- MANOVA is an extension of ANOVA in which main effects and interactions are assessed on a linear combination of DVs
- It tests whether there are statistically significant mean differences among groups on a combination of DVs

Thus in a nutshell, Multivariate analysis of variance (MANOVA) is a statistical test procedure for comparing multivariate (population) means of several groups. Unlike ANOVA, it uses the variance-covariance between variables in testing the statistical significance of the mean differences.

It is a generalized form of univariate analysis of variance (ANOVA). It is used when there are two or more dependent variables. It helps to answer the following questions:-

1. do changes in the independent variable(s) have significant effects on the dependent variables;
2. what are the interactions among the dependent variables and
3. what are the interactions among the independent variables. Statistical reports however will provide individual p-values for each dependent variable, indicating whether differences and interactions are statistically significant.

What a multivariate analysis of variance does???

Like an ANOVA, MANOVA examines the degree of variance within the independent variables and determines whether it is smaller than the degree of variance between the independent variables. If the within subjects variance is smaller than the between subjects variance it means the independent variable has had a significant effect on the dependent variables.

The main differences between MANOVAs and ANOVAs:- Multivariate analysis of variance (MANOVA), and analysis of variance (ANOVA) tests are statistical methods for analyzing the difference in means between variables. The MANOVA and ANOVA tests are similar in nature to one another, because they work on the same assumptions; however, there are some key advantages to using a MANOVA over an ANOVA test.

1. The first is that MANOVAs are able to take into account multiple independent and multiple dependent variables within the same model, permitting greater complexity.

2. Secondly, rather than using the F value as the indicator of significance a number of multivariate measures (Wilks' lambda, Pillai's trace, Hotelling trace and Roy's largest root) are used.

3. The MANOVA can measure multiple dependent variables, while the ANOVA only allows for one. The ability to measure the effects of an independent variable on multiple dependent variables is useful for comparing the effect of the independent variable in different settings. You would need to run multiple ANOVA tests to measure the same number of things that one MANOVA does.

Simultaneous Testing

Because the MANOVA tests multiple dependent variables at once, you're testing the effects of the independent variables simultaneously. Running multiple ANOVA tests on each variable not only takes more time, but increases the risk of type I statistical errors. A type I error occurs when a statistical test rejects a null hypothesis when it is true. For example, if your null hypothesis is "students who study have higher test scores than students who don't study," then a type one error would cause your results to reject that statement, even though your data actually supported it.

Multiple Dependent Variables

MANOVA deals with the multiple dependent variables by combining them in a linear manner to produce a combination which best separates the independent variable groups. An ANOVA is then performed on the newly developed dependent variable.

In MANOVAs the independent variables relevant to each main effect are weighted to give them priority in the calculations performed. In interactions the independent variables are equally weighted to determine whether or not they have an additive effect in terms of the combined variance they account for in the dependent variable/s.

The main effects of the independent variables and of the interactions are examined with "all else held constant". The effect of each of the independent variables is tested separately. Any multiple interactions are tested separately from one another and from any significant main effects. Assuming there are equal sample sizes both in the main effects and the interactions, each test performed will be independent of the next or previous calculation (except for the error term which is calculated across the independent variables).

Finding Effect

The MANOVA also increases your chance of finding an effect that an independent variable has. When you're measuring the independent variable's effect on multiple dependent variables, you may find that there is a significant influence on one of the dependent variables, but not the others. Using an ANOVA, you would have only been testing one of the dependent variables.

Disadvantages

Although MANOVA tests have significant advantages over the ANOVA, there are also some key disadvantages. The test is more complex to run than a single ANOVA, and your results can be more ambiguous. For example, if you find that an independent variable affects multiple dependent variables, you can't tell for certain whether or not it truly was the independent variable, or the multiple dependent variables having an affect on each other. Because ANOVA tests only have one dependent variable, the results are clearer.

There are two aspects of MANOVAs which are left to researchers:

1. Firstly, they decide which variables are placed in the MANOVA. Variables are included in order to address a particular research question or hypothesis, and the best combination of dependent variables is one in which they are not correlated with one another, as explained above.

2. Second, the researcher has to interpret a significant result. A statistical main effect of an independent variable implies that the independent variable groups are significantly different in terms of their scores on the dependent variable. (But this does not establish that the independent variable has caused the changes in the dependent variable. In a study which was poorly designed, differences in dependent variable scores may be the result of extraneous, uncontrolled or confounding variables.)

To tease out higher level interactions in MANOVA, smaller ANOVA models which include only the independent variables which were significant can be used in separate analyses and followed by post hoc tests. Post hoc and preplanned comparisons compare all the possible paired combinations of the independent variable groups e.g. for three ethnic groups of white, African and Asian the comparisons would be: white v African, white v Asian, African v Asian. The most frequently used preplanned and post hoc tests are Least Squares Difference (LSD), Scheffe, Bonferroni, and Tukey. The tests will give the mean difference between each group and a p value to indicate whether the two groups differ significantly.

MANOVA PROCEDURES

- Find a set of weights for the dependent variables and the predictor variables that maximises the correlation between the two.
- MANOVA F tests – Wilk's Lambda – Pillai's Trace
  – Hotellings
  – Roy's Largest Root

1. simultaneously explore relationship between several categorical independent variables and two or more metric variables
2. Brewer(1996) – supervisor & subordinate gender affects the nature of performance feedback toward subordinates (2x2x2 matrix)

ANOVA

- One-way ANOVA tests the equality of group means
  - Assumptions: independent observations; normality; homogeneity of variance
- Two-way ANOVA tests three hypotheses simultaneously:
  - Test the interaction of the levels of the two independent variables
    - Interaction occurs when the effects of one factor depends on the different levels of the second factor
  - Test the two independent variable separately
  - Interaction is significant means the two IVs in combination result in a significant effect on the DV, thus, it does not make sense to interpret the main effects.
  - Assumptions: the same as One-way ANOVA
  - Example: the impact of gender (sex) and age (agecat4) on income (rincome_2)

- Explore (omitted)
    - Analysis>GLM>univariate
    - Click model>click Full factorial>Cont.
    - Click Options>Click Descriptive Stat; Estimates of effect size; Homogeneity test
    - Click Post Hoc>click LSD; Bonferroni; Scheffe; Cont.
    - Click Plots>put one IV into Horizontal and the other into Separate line

ANCOVA

- Idea: the difference on a DV often does not just depend on one or two IVs, it may depend on other measurement variables. ANCOVA takes into account of such dependency.
    - i.e. it removes the effect of one or more covariates
- Assumptions: in addition to the regular ANOVA assumptions, we need:
    - Linear relationship between DV and covariates
    - The slope for the regression line is the same for each group
    - The covariates are reliable and is measured without error
    - Homogeneity of slopes = homogeneity of regression = there is interaction between IVs and the covariate
        - If the interaction between covariate and IVs are significant, ANCOVA should not be conducted
- Example: determine if hours worked per week (hrs2) is different by gender (sex) and for those satisfied or dissatisfied with their job (satjob2), after adjusted to their income (or equalized to their income)
    - Analysis>GLM>Univariate
        - Move hrs2 into DV box; move sex and satjob2 into Fixed Factor box; move rincome_2 into Covariate box
    - Click at Model>Custom
        - Highlight all variables and move it to the Model box
        - Make sure the Interaction option is selected
    - Click at Option
        - Move sex and satjob2 into Display Means box

- Click Descriptive Stat.; Estimates of effect size; and Homogeneity tests
- This tests the homogeneity of regression slopes
    - If there is no interaction found by the previous step, then repeat the previous step except click at Model>Factorial instead of Model>Custom

MANOVA

- Characteristics
    - Similar to ANOVA
    - Multiple DVs
    - The DVs are correlated and linear combination makes sense
    - It tests whether mean differences among k groups on a combination of DVs are likely to have occurred by chance
    - The idea of MANOVA is to find a linear combination that separates the groups 'optimally', and perform ANOVA on the linear combination

Advantages

- The chance of discovering what actually changed as a result of the different treatment increases
- May reveal differences not shown in separate ANOVAs
- Without inflation of type one error
- The use of multiple ANOVAs ignores some very important info (the fact that the DVs are correlated)

Disadvantages

- More complicated
- ANOVA is often more powerful

Assumptions:

- Independent random samples
- Multivariate normal distribution in each group
- Homogeneity of covariance matrix
- Linear relationship among DVs

- Steps to carry out MANOVA
    - Check for assumptions
    - If MANOVA is not significant, stop
    - If MANOVA is significant, carry out univariate ANOVA
    - If univariate ANOVA is significant, do Post Hoc
- If homoscedasticity, use Wilks Lambda, if not, use Pillai's Trace. In general, all 4 statistics should be similar.
- Example: An experiment looking at the memory effects of different instructions: 3 groups of human subjects learned nonsense syllables as they were presented and were administered two memory tests: recall and recognition. The first group of subjects was instructed to like or dislike the syllables as they were presented (to generate affect). A second group was instructed that they will be tested (induce anxiety?). The 3rd group was told to count the syllable as they were presented (interference). The objective is to access group differences in memory
- How to do it?
    - File>Open Data
        - Open the file As9.por in Instruct>Zhang Multivariate Short Course folder
    - Analyze>GLM>Multivariate
        - Move recall and recog into Dependent Variable box; move group into Fixed Factors box
        - Click at Options; move group into Display means box (this will display the marginal means predicted by the model, these means may be different than the observed means if there are covariates or the model is not factorial); Compare main effect box is for testing the every pair of the estimated marginal means for the selected factors.
        - Click at Estimates of effect size and Homogeneity of variance
- Push buttons:
    - Plots: create a profile plot for each DV displaying group means
    - Post Hoc: Post Hoc tests for marginal means
    - Save: save predicted values, etc.
    - Contrast: perform planned comparisons

        - Model: specify the model
        - Options:
            - Display Means for: display the estimated means predicted by the model
            - Compare main effects: test for significant difference between every pair of estimated marginal means for each of the main effects
        - Observed power: produce a statistical power analysis for your study
        - Parameter estimate: check this when you need a predictive model
        - Spread vs. level plot: visual display of homogeneity of variance
    - Example 2: Check for the impact of job satisfaction (satjob) and gender (sex) on income (rincome_2) and education (educ) (in gssft.sav)
        - Screen data: transform educ to educ2 to eliminate cases with '6 or less'
        - Check for assumptions: explore
        - MANOVA
- MANCOVA
- Objective: Test for mean differences among groups for a linear combination of DVs after adjusted for the covariate.
- Example: to test if there are differences in productivity (measured by income and hours worked) for individuals in different age groups after adjusted for the education level
- Assumptions: similar to ANCOVA
- SPSS how to:
    - Analysis>GLM>Multivariate
        - Move rincome_2 and educ2 to DV box; move sex and satjob into IV box; move age to Covariate box
        - Check for homogeneity of regression
            - Click at Model>Custom; Highlight all variables and move them to Model box
        - If the covariate-IVs interaction is not significant, repeat the process but select the Full under model

- Repeated Measure analysis
- Objective: test for significant differences in means when the same observation appears in multiple levels of a factor
- Examples of repeated measure studies:
    - Marketing – compare customer's ratings on 4 different brands
    - Medicine – compare test results before, immediately after, and six months after a procedure
    - Education – compare performance test scores before and after an intervention program
- The logic of repeated measure: SPSS performs repeated measure ANOVA by computing contrasts (differences) across the repeated measures factor's levels for each subject, then testing if the means of the contrasts are significantly different from 0; any between subject tests are based on the means of the subjects.
- Assumptions:
    - Independent observations
    - Normality
    - Homogeneity of variances
    - Sphericity: if two or more contrasts are to be pooled (the test of main effect is based on this pooling), then the contrasts should be equally weighted and uncorrelated (equal variances and uncorrelated contrasts); this assumption is equivalent to the covariance matrix is diagonal and the diagonal elements are the same)
- Example 1: A study in which 5 subjects were tested in each of 4 drug conditions
- Open data file:
    - File>Open...Data; select Repmeas1.por
- SPSS repeated measure procedure:
    - Analyze>GLM>Repeated Measure
        - Within-Subject Factor Name (the name of the repeated measure factor): a repeated measure factor is expressed as a set of variables
            - Replace factor1 with Drug

        - Number of levels: the number of repeated measurements
            - Type 4
    - The Measure pushbutton for two functions
        - For multiple dependent measures (e.g. we recorded 4 measures of physiological stress under each of the drug conditions)
        - To label the factor levels
            - Click Measure; type memory in Measure name box; click add
        - Click Define: here we link the repeated measure factor level to variable names; define between subject factors and covariates
        - Move drug1 – drug 4 to the Within-Subject box
            - Select a variable by the up and down button
        - Model button: by default a complete model
        - Contrast button: specify particular contrasts
        - Plot button: create profile plots that graph factor level estimated marginal means for up to 3 factors at a time
        - Post Hoc: provide Post Hoc tests for between subject factors
        - Save button: allow you to save predicted values, residuals, etc.
        - Options: similar to MANOVA
            - Click Descriptive; click at Transformation Matrix (it provides the contrasts)
- Interpret the results
    - Look at the descriptive statistics
    - Look at the test for Sphericity
        - If Sphericity is significant, use the Multivariate results (test on the contrasts). It tests whether all of the contrast variables are zero in the population
        - If Sphericity is not significant, use the Sphericity Assumed result

    - Look at the tests for within subject contrasts: it tests the linear trend; the quadratic trend...

- It may not make sense in some applications, as in this example (but it makes sense in terms of time and dosage)
    - Transformation matrix provides info on what are linear contrast, etc.
        - The first table is for the average across the repeated measure factor (here they are all .5, it means each variable is weighted equally, normalization requires that the square of the sums equals to 1)
        - The second table defines the corresponding repeated measure factor
            - Linear – increase by a constant, etc.
            - Linear and quadratic is orthogonal, etc.
        - Having concluded there are memory differences due to drug condition, we want to know which condition differ to which others
        - Repeat the analysis, except under Option button, move 'drug' into Display Means, click at Compare Main effects and select Bonferroni adjustment
            - Transformation Coefficients (M Matrix): it shows how the variables are created for comparison. Here, we compare the drug conditions, so the M matrix is an identity matrix
        - Suppose we want to test each adjacent pair of means: drug1 vs. drug2; drug2 vs. drug3; drug3 vs. drug 4:
            - Repeated measure>Define>Contrast>Select Repeated
- Example 2: A marketing experiment was devised to evaluate whether viewing a commercial produces improved ratings for a specific brand. Ratings on 3 brands were obtained from objects before and after viewing the commercial. Since the hope was that the commercial would improve ratings of only one brand (A), researchers expected a significant brand by pre-post commercial interaction. There are two between-subjects factors: sex and brand used by the subject

- SPSS how to:
    - Analyze>GLM>Repeated Measures
        - Replace factor1 with prepost in the Within-Subject Factor box; type 2 in the Number of level box; click add
        - Type brand in the Within-Subject Factor box; type 3 in the Number of level box; click add
        - Click measure; type measure in Measure Name box; click add
        - Note: SPSS expects 2 between-subject factors
        - Click Define button; move the appropriate variable into place; move sex and user into Between-Subject Factor box
        - Click Options button; move sex, user, prepost and brand into the Display means box
        - Click Homogeneity tests and descriptive boxes
        - Click Plot; move user into Horizontal Axis box and brand into Separate Lines box
        - Click continue; OK

One Way Manova in SPSS

Introduction

The one-way multivariate analysis of variance (one-way MANOVA) is used to determine whether there are any differences between independent groups on more than one continuous dependent variable. In this regard, it differs from a one-way ANOVA, which only measures one dependent variable. For example, one-way MANOVA is used to understand whether there were differences in the perceptions of attractiveness and intelligence of drug users in movies (i.e., the two dependent variables are "perceptions of attractiveness" and "perceptions of intelligence", whilst the independent variable is "drug users in movies", which has three independent groups: "non-user", "experimenter" and "regular user"). Alternately, one-way MANOVA is used to understand whether there were differences in students' short-term and long-term recall of facts based on three different lengths of lecture (i.e., the two dependent variables are "short-term memory recall" and "long-term memory recall", whilst the independent variable is "lecture duration", which has four independent groups: "30 minutes", "60 minutes", "90 minutes" and "120 minutes").

It is important to realise that the one-way MANOVA is an omnibus test statistic and cannot notify which specific groups were significantly different from each other; it only report at least two groups were different. While there are three, four, five or more groups in the study design, determining which of these groups differ from each other is important. This can be done by using a post-hoc test.

Assumptions

When you choose to analyse your data using a one-way MANOVA, part of the process involves checking to make sure that the data you want to analyse can actually be analysed using a one-way MANOVA. You need to do this because it is only appropriate to use a one-way MANOVA if your data "passes" nine assumptions that are required for a one-way MANOVA to give you a valid result. Do not be surprised if, when analysing your own data using SPSS, one or more of these assumptions is violated (i.e., is not met). This is not uncommon when working with real-world data. However, even when your data fails certain assumptions, there is often a solution to overcome this.

In practice, checking for these nine assumptions adds some more time to your analysis, requiring you to work through additional procedures in SPSS when performing your analysis, as well as thinking a little bit more about your data. These nine assumptions are presented below:

1. Two or more dependent variables should be measured at the interval or ratio level (i.e., they are continuous).

2. Independent variable should consist of two or more categorical, independent groups. It should have independence of observations, which means that there is no relationship between the observations in each group or between the groups themselves. For example, there must be different participants in each group with no participant being in more than one group. This is more of a study design issue than something you can test for, but it is an important assumption of the one-way MANOVA.

3. Adequate sample size is required. Although the larger your sample size, the better; for MANOVA you need to have more cases in each group than the number of dependent variables you are analysing.

4. There are no univariate or multivariate outliers. First, there can be no (univariate) outliers in each group of the independent variable for any of the dependent variables. This is a similar assumption to the one-way ANOVA, but for each dependent variable that you have in your MANOVA analysis. Univariate outliers are often just called outliers and are the same type of outliers you would have come across if t-tests or ANOVAs were conducted. It also refers to them as univariate in this guide to distinguish them from multivariate outliers. Multivariate outliers are cases which have an unusual combination of scores on the dependent variables. In our enhanced one-way MANOVA guide, we show you how to: (1) detect univariate outliers using boxplots, which you can do using SPSS, and discuss some of the options you have in order to deal with outliers; and (2) check for multivariate outliers using a measure called Mahalanobis distance, which you can also do using SPSS, and discuss what you should do if you have any.

5. There is multivariate normality. Unfortunately, multivariate normality is a particularly tricky assumption to test for and cannot be directly tested in SPSS. Instead, normality of each of the dependent variables for each of the groups of the independent variable is often used in its place as a best 'guess' as to whether there is multivariate normality. It can test for this using the Shapiro-Wilk test of normality, which is easily tested for using SPSS. In addition to showing you how to do this in our enhanced one-way MANOVA guide, we also explain what you can do if your data fails this assumption.

6. There is a linear relationship between each pair of dependent variables for each group of the independent variable. If the variables are not linearly related, the power of the test is reduced. It tests for this assumption by plotting a scatterplot matrix for each group of the independent variable. In order to do this, you will need to split your data file in SPSS before generating the scatterplot matrices.
7. There is a homogeneity of variance-covariance matrices. You can test this assumption in SPSS using Box's M test of equality of covariance. If your data fails this assumption, you may also need to use SPSS to carry out Levene's test of homogeneity of variance to determine where the problem may lie. We show you how to carry out these tests using SPSS in our enhanced one-way MANOVA guide, as well as discuss how to deal with situations where your data fails this assumption.
8. There is no multicollinearity. Ideally, you want your dependent variables to be moderately correlated with each other. If the correlations are low, you might be better off running separate one-way ANOVAs, and if the correlation(s) are too high (greater than 0.9), you could have multicollinearity. This is problematic for MANOVA and needs to be screened out. Whilst there are many different methods to test for this assumption, in our enhanced one-way MANOVA guide, we take you through one of the most straightforward methods using SPSS, and explain what you can do if your data fails this assumption.

- Check assumptions #5, #6, #7, #8 and #9 using SPSS.
- Before doing this, it is necessary to make sure that your data meets assumptions #1, #2, #3 and #4.

In the section, Procedure, we illustrate the SPSS procedure to perform a one-way MANOVA assuming that no assumptions have been violated. First, we set out the example we use to explain the one-way MANOVA procedure in SPSS.

Example

The pupils at a high Food & Grocery [FG2] come from three different primary Food & Grocery [FG2]s. The headteacher wanted to know whether there were academic differences between the pupils from the three different primary Food & Grocery [FG2]s. As such, she randomly selected 20 pupils from Food & Grocery [FG2] A, 20 pupils from Food & Grocery [FG2] B and 20 pupils from Food & Grocery [FG2] C, and measured their academic performance as assessed by the marks they received for their end-of-year English and Maths exams. Therefore, the two dependent variables were "English score" and "Maths score", whilst the independent variable was "Food & Grocery [FG2]", which consisted of three categories: "Food & Grocery [FG2] A", "Food & Grocery [FG2] B" and "Food & Grocery [FG2] C".

Setup in SPSS

In SPSS, we separated the groups for analysis by creating a grouping variable called Food & Grocery [FG2] (i.e., the independent variable), and gave the three categories of the independent variable the labels "Food & Grocery [FG2] A", "Food & Grocery [FG2] B" and "Food & Grocery [FG2] C". The two dependent variables were labelled English_Score and Maths_Score, respectively.

Test Procedure in SPSS

Following are the steps to analyse data using one-way MANOVA in SPSS.

Step - 1: Click Analyze > General Linear Model > Multivariate as shown in the figure -1, the Multivariate dialogue box will appear (as given in the figure - 2).

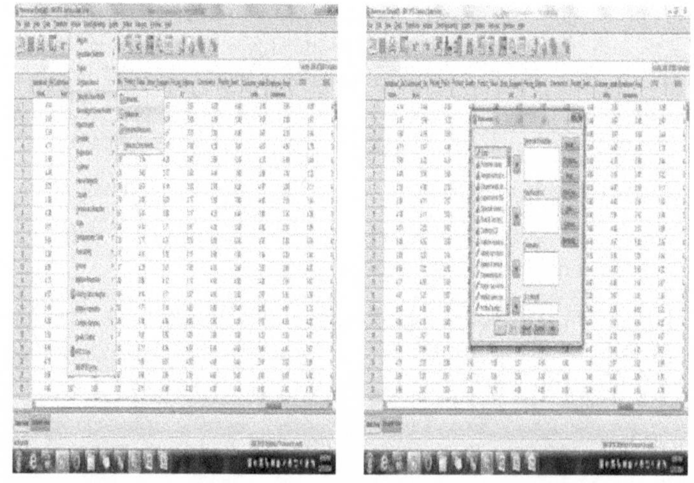

Figure – 1 **Figure – 2**

Step - 2: Transfer the independent variable, Food and Grocery, into the Fixed Factor(s): box and transfer the dependent variables, satisfaction level of organised outlet and satisfaction level of unorganised outlet, into the Dependent Variables: box by drag-and-dropping the variables into their respective boxes or by using the SPSS Right Arrow Button. The result is shown below in figure -3

Note: For this analysis, no need to use the Covariate(s): box (used for MANCOVA) or the WLS Weight: box.

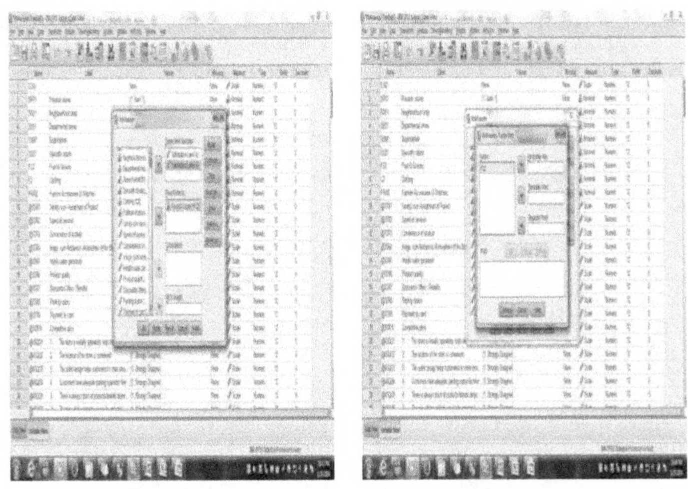

Figure – 3 **Figure – 4**

Step - 3: In Multivariate dialogue box, Click on the SPSS Plots Button, the Multivariate: Profile Plots dialogue box will appear (as given in figure -4)

Step - 4: Transfer the independent variable, Food & Grocery i.e. [FG2], into the Horizontal Axis: box, as shown in figure -5 and Click the SPSS Add Button and that "FG2" has been added to the Plots: box, as shown in figure -6 and Click the Continue button to return to Multivariate dialogue box.

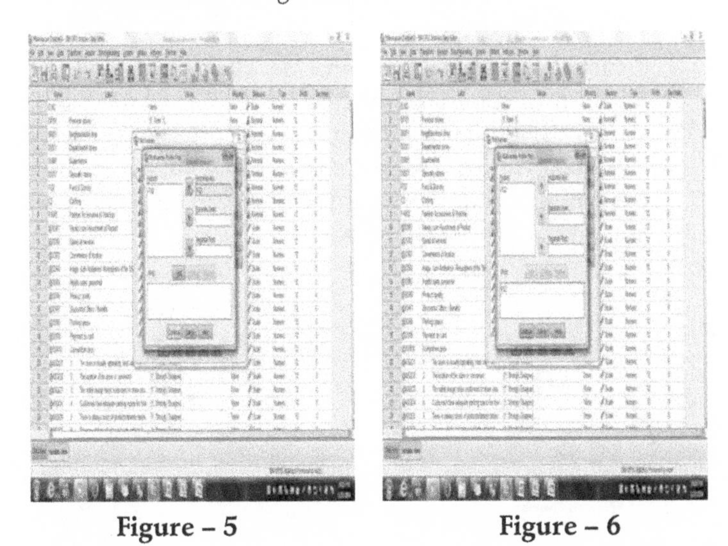

Figure – 5 **Figure – 6**

Step - 5: In Multivariate dialogue box, Click the Posthoc button, the Multivariate: Post Hoc Multiple Comparisons for Observed dialogue box will appear, as shown in figure -7:

Step - 6: Transfer the independent variable, FG2, into the Post Hoc Tests for: box and select the Tukey checkbox in the Equal Variances Assumed- area, as shown in figure -8 and click continue button to return to Multivariate dialogue box.

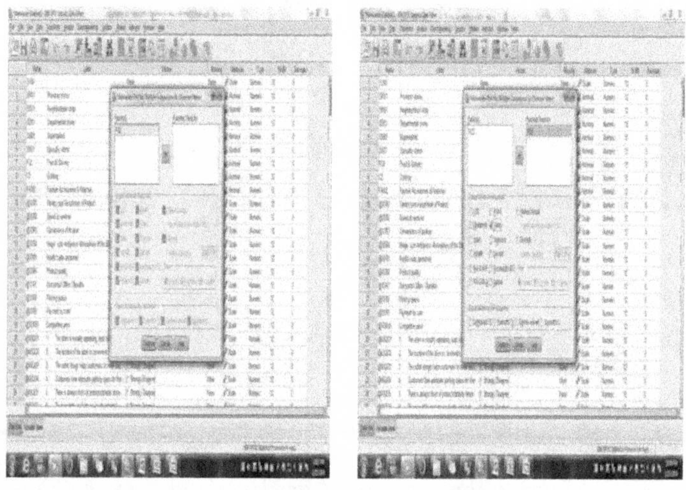

Figure – 7 **Figure – 8**

Step - 7: Click the SPSS Options Button in Multivariate dialogue box, the Multivariate: Options dialogue box, as shown in figure -9:

Step - 8: Transfer the independent variable, "FG2", from the Factor(s) and Factor Interactions: box into the Display Means for: box. Select the Descriptive statistics, Estimates of effect size and Observed power checkboxes in the -Display- area as shown in figure -10 and Click the continue button to return to Multivariate dialogue box.

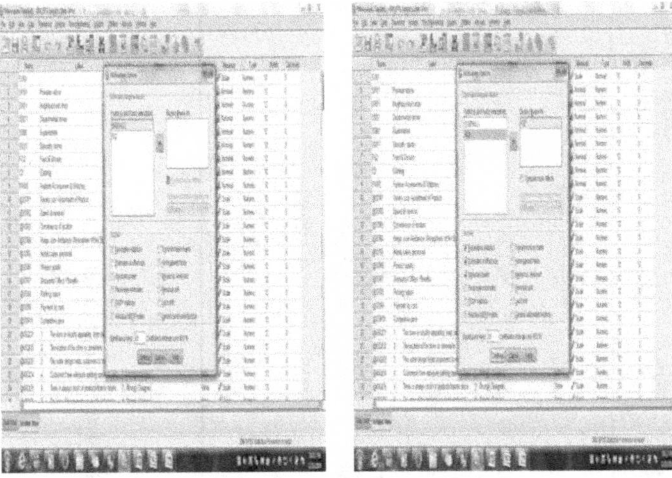

Figure – 9 **Figure – 10**

Step - 9: Click the OK button to generate the output.

SPSS Output of the One-Way MANOVA

Table 1 General Linear Model

Between-Subjects Factors			
		Value Label	**N**
Food & Grocery	1	Neighbourhood Shop	218
	2	Branded Showroom	124
	3	Specialty Store	174

Table 2 Descriptive Statistics

	Food & Grocery	**Mean**	**Std. Deviation**	**N**
Satisfaction Level of Customer with the Organised Retail outlet	Neighbourhood Shop	13.9495	2.34024	218
	Branded Showroom	14.1694	2.06292	124
	Specialty Store	14.5805	1.82844	174
	Total	14.2151	2.12663	516
Satisfaction Level of Customer with the Unorganised Retail outlet	Neighbourhood Shop	12.8899	2.11355	218
	Branded Showroom	13.5726	2.00476	124
	Specialty Store	13.8391	1.80477	174
	Total	13.3740	2.02931	516

Table 3 Multivariate Tests[a]

Effect		Value	F	Hypothesis df	Error df	Sig.	Partial Eta Squared	Noncent. Parameter	Observed Power[d]
Intercept	Pillai's Trace	.984	16114.505[b]	2.000	512.000	.000	.984	32229.011	1.000
	Wilks' Lambda	.016	16114.505[b]	2.000	512.000	.000	.984	32229.011	1.000
	Hotelling's Trace	62.947	16114.505[b]	2.000	512.000	.000	.984	32229.011	1.000
	Roy's Largest Root	62.947	16114.505[b]	2.000	512.000	.000	.984	32229.011	1.000
FG2	Pillai's Trace	.048	6.350	4.000	1026.000	.000	.024	25.401	.990
	Wilks' Lambda	.952	6.403[b]	4.000	1024.000	.000	.024	25.614	.991
	Hotelling's Trace	.051	6.456	4.000	1022.000	.000	.025	25.826	.991
	Roy's Largest Root	.048	12.362[c]	2.000	513.000	.000	.046	24.723	.996

a. Design: Intercept + FG2

b. Exact statistic

c. The statistic is an upper bound on F that yields a lower bound on the significance level.

d. Computed using alpha = .05

Table 4 Tests of Between-Subjects Effects

Source	Dependent Variable	Type III Sum of Squares	df	Mean Square	F	Sig.	Partial Eta Squared	Noncent. Parameter	Observed Power[c]
Corrected Model	Satisfaction Level of Customer with the Organised Retail outlet	38.860[a]	2	19.430	4.352	.013	.017	8.704	.753
	Satisfaction Level of Customer with the Unorganised Retail outlet	93.613[b]	2	46.807	11.845	.000	.044	23.690	.995
Intercept	Satisfaction Level of Customer with the Organised Retail outlet	99095.329	1	99095.329	22196.544	.000	.977	22196.544	1.000
	Satisfaction Level of Customer with the Unorganised Retail outlet	88278.400	1	88278.400	22339.604	.000	.978	22339.604	1.000

Source	Dependent Variable	Type III Sum of Squares	df	Mean Square	F	Sig.	Partial Eta Squared	Noncent. Parameter	Observed Power[c]
FG2	Satisfaction Level of Customer with the Organised Retail outlet	38.860	2	19.430	4.352	.013	.017	8.704	.753
	Satisfaction Level of Customer with the Unorganised Retail outlet	93.613	2	46.807	11.845	.000	.044	23.690	.995
Error	Satisfaction Level of Customer with the Organised Retail outlet	2290.262	513	4.464					
	Satisfaction Level of Customer with the Unorganised Retail outlet	2027.199	513	3.952					
Total	Satisfaction Level of Customer with the Organised Retail outlet	106597.000	516						
	Satisfaction Level of Customer with the Unorganised Retail outlet	94415.000	516						

Source	Dependent Variable	Type III Sum of Squares	df	Mean Square	F	Sig.	Partial Eta Squared	Noncent. Parameter	Observed Power[c]
Corrected Total	Satisfaction Level of Customer with the Organised Retail outlet	2329.122	515						
	Satisfaction Level of Customer with the Unorganised Retail outlet	2120.812	515						

a. R Squared = .017 (Adjusted R Squared = .013) **b.** R Squared = .044 (Adjusted R Squared = .040) **c.** Computed using alpha = .05

Table 5 Estimated Marginal Means

Food & Grocery					
Dependent Variable	**Food & Grocery**	**Mean**	**Std. Error**	**95% Confidence Interval**	
				Lower Bound	**Upper Bound**
Satisfaction Level of Customer with the Organised Retail outlet	Neighbourhood Shop	13.950	.143	13.668	14.231
	Branded Showroom	14.169	.190	13.797	14.542
	Specialty Store	14.580	.160	14.266	14.895
Satisfaction Level of Customer with the Unorganised Retail outlet	Neighbourhood Shop	12.890	.135	12.625	13.154
	Branded Showroom	13.573	.179	13.222	13.923
	Specialty Store	13.839	.151	13.543	14.135

Table 6 Post Hoc Test
Food & Grocery

Multiple Comparisons
Tukey HSD

Dependent Variable	(I) Food & Grocery	(J) Food & Grocery	Mean Difference (I-J)	Std. Error	Sig.	95% Confidence Interval	
						Lower Bound	Upper Bound
Satisfaction Level of Customer with the Organised Retail outlet	Neighbourhood Shop	Branded Showroom	-.2198	.23766	.625	-.7784	.3388
		Specialty Store	-.6309*	.21479	.010	-1.1358	-.1260
	Branded Showroom	Neighbourhood Shop	.2198	.23766	.625	-.3388	.7784
		Specialty Store	-.4111	.24832	.224	-.9948	.1726
	Specialty Store	Neighbourhood Shop	.6309*	.21479	.010	.1260	1.1358
		Branded Showroom	.4111	.24832	.224	-.1726	.9948
Satisfaction Level of Customer with the Unorganised Retail outlet	Neighbourhood Shop	Branded Showroom	-.6827*	.22360	.007	-1.2082	-.1571
		Specialty Store	-.9492*	.20208	.000	-1.4242	-.4742
	Branded Showroom	Neighbourhood Shop	.6827*	.22360	.007	.1571	1.2082
		Specialty Store	-.2665	.23362	.489	-.8156	.2826
	Specialty Store	Neighbourhood Shop	.9492*	.20208	.000	.4742	1.4242
		Branded Showroom	.2665	.23362	.489	-.2826	.8156

Based on observed means.

The error term is Mean Square(Error) = 3.952.

*. The mean difference is significant at the .05 level.

Table 7 Homogeneous Subjects

Satisfaction Level of Customer with the Organised Retail outlet

Tukey HSD

Food & Grocery	**N**	**Subset**	
		1	**2**
Neighbourhood Shop	218	13.9495	
Branded Showroom	124	14.1694	14.1694
Specialty Store	174		14.5805
Sig.		.616	.185

Means for groups in homogeneous subsets are displayed.
Based on observed means.
The error term is Mean Square(Error) = 4.464.
a. Uses Harmonic Mean Sample Size = 163.054.
b. Alpha = .05.

Table -8

Satisfaction Level of Customer with the Unorganised Retail outlet

Tukey HSD

Food & Grocery	**N**	**Subset**	
		1	**2**
Neighbourhood Shop	218	12.8899	
Branded Showroom	124		13.5726
Specialty Store	174		13.8391
Sig.		1.000	.447

Means for groups in homogeneous subsets are displayed.
Based on observed means.
The error term is Mean Square(Error) = 3.952.
a. Uses Harmonic Mean Sample Size = 163.054.
b. Alpha = .05.

Profile Plots

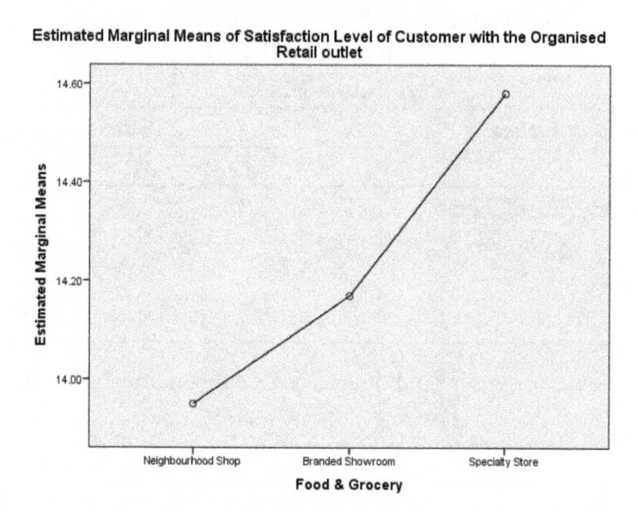

Figure 11 Satisfaction Level of Customer with the Organised Retail outlet

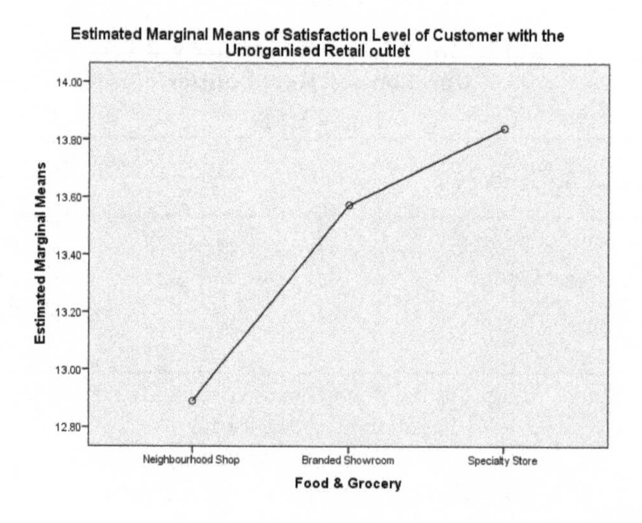

Figure 12 Satisfaction Level of Customer with the Unorganised Retail outlet

RESULTS & DISCUSSION

The SPSS Produces different tables in its one way MANOVA, which includes General linear model, descriptive statistics, multivariate tests, test of between subjects effect, estimated marginal means, post hoc test multiple comparison, homogeneous subjects and profile plots. To understand the one-way MANOVA the results are discussed.

Descriptive Statistics

The first important one is the Descriptive Statistics table shown below. This table is very useful as it provides the mean and standard deviation for the two different dependent variables, which have been split by the independent variable. In addition, the table provides "Total" rows, which allows means and standard deviations for groups only split by the dependent variable to be known.

Multivariate Tests

The Multivariate Tests table is where we find the actual result of the one-way MANOVA. You need to look at the second Effect, labelled "Food & Grocery [FG2]", and the Wilks' Lambda row (highlighted in red). To determine whether the one-way MANOVA was statistically significant you need to look at the "Sig." column. We can see from the table that we have a "Sig." value of .000, which means $p < .0005$. Therefore, we can conclude that this Food & Grocery [FG2]'s pupils academic performance was significantly dependent on which prior Food & Grocery [FG2] they had attended ($p < .0005$).

Reporting the Result (without follow-up tests)

Generally, there was a statistically significant difference in academic performance based on a pupil's prior Food & Grocery [FG2], $F(4, 112) = 13.74$, $p < .0005$; Wilk's $\Lambda = 0.450$, partial $\eta 2 = .33$.

If you had not achieved a statistically significant result, you would not perform any further follow-up tests. However, as our case shows that we did, we will continue with further tests.

Univariate ANOVAs

To determine how the dependent variables differ for the independent variable, we need to look at the Tests of Between-Subjects Effects table (highlighted in red):

We can see from this table that prior Food & Grocery [FG2]ing has a statistically significant effect on both English ($F(2, 57) = 18.11$; $p < .0005$; partial $\eta 2 = .39$) and Maths scores ($F(2, 57) = 14.30$; $p < .0005$; partial $\eta 2 = .33$). It is important to note that you should make an alpha correction to account for multiple ANOVAs being run, such as a Bonferroni correction. As such, in this case, we accept statistical significance at $p < .025$.

Multiple Comparisons

The significant ANOVAs with Tukey's HSD post-hoc tests, as shown below in the Multiple Comparisons table:

The table above shows that mean scores for English were statistically significantly different between Food & Grocery [FG2] A and Food & Grocery [FG2] B ($p < .0005$), and Food & Grocery [FG2] A and Food & Grocery [FG2] C ($p < .0005$), but not between Food & Grocery [FG2] B and Food & Grocery [FG2] C ($p = .897$). Mean maths scores were statistically significantly different between Food & Grocery [FG2] A and Food & Grocery [FG2] C ($p < .0005$), and Food & Grocery [FG2] B and Food & Grocery [FG2] C ($p = .001$), but not between Food & Grocery [FG2] A and Food & Grocery [FG2] B ($p = .443$). These differences can be easily visualised by the plots generated by this procedure.

Chapter - 4

Multiple Regression Analysis

Learning Objectives

This chapter helps to understand the following

- Multiple Regression Analysis using SPSS
- Procedure to run Multiple Regression Analysis using SPSS
- Results and Discussion of Multiple Regression Analysis Output

Introduction

Regression analysis is a way of predicting an outcome variable from one predictor variable (simple regression) or several predictor variables (multiple regression). This tool is incredibly useful because it allows us to go a step beyond the data that we collected.

Data: The dependent and independent variables should be quantitative. Categorical variables, such as religion, major field of study, or region of residence, need to be recoded to binary (dummy) variables or other types of contrast variables.

ASSUMPTIONS

To draw conclusions about a population based on a regression analysis done on a sample, several assumptions must be true (see Berry, 1993)

Variable types

All predictor variables must be quantitative or categorical (with two categories), and the outcome variable must be quantitative, continuous and unbounded. By quantitative I mean that they should be measured at the interval level and by unbounded I mean that there should be no constraints on the variability of the outcome. If the outcome is a measure ranging from 1 to 10 yet the data collected vary between 3 and 7, then these data are constrained.

Non-zero variance

The predictors should have some variation in value (i.e. they do not have variances of 0).

No perfect Multicollinearity

There should be no perfect linear relationship between two or more of the predictors. So, the predictor variables should not correlate too highly.

Predictors are uncorrelated with external variables

External variables are variables that haven't been included in the regression model which influence the outcome variable. These variables can be thought of as similar to the 'third variable' that was discussed with reference to correlation. This assumption means that there should be no external variables that correlate with any of the variables included in the regression model. Obviously, if external variables do correlate with the predictors, then the conclusions we draw from the model become unreliable (because other variables exist that can predict the outcome just as well).

Homoscedasticity

At each level of the predictor variable(s), the variance of the residual terms should be constant. This just means that the residuals at each level of the predictor(s) should have the same variance (homoscedasticity); when the variances are very unequal there is said to be heteroscedasticity.

Independent errors

For any two observations the residual terms should be uncorrelated (or independent). This eventuality is sometimes described as a lack of autocorrelation. This assumption can be tested with the Durbin–Watson test, which tests for serial correlations between errors. Specifically, it tests whether adjacent residuals are correlated. The test statistic can vary between 0 and 4 with a value of 2 meaning that the residuals are uncorrelated. A value greater than 2 indicates a negative correlation between adjacent residuals, whereas a value below 2 indicates a positive correlation. The size of the Durbin–Watson statistic depends upon the number of predictors in the model and the number of observations. For accuracy, you should look up the exact acceptable values in Durbin and Watson's

(1951) original paper. As a very conservative rule of thumb, values less than 1 or greater than 3 are definitely cause for concern; however, values closer to 2 may still be problematic depending on your sample and model.

Normally distributed errors

It is assumed that the residuals in the model are random, normally distributed variables with a mean of 0. This assumption simply means that the differences between the model and the observed data are most frequently zero or very close to zero, and that differences much greater than zero happen only occasionally.

Independence

It is assumed that all of the values of the outcome variable are independent (in other words, each value of the outcome variable comes from a separate entity).

Linearity

The mean values of the outcome variable for each increment of the predictor(s) lie along a straight line. In plain English this means that it is assumed that the relationship we are modelling is a linear one. If we model a non-linear relationship using a linear model then this obviously limits the generalizability of the findings.

PROCEDURE

To open the Linear Regression dialog box, from the menus choose

Step 1: Open the SPSS data file.

Step 2: From the SPSS menu choose Analyse> Regression > Linear as given in figure-1.

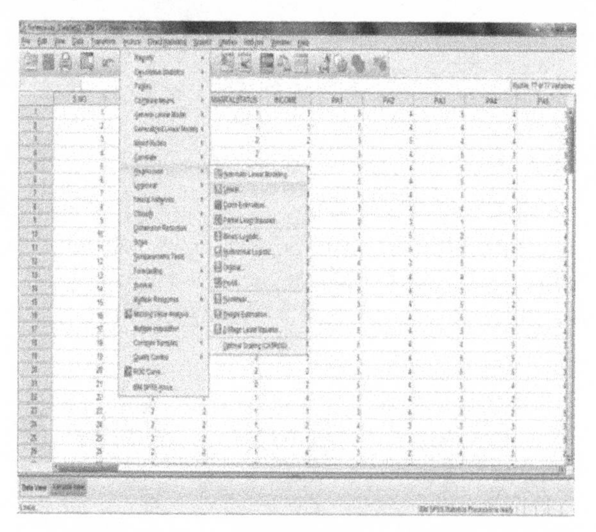

Figure – 1

The **Linear Regression** dialogue box will be shown as below (Figure -2). It helps to select the criterion (dependent) and the predictor (independent) variables.

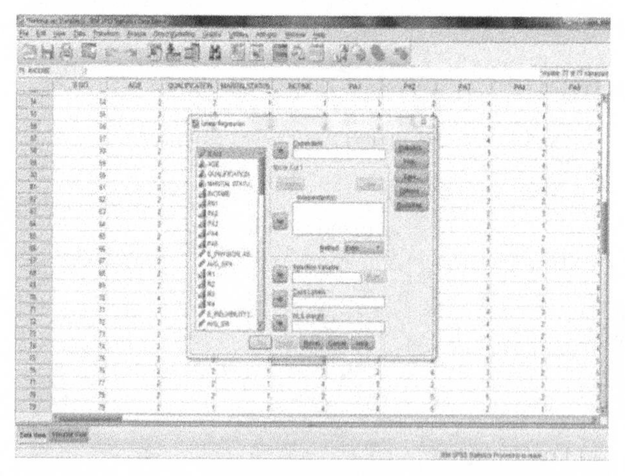

Figure – 2

Step 3: Select the Criterion (or dependent) variable and move it into the **Dependent** box. Select the predictor (or independent) variables and move them into **Independent(s)** box. specify more than one list, or "block" of variables, using the *Next* and *Previous* buttons to display the different lists. Up to nine blocks can be specified in figure - 4.

- **Method** allows selecting a method for including / excluding independent variables in a model.
- **Enter** option allows All variables in the block are entered into the equation as a group.
- **Stepwise**: Selection of variables within the block proceeds by steps. At each step, variables already in the equation are evaluated according to the selection criteria for removal, then variables not in the equation are evaluated for entry. This process repeats until no variable in the block is eligible for entry or removal.
- **Remove**: All variables in the block that are already in the equation are removed as a group.
- **Backward**: All variables in the block that are already in the equation are evaluated according to the selection criteria for removal. Those eligible are removed one at a time until no more are eligible.
- **Forward**: All variables in the block that are not in the equation are evaluated according to the selection criteria for entry. Those eligible are entered one at the time until no more are eligible.

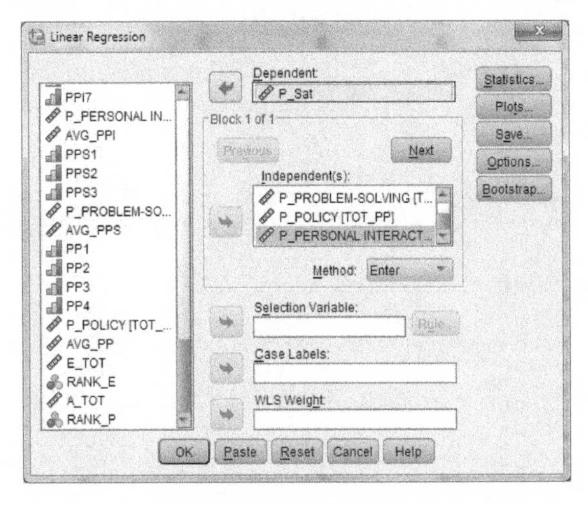

Figure – 4

Step 4: Click on **'Statistics...'** button to request optional statistical output including regression coefficients, descriptives, model fit statistics, etc (Figure -5).

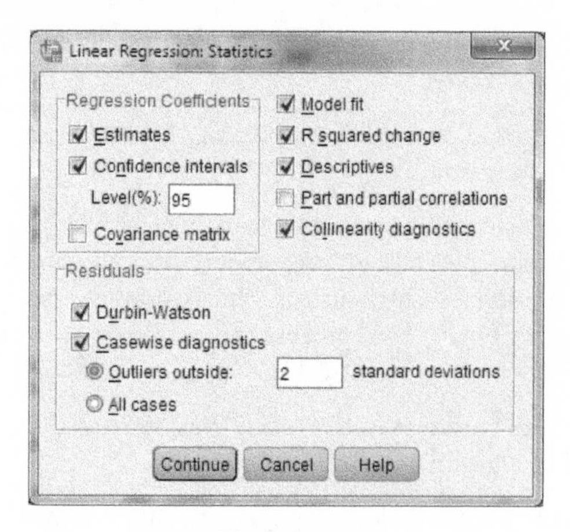

Figure – 5

Step 5: Click on **'Plots...'** to request optional plots, including scatter plots, histograms, normal probability plots and outlier plots. On the left-hand side of the dialog box is a list of several variables (Figure -6).

DEPENDNT (*the outcome variable*).

***ZPRED** (*the standardized predicted values* of the dependent variable based on the model). These values are standardized forms of the values predicted by the model.

***ZRESID** (*the standardized residuals*, or errors). These values are the standardized differences between the observed data and the values that the model predicts).

***DRESID** (*the deleted residuals*). See section 7.6.1.1 for details.

***ADJPRED** (*the adjusted predicted values*). See section 7.6.1.1 for details.

***SRESID** (*the Studentized residual*). See section 7.6.1.1 for details.

***SDRESID** (*the Studentized deleted residual*). This value is the deleted residual divided by its standard deviation.

The variables listed in this dialog box all come under the general heading of residuals.For a basic analysis it is worth plotting *ZRESID (*Y*-axis) against *ZPRED (*X*-axis), because this plot is useful to determine whether the assumptions of random errors and homoscedasticity have been met. A plot of *SRESID (*y*-axis) against *ZPRED (*x*-axis) will show up any heteroscedasticity also. Although often these two plots are virtually identical, the latter is more sensitive on a case-by-case basis. To create these plots simply select a variable from the list, and transfer it to the space labelled either *x* or *y* (which refer to the axes). When you have selected two variables for the first plot you can specify a new plot by clicking on next button. This process clears the spaces in which variables are specified. If you click on next button and would like to return to the plot that you last specified, then simply click on previous button. You can specify up to nine plots.

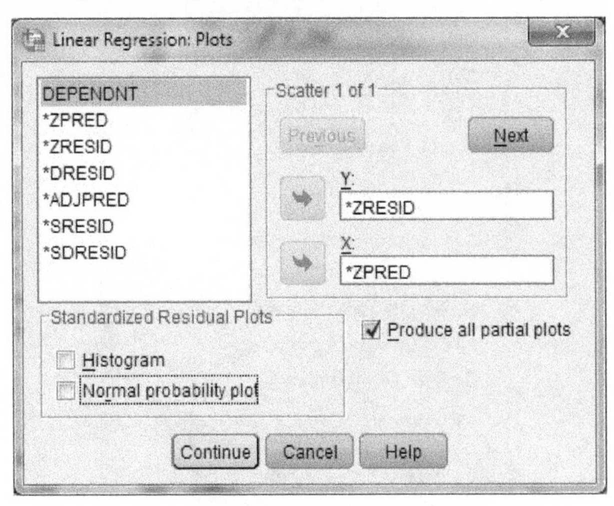

Figure – 6

Step 6: The **'Save...'** button allows to save predicted values, residuals, and related measures as new variables, which are added to the working data file. A table in the output shows the name of each new variable and its content.

Figure – 7

Step 7: The **'Options...'** button allows to control the criteria by which variables are chosen for entry or removal from the regression model, to suppress the constant term, and to control the handling of missing values.

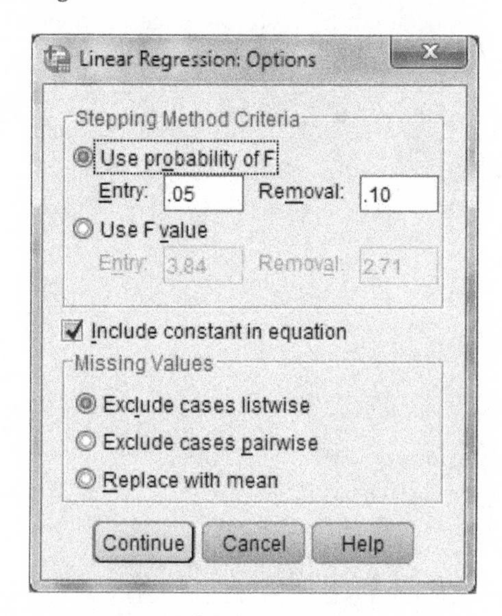

Figure – 8

Step – 8: Click continue tab in the linear option dialogue box, to return to Regression Dialogue box. Then Click Ok to generate output.

SPSS OUTPUT FOR MULTIPLE REGRESSION ANALYSIS USING ENTER METHOD

The following table produced by the Descriptives option.

Descriptive Statistics

	Mean	Std. Deviation	N
P_Sat	2.59	1.245	41
P_Physical Aspects	16.46	3.257	41
P_Reliability	13.27	2.721	41
P_Problem-Solving	10.02	2.329	41
P_Policy	13.68	3.350	41
P_Personal Interaction	21.29	4.849	41

Correlations

The below table gives details of the correlation between each pair of variables. We do not want strong correlations between the criterion and the predictor variables. The values here are acceptable.

Correlations

		P_Sat	P_ PHYSICAL ASPECTS	P_ RELI ABILITY	P_ PROBLEM - SOLVING	P_ POLICY	P_ PERSONAL INTER ACTION
Pearson Correla tion	P_Sat	1.000	.332	.440	.297	.675	.398
	P_Physical Aspects	.332	1.000	.420	.289	.419	.352
	P_ Reliability	.440	.420	1.000	.382	.542	.521
	P_ Problem-Solving	.297	.289	.382	1.000	.488	.502
	P_Policy	.675	.419	.542	.488	1.000	.595
	P_Personal Interaction	.398	.352	.521	.502	.595	1.000
Sig. (1-t ailed)	P_Sat	.	.017	.002	.030	.000	.005
	P_Physical Aspects	.017	.	.003	.034	.003	.012
	P_ Reliability	.002	.003	.	.007	.000	.000
	P_ Problem-Solving	.030	.034	.007	.	.001	.000
	P_Policy	.000	.003	.000	.001	.	.000
	P_Personal Interaction	.005	.012	.000	.000	.000	.
N	P_Sat	41	41	41	41	41	41
	P_Physical Aspects	41	41	41	41	41	41
	P_ Reliability	41	41	41	41	41	41
	P_ Problem-Solving	41	41	41	41	41	41
	P_Policy	41	41	41	41	41	41
	P_Personal Interaction	41	41	41	41	41	41

Variables Entered/Removed

The following table explains about the predictor variables and the method used. Here we can see that all of our predictor variables were entered simultaneously (because we selected the Enter method.

Variables Entered/Removed[a]

Model	Variables Entered	Variables Removed	Method
1	P_Personal Interaction, P_Physical Aspects, P_Problem-Solving, P_ Reliability, P_Policy[b]	.	Enter

a. Dependent Variable: P_Sat **b.** All requested variables entered.

Model Summary

The below model summary table is important. The Adjusted R Square value tells us that our model accounts for 39.2% of variance in the satisfaction.

Model Summary[b]

Model	R	R Square	Adjusted R Square	Std. Error of the Estimate	Change Statistics					Durbin-Watson
					R Square Change	F Change	df1	df2	Sig. F Change	
1	.684[a]	.468	.392	.970	.468	6.165	5	35	.000	2.206

a. Predictors: (Constant), P_PERSONAL INTERACTION, P_PHYSICAL ASPECTS, P_PROBLEM-SOLVING, P_RELIABILITY, P_POLICY **b.** Dependent Variable: P_Sat

ANOVA

This table reports an ANOVA, which assesses the overall significance of the model. As p value < 0.05, the model is significant.

ANOVA[a]

	Model	Sum of Squares	df	Mean Square	F	Sig.
1	Regression	29.011	5	5.802	6.165	.000[b]
	Residual	32.941	35	.941		
	Total	61.951	40			

a. Dependent Variable: P_Sat

b. Predictors: (Constant), P_PERSONAL INTERACTION, P_PHYSICAL ASPECTS, P_PROBLEM-SOLVING, P_RELIABILITY, P_POLICY

Coefficients[a]

The Standardized Beta Coefficients give a measure of the contribution of each variable to the model. A large value indicates that a unit change in this predictor variable has a large effect on the criterion variable. The t and Sig (p) values give a rough indication of the impact of each predictor variable – a big absolute t value and small p value suggests that a predictor variable is having a large impact on the criterion variable.

Coefficients[a]

Model		Unstandardized Coefficients		Standardized Coefficients	t	Sig.	95.0% Confidence Interval for B		Collinearity Statistics	
		B	Std. Error	Beta			Lower Bound	Upper Bound	Tolerance	VIF
1	(Constant)	-1.159	.985		-1.177	.247	-3.158	.840		
	P_Physical Aspects	.016	.054	.043	.306	.761	-.093	.126	.765	1.306
	P_Reliability	.051	.072	.110	.700	.488	-.096	.197	.611	1.637
	P_Problem-Solving	-.030	.079	-.057	-.381	.705	-.192	.131	.687	1.456
	P_Policy	.239	.063	.642	3.767	.001	.110	.367	.523	1.914
	P_Personal Interaction	-.007	.043	-.029	-.174	.863	-.094	.080	.545	1.833

a. Dependent Variable: P_Sat

Collinearity Diagnostics

The tolerance values are a measure of the correlation between the predictor variables and can vary between 0 and 1. The closer to zero the tolerance value is for a variable, the stronger the relationship between this and the other predictor variables. SPSS will not include a predictor variable in a model if it has a tolerance of less that 0.0001. VIF is an alternative measure of collinearity (in fact it is the reciprocal of tolerance) in which a large value indicates a strong relationship between predictor variable

Collinearity Diagnostics[a]

Model	Dimension	Eigenvalue	Condition Index	Variance Proportions					
				(Constant)	P_ Physical Aspects	P_ Reliability	P_ Problem-Solving	P_ Policy	P_ Personal Interaction
1	1	5.877	1.000	.00	.00	.00	.00	.00	.00
	2	.036	12.849	.11	.28	.01	.20	.10	.09
	3	.030	14.113	.07	.00	.06	.61	.26	.04
	4	.022	16.401	.05	.22	.17	.05	.41	.34
	5	.019	17.502	.00	.18	.58	.04	.02	.52
	6	.016	18.954	.77	.32	.17	.09	.21	.01

a. Dependent Variable: P_Sat

Residuals Statistics[a]

	Minimum	Maximum	Mean	Std. Deviation	N
Predicted Value	.22	4.31	2.59	.852	41
Std. Predicted Value	-2.776	2.021	.000	1.000	41
Standard Error of Predicted Value	.190	.609	.355	.110	41
Adjusted Predicted Value	-.20	4.20	2.55	.878	41
Residual	-1.906	1.391	.000	.907	41
Std. Residual	-1.965	1.434	.000	.935	41
Stud. Residual	-2.042	1.556	.017	1.009	41
Deleted Residual	-2.059	1.861	.036	1.061	41
Stud. Deleted Residual	-2.144	1.589	.015	1.023	41
Mahal. Distance	.558	14.762	4.878	3.669	41
Cook's Distance	.000	.210	.029	.040	41
Centered Leverage Value	.014	.369	.122	.092	41

a. Dependent Variable: P_Sat

Residuals Statistics table which displays descriptive statistics for predicted values, adjusted predicted values, and residual values. Residuals are the differences between the actual values of our outcome y and the predicted values of our outcome y based on the model we have specified. The table also produces descriptive summary statistics for measures of multivariate distance and leverage; which allow us to get an idea of whether or not we have outliers or influential data points.

Chapter - 5

Binary Logistic Regression

Learning objectives

This chapter helps to understand the following

- Binary Logistic Regression using SPSS
- Procedure to run Binary Logistic Regression using SPSS
- Results and Discussion of Binary Logistic Regression Output

Introduction

Logistic regression analysis is used when the dependent variable is categorical in nature, the independent variable may be qualitative, dichotomous or both. The categorical dep000 endent variable could consist of more than two categories called as polytomous or multi nominal variable. In this logistic regression designs with a dichotomous dependent variable is discussed. Many research studies in the social and behavioral science investigates dependent variables of a dichotomous in nature, i.e. Yes or no responses to diverse questions about behavior e.g. Employed, training required, marital status or binary responses such as Agree-Disagree, Success-Failure, Presence-Absence and Pro-Con.

The use of logistic regression is increasing because of the availability of sophisticated statistical software and high speed computer. Logistic regression has expanded from its origins in biomedical research further expanded to business, finance, biology etc. The assumption of logistic regression are followed as

- There must be an absence of perfect multicollinearity
- There must be no specification errors.
- The independent variable must be measures at the summative response scale, interval or ratio level.

- Dichotomous variable also allowed
- Logistic regression requires larger samples than does linear regression for valid interpretation of the results.

In this chapter, example of logistic regression will be presented. The example of logistic regression with a single dichotomous independent variable is location and level of job involvement with a dichotomous dependent variable of identifying occupational stress is (yes or no).

Test Procedure in SPSS

Following are the steps to analyse data using Factor Analysis in SPSS.

Step -1: Click Analyze > Regression > Binary Logistic as shown in the figure -1, the Logistic Regression dialogue box will appear (as given in the figure - 2).

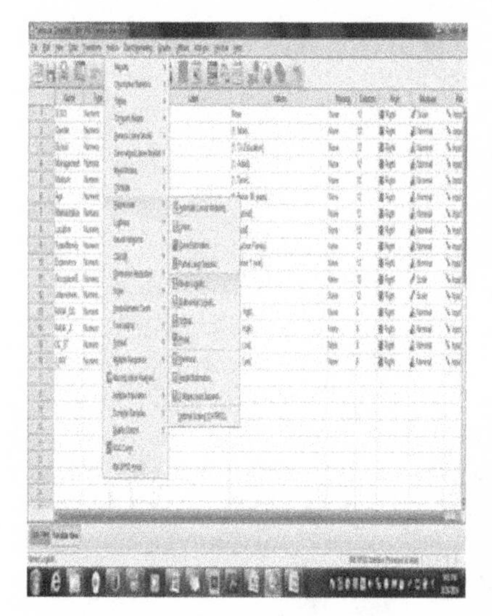

Figure – 1

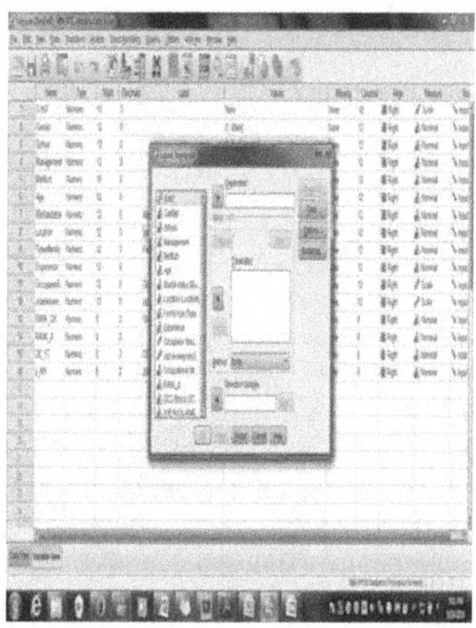

Figure – 2

Step - 2: From the Logistic Regression dialogue box, Select the required variable and transfer to dependent box and covariate box as given in figure -3 by using the arrow tab. In this example, occupational stress is given in dependent tab and location, job involvement in covariates tab.

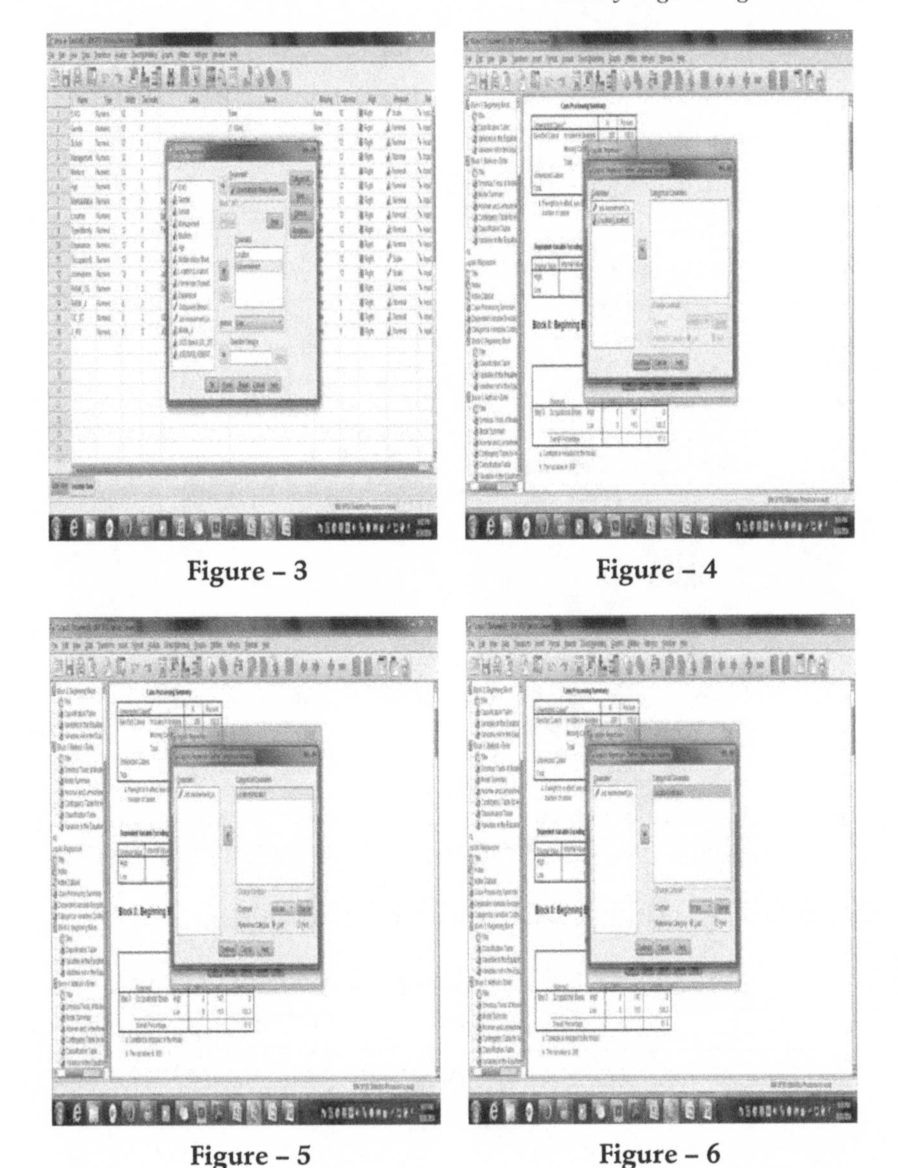

Figure – 3

Figure – 4

Figure – 5

Figure – 6

Step -3: In Logistic regression dialogue box, click on categorical tab, a dialogue box will appear as given in figure -4. Transfer location variable in the categorical covariates as given in figure -5. Then in change contrast indicator column using the drop down box, select simple as given in figure - 6 and click on continue tab to return to logistic regression box.

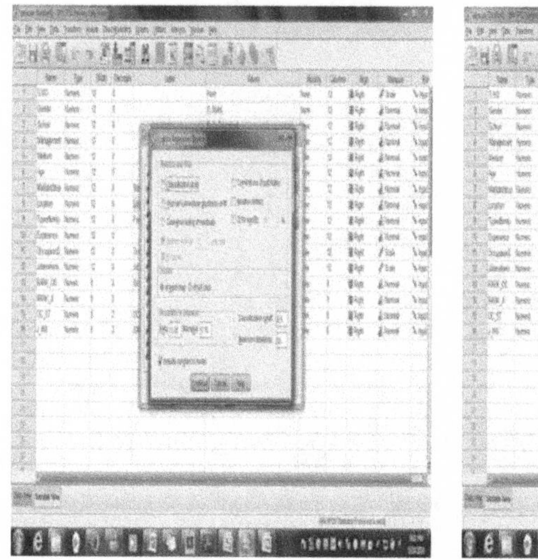

Figure – 7

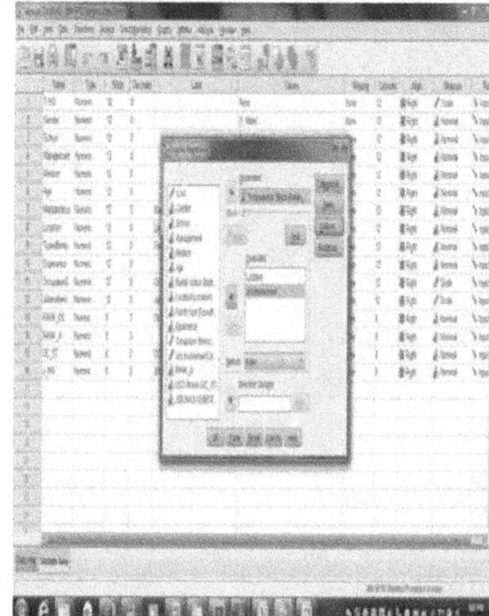

Figure – 8

Step – 6: In the logistic regression dialogue box, click on options tab, a dialogue box will appear as given in figure -7. In the box select Hosmer – Lemeshow goodness of fit and CI for exp(B), in display column, at each step and include constant model as default and click on continue.

Step – 7: In logistic regression dialogue box, Method column - Enter by default. Click on Ok tab to generate output as given in figure -8.

SPSS OUTPUT OF LOGISTIC REGRESSION

Table 1 Case Processing Summary

Unweighted Cases[a]		N	Percent
Selected Cases	Included in Analysis	300	100.0
	Missing Cases	0	.0
	Total	300	100.0
Unselected Cases		0	.0
Total		300	100.0

a. If weight is in effect, see classification table for the total number of cases.

Table 2 Dependent Variable Encoding

Original Value	Internal Value
High	0
Low	1

Table 3 Categorical Variables Codings

		Frequency	Parameter coding
			(1)
Location	Rural	188	1.000
	Urban	112	.000

Block 0: Beginning Block

Table 4 Classification Table[a,b]

Observed			Predicted		
			Occupational Stress		Percentage Correct
			High	Low	
Step 0	Occupational Stress	High	0	147	.0
		Low	0	153	100.0
	Overall Percentage				51.0

a. Constant is included in the model.

b. The cut value is .500

Table 5 Variables in the Equation

		B	S.E.	Wald	df	Sig.	Exp(B)
Step 0	Constant	.040	.115	.120	1	.729	1.041

Table 6 Variables not in the Equation

			Score	df	Sig.
Step 0	Variables	Location	3.540	1	.060
		Job involvement	71.418	1	.000
	Overall Statistics		71.805	2	.000

Block 1: Method = Enter

Table 7 Omnibus Tests of Model Coefficients

		Chi-square	df	Sig.
Step 1	Step	87.330	2	.000
	Block	87.330	2	.000
	Model	87.330	2	.000

Table 8 Model Summary

Step	-2 Log likelihood	Cox & Snell R Square	Nagelkerke R Square
1	328.438[a]	.253	.337

a. Estimation terminated at iteration number 5 because parameter estimates changed by less than .001.

Table 9 Hosmer and Lemeshow Test

Step	Chi-square	df	Sig.
1	8.685	8	.370

Table 10 Contingency Table for Hosmer and Lemeshow Test

		Occupational Stress = High		Occupational Stress = Low		Total
		Observed	Expected	Observed	Expected	
Step 1	1	28	29.531	3	1.469	31
	2	26	26.020	5	4.980	31
	3	25	21.295	7	10.705	32
	4	18	16.329	12	13.671	30
	5	16	13.624	15	17.376	31
	6	8	9.822	19	17.178	27
	7	6	10.579	26	21.421	32
	8	8	9.218	23	21.782	31
	9	8	7.192	23	23.808	31
	10	4	3.389	20	20.611	24

Table 11 Classification Table[a]

Observed			Predicted		
			Occupational Stress		Percentage Correct
			High	Low	
Step 1	Occupational Stress	High	96	51	65.3
		Low	26	127	83.0
	Overall Percentage				74.3

a. The cut value is .500

Table 12 Variables in the Equation

		B	S.E.	Wald	df	Sig.	Exp(B)	95% C.I.for EXP(B)	
								Lower	Upper
Step 1[a]	Location	.154	.274	.317	1	.574	1.167	.682	1.998
	Jobinvo lvement	-.108	.015	50.190	1	.000	.897	.871	.925
	Constant	11.197	1.670	44.954	1	.000	72878.109		

a. Variable(s) entered on step 1: Location, Jobinvolvement.

Results & Discussion

Case Processing Summary

The Case Processing Summary table displays information on the number of cases in the analysis. The case processing summary table seen in Table-1 is the first piece of output provided by SPSS. It provides information on the number of cases being analyzed. This is useful as a check to make sure that you have the expected number of cases in the analysis. In this instance, all 300 of the cases are included. This table also provides information on missing cases because researcher should be aware of situations where they have unexpected loss of data.

Dependent Variable Encoding

The Dependent Variable Encoding shows the "internal recoding" of the binary outcome variable. SPSS does this recoding because not all researchers may assign codes of 0 and 1 to the outcomes. The software thus assigns the lower code (whatever was used by the researchers) a value of 0 and the higher code a value of 1; IBM SPSS ignored the fact that had already anticipated the need to code as 0 and 1, and so it performed its recoding anyway. The next part of the logistic output is the dependent variable encoding table also shown in table -2. It indicates whether SPSS had to temporarily recode the outcome variable to conform to the "High - 0" or "Low -1" coding scheme.

Categorical Variables Codings

The categorical variables codings obtain a count of high and low in the Categorical Variables Codings table together with their codes. The codes under Parameter coding reflect our specification of the reference group; the category shown as 1.000 (Rural) is treated as the reference category in the analysis as given in table -3.

Block 0: Beginning Block

The results for the intercept-only model, computed with only the constant in the equation but with none of the predictor variables; it is called the Step 0 model (the model as of Block 0), as there are no predictor variables in the equation yet.

Classification Table

The Classification Table simply provides counts of the number of cases in each binary outcome. It is a prediction table, with the Observed cases in the rows and the Predicted group membership represented by the columns. With only the intercept in the model, our prediction is based exclusively on the frequencies in that table: 147 cases are high in occupational stress and 153 cases are low in occupational stress. Thus, if there is additional information, our best single guess is that a teacher participant has occupation stress and the classification (predictions) would be correct 51% of the time. It correctly predicts the occupational stress, but those who are affected by occupational stress. The predicted 51% is the proportion of cases that is affected by occupational stress.

Variables in the Equation and Variables Not in the Equation

The Variables in the Equation in the intercept-only model have no predictor variables, so the only factor in the model is the intercept (shown as constant). The odds ratio, shown as Exp(B), has a value of 1.041. This is because 153 is 1.971 times which is nearer to 147 i.e. the teachers who participated in the study are with occupational stress. The odds ratio informs us that a random program participant is 1.041 times are with less occupational stress. The bottom table labeled as Variables not in the Equation reminds us that these predictor variables have yet to be entered into the model. The Wald test is analogous to the t test. The sig. value is the constant by itself and does not significantly improve prediction.

Block 1: Method = Enter

This is Block 1 or Step 1 and is our only step because we entered both of the predictors together. The Omnibus Tests of Model Coefficients contains the model chi-square, a statistical test of the null hypothesis that all the predictor coefficients are zero. It is equivalent to the overall F test in linear regression.

Omnibus Tests of Model Coefficients

The Model chi-square value (in the last row) is 87.330, and with 2 degrees of freedom (there are two predictors in the model), we have a statistically significant amount of prediction ($p<.001$). In this example, the null hypothesis is rejected because the significance is less than .05 i.e. .000.

Model Summary

The Model Summary table provides three indexes of how well the logistic regression model fits the data. With all the variables in the model, the goodness-of-fit −2 Log likelihood statistic is 328.438. It usually cannot interpret this statistic directly but use it to compare different logistic models. The Cox and Snell pseudo R2 is .253 and the Nagelkerke pseudo R Square, which is always the higher of the two, is .337. On the basis of the Nagelkerke pseudo R Square, the model account for 33.7% of the variance associated with occupational stress, which is explained by the predictor variables.

Hosmer and Lemeshow Test

A non-significant chi square means that the predicted probabilities match the observed probabilities. This is what most researchers strive to achieve. The Hosmer and Lemeshow Test provides formal test assessing whether the predicted probabilities match the observed probabilities. We hope to get non-significant p value for this test because the goal of the research is to derive predictors that will accurately predict the actual probabilities. In this example, the goodness-of-fit statistic is 8.685; it is tested as a chi-square value and is associated with a p value of .370, indicating an acceptable match between predicted and observed probabilities.

Contingency Table for Hosmer and Lemeshow Test

This is a more detailed assessment of the Hosmer and Lemeshow test. Note the close match between the observed and the expected values for each group. The Contingency Table for Hosmer and Lemeshow Test, shown in the table -10 demonstrates more details of the Hosmer and Lemeshow test. This output has divided the data into 10 groups based on the outcome variable. These groups are defined by increasing level of occupational stress (called "steps" in the table). For example, the first group (Step 1) represents those students least likely to graduate. The observed frequencies were that 28 cases with high occupational stress and 3 cases are with low level of occupational stress. The observed and the expected frequencies (based on the prediction model) match reasonably well for all of the steps and is a desirable result.

Classification Table

The overall Predictive accuracy is 74.3%, in that High level of Occupational stress is 65.3% and low level of occupation stress is 83%. To recall the block 0 predicted level, it is 51% but now it is improved.

Variables in the Equation

The fine-tuned results of the analysis are presented in the **Variables in the Equation** table shown in Figure 30.10. The table presents for each predictor the raw score partial regression coefficient and its standard error (**S.E.**). These coefficients indicate the amount of change expected in the log odds when there is a one-unit change in the predictor variable, with all the other variables in the model held constant. A coefficient close to 0 suggests that there is no change in the outcome variable associated with the predictor variable.

The **Sig**. column represents the *p* value for testing whether a predictor is significantly associated with graduation controlling for the other predictor(s). The logistic coefficients can be used in a manner similar to linear regression coefficients to generate predicted values. In this example, the model is as follows:

Stress = 11.197 = + .154 (location) - .108 (Job Involvement)

The **Exp(B)** column provides the odds ratios associated with each predictor (adjusting for the other predictor), with the 95% confidence interval associated with each provided in the final two columns. The adjusted odds ratio for **location** is **1.167**, with a 95% confidence interval of 1.998–.682. This odds ratio indicates that in this sample, the odds of urban (because they were the focal group) stress are **1.167** times the odds of occupational stress, controlling for **Job Involvement**.

The adjusted odds ratio for **Job Involvement** is **.897**, with a confidence interval of .925–**.871**. This is a quantitatively measured variable, and so we interpret this odds ratio of **.897** to mean that an increase of 1 in the **job involvement** measure increases the level of stress over the odds for occupational stress by **.897** times, controlling for **location**.

It is concluded that influence of location in job involvement is weak in connection to occupational stress of the teachers at work place.

Chapter - 6

Factor Analysis

Learning Objectives

This chapter helps to understand the following

- Introduction to Factor Analysis
- Meaning of Factor Analysis
- Evolution of Factor Analysis
- Statistics Associated with Factor Analysis
- Steps Involved in Conducting Factor Analysis
- Three Important Measures Used in Factor Analysis
- Types of Factor Analysis
- COMMON FACTOR ANALYSIS Vs COMPONENT FACTOR ANALYSIS
- Factor Rotation
- Interpretation of Factors
- Advantage and Disadvantage of Factor analysis
- Limitations of Factor analysis

INTRODUCTION TO FACTOR ANALYSIS

Factor analysis is a general name which denotes a class of procedures primarily used for reducing and summarizing data. In marketing research, there might be a large number of variables, most of which are correlated and which must be reduced to a manageable level. Relationships among sets of many interrelated variables are examined and represented in terms of a few underlying factors. For example, store image may be measured by asking respondents to evaluate stores on a series of items on a sematic differential scale. The item evaluations may then be analysed to determine the factors underlying store image.

In analysis of variance, multiple regression, and discriminant analysis, one variable is considered as the dependent or criterion variable, and the others as independent or predictor variables. However, no such distinction is made in factor analysis. Rather, factor analysis is an interdependence technique in that an entire set of interdependent relationships is examined.

MEANING

Factor analysis is a multi-variate analysis procedure that attempts to identify any underlying "factors" that are responsible for the co-variation among a group independent variables. The goals of a factor analysis are typically to reduce the number of variables used to explain a relationship or to determine which variables show a relationship. Like a regression model, a factor is a linear combination of a group of variables (items) combined to represent a scale measure of a concept. To successfully use a factor analysis, the variables must represent indicators of some common underlying dimension or concept such that they can be grouped together theoretically as well as mathematically. For example, the variables income, dollars in savings, and home value might be grouped together to represent the concept of the economic status of research subjects.

DEFINITION

Factor analysis is a class of procedures primarily used for summarization and data reduction. Factor analysis is an interdependence technique in that an entire set of interdependent relationships is examined. Factors are defined as an underlying dimension that explains the correlation among a set of variables.

EVOLUTION OF FACTOR ANALYSIS

Charles Spearman first used the factor analysis as a technique of indirect measurement. When they test human personality and intelligence, a set of questions and tests are developed. They believe that a person who is given this set of questions and tests would respond on the basis of some structure that exists in his mind. Thus, his responses would form a certain pattern. This approach is based on the assumption that the underlying structure in answering the questions would be the same in the case of different respondents.

Even though it is in the field of psychology that factor analysis has its beginning, it has since been applied to problems in different areas including marketing. Its use has become far more frequent as a result of the introduction of specialized software packages such as SPSS, SAS, etc.

STATISTICS ASSOCIATED WITH FACTOR ANALYSIS

The key statistics associated with factor analysis are as follows:

Bartlett's test of sphericity

It is a test statistic used to examine the hypothesis that the variables are uncorrelated in the population. In other words, the population correlation matrix is an indentity matrix; each variable correlates perfectly with itself ($r=1$) but has no correlation with the other variables ($r=0$).

Correlation matrix

It is a lower triangle matrix showing the simple correlations, r, between all possible pairs of variables included in the analysis. The diagonal elements, which are all 1, are omitted.

Communality

It is the amount of variance a variable shares with all the other variables being considered. This is also the proportion of variance explained by the common factors.

Eigenvalue

It represents the total variance explained by each factor.

Factor loadings

It is a simple correlation between the variables and the factors.

Factor loading plot

It is a plot of the original variables using the factor loadings as coordinates.

Factor matrix

It contains the factor loadings of all the variables on all the factors extracted.

Factor scores

It is composite scores estimated for each respondent on the derived factors.

Kaiser-Meyer-Olkin (KMO)

It is a measure of sampling adequacy. An index used to examine the appropriateness of factor analysis. High values (between 0.5 and 1.0) indicate the factor analysis is appropriate. Values below 0.5 imply that the factor analysis may not be appropriate.

Percentage of variance

It is the Percentage of the total variance attributed to each factor and is the percentage or proportion of the common variance (defined by the sum of communality estimates) that is explained by successive factors. For example, if you set the cutting line at 75 percent of the common variance (PROPORTION=.75 or PERCENT=75), then factors will be extracted until the sum of eigenvalues for the retained factors exceeds 75 percent of the common variance, defined as the sum of initial communality estimates.

Residuals

If the factors are doing a good job in explaining the correlations among the original variables, we expect the predicted correlation matrix **R*** to closely approximate the input correlation matrix. In other words, we expect the residual matrix **R** - **R*** to approximate a null matrix. The RESIDUAL (or RES) option in the PROC FACTOR statement prints the residual correlation matrix and the partial correlation matrix (correlation between variables after the factors are partialled out or statistically controlled). If the residual correlations or partial correlations are relatively large (> 0.1), then either the factors are not doing a good job explaining the data or we may need to extract more factors to more closely explain the correlations.

Scree plot

Sometimes plotting the eigenvalues against the corresponding factor numbers gives insight into the maximum number of factors to extract. The SCREE option in the PROC FACTOR statement produces a scree plot that illustrates the rate of change in the magnitude of the eigenvalues for the factors. The rate of decline tends to be fast for the first few factors but then levels off. The "elbow", or the point at which the curve bends, is considered to indicate the maximum number of factors to extract.

STEPS INVOLVED IN CONDUCTING THE FACTOR ANALYSIS

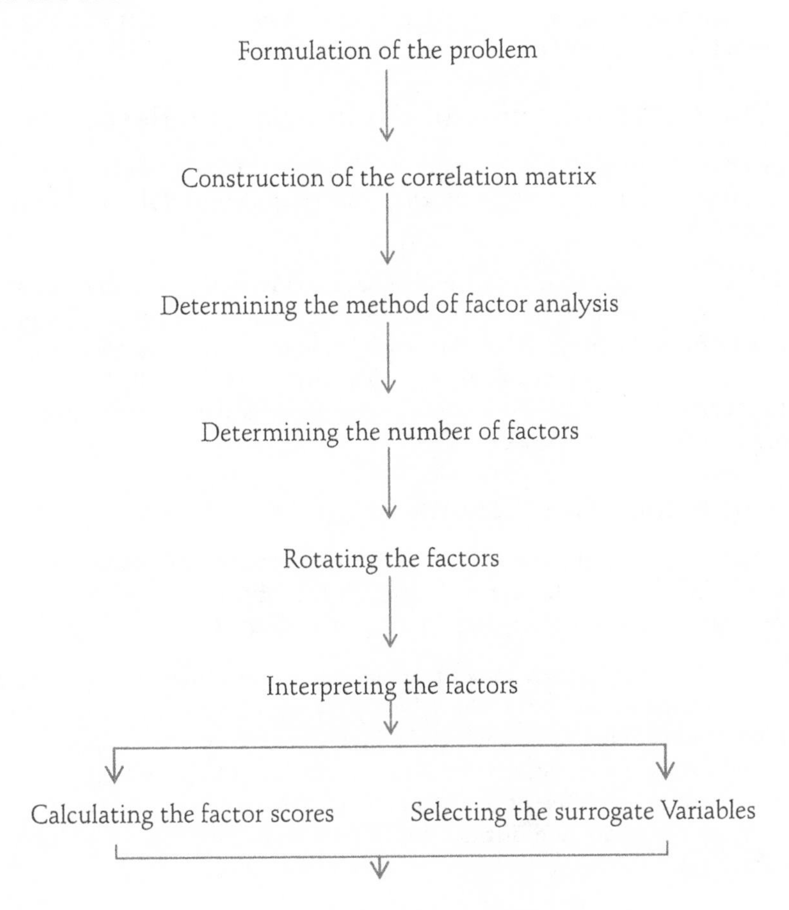

THREE IMPORTANT MEASURES USED IN FACTOR ANALYSIS

1. Variance
2. Standardized scores of an individual's responses
3. Correlation coefficient

1. Variance:

A factor analysis is somewhat like regression analysis in that it tries to "best fit" factors to a scatter diagram of the data in such a way that the factors explain the variance associated with the responses to each statement.

2. Standardized scores of An Individual's Responses:

To facilitate comparisons of the responses from such different scales, researchers standardize all of the answers from all of the respondents on all statements and questions.

Thus, an individual's standardized score is nothing more than an actual response measured in terms of the number of standard deviations (+ or -) it lies away from the mean. Therefore, each standard score is likely to be a value somewhere in the range of +3.00 and -3.00, with +3.00 typically being equated to the "agree very strongly" response and -3.00 typically being equated to the "disagree very strongly" response.

3. Correlation Coefficient:

The third measure used is the correlation coefficient associated with the standardized scores of the responses to each pair of statements. The matrix of correlation coefficients is very important part of factor analysis.

The factor analysis searches through a large set of data to locate two or more sets of statements, which have highly correlated responses. The responses to the statements in one set will all be highly correlated with each other, but they will also be quite uncorrelated with the responses to the statements in the other sets. Since the different sets of statements are relatively uncorrelated with each other, a separate and distinct factor relative to motorcycles is associated with each set.

TYPES OF FACTOR ANALYSIS

A factor analysis is mainly used for interpretation of data and in analyzing the underlying relationships between variable and other underlying factors that may determine consumer behavior. Instead of grouping responses and response types, factor analysis segregates the variable and groups these according to their co-relevance.

There are mainly three types of factor analysis that are used for different kinds of market research and analysis.

- Exploratory factor analysis
- Confirmatory factor analysis
- Structural equation modeling

1. **Exploratory factor analysis** is used to measure the underlying factors that affect the variables in a data structure without setting any predefined structure to the outcome.
2. **Confirmatory factor analysis** on the other hand is used as tool in market research and analysis to reconfirm the effects and correlation of an existing set of predetermined factors and variables that affect these factors. Structural equation modeling hypothesizes a relationship between a set of variables and factors and tests these casual relationships on the linear equation model.
3. **Structural equation modeling** can be used for exploratory and confirmatory modeling alike, and hence it can be used for confirming results as well as testing hypotheses.

DETERMINING THE NUMBER OF FACTORS

It is possible to compute as many principal components as there are variables, but doing so, no parsimony is gained. In order to summarize the information contained in the original variables, a smaller number of factors should be extracted.

The following are the various procedures that have been suggested for determining the number of factors,

- A Priori Determination
- Determination Based on Eigenvalues
- Determination Based on Scree Plot
- Determination Based on Percentage of Variance
- Determination Based on Split-Half Reliability
- Determination Based on Significance Test

ALGORITHM

The Factor analysis operation results in a linear transformation of a set of (satellite) raster maps. The map values are transformed into raster map values, called factors.

The following steps apply:

1. Compute the correlation matrix of the input raster maps.
2. Based on the eigenvectors and eigenvalues of the correlation matrix compute the orientation of the factors.
3. The vectors in each column are normalized. The normalized values are the factor analysis coefficients.

Look at two input raster maps the factor analysis coefficients matrix which looks like:

$$\begin{pmatrix} \alpha1,\alpha2 \\ \beta1,\beta2 \end{pmatrix}$$

where:

a_1 and a_2 = coefficients of the first factor (factor loadings)

b_1 and b_2= coefficients of the second factor (factor loadings)

Compute an MxM matrix of factor analysis coefficients is if M bands are used,

4. A map list is created which contains an expression with which the transformed raster maps (factors) can be defined and calculated.

When the map list is opened, the pixel values of the input maps are transformed into the new raster maps (factors):

$$X=\alpha1X+\alpha2Y$$

$$Y=\beta1X+\beta2Y$$

where:

x and y = spectral values in the first and second factor (output maps)

X and Y = spectral values in the two input raster maps

a_1 and a_2 = coefficients of the first factor

b_1 and b_2 = coefficients of the second factor.

COMMON FACTOR ANALYSIS Vs COMPONENT FACTOR ANALYSIS

Factor analysis as a generic term includes *principal component analysis*. While the two techniques are functionally very similar and are used for the same purpose (data reduction), they are quite different in terms of underlying assumptions.

The term "common" in *common factor analysis* describes the variance that is analyzed. It is assumed that the variance of a single variable can be decomposed into common variance that is shared by other variables included in the model, and unique variance that is unique to a particular variable and includes the error component. Common factor analysis (CFA) analyzes only the *common* variance of the observed variables; principal component analysis considers the *total* variance and makes no distinction between common and unique variance.

The selection of one technique over the other is based upon several criteria. First of all, what is the objective of the analysis? Common factor analysis and principal component analysis are similar in the sense that the purpose of both is to reduce the original variables into fewer composite variables, called *factors* or *principal components*. However, they are distinct in the sense that the obtained composite variables serve different purposes. In common factor analysis, a small number of factors are extracted to account for the intercorrelations among the observed variables--to identify the latent dimensions that explain why the variables are correlated with each other. In principal component analysis, the objective is to account for the maximum portion of the variance present in the original set of variables with a minimum number of composite variables called principal components.

Secondly, what are the assumptions about the variance in the original variables? If the observed variables are measured relatively error free, (for example, age, years of education, or number of family members), or if it is assumed that the error and specific variance represent a small portion of the total variance in the original set of the variables, then principal component analysis is appropriate. But if the observed variables are only indicators of the latent constructs to be measured (such as test scores or responses to attitude scales), or if the error (unique) variance represents a significant portion of the total variance, then the appropriate technique to select is common factor analysis.

THE FACTOR MODEL

In application, there are not one but several factor models which differ in significant respects. A model which is most often applied in psychology is called *common factor analysis*. Indeed, psychologists usually used to reserve the term "factor analysis" for just this model. Common factor analysis is one which is concerned with defining the patterns of common variation among a set of variables. Variation unique to a variable is being ignored. In contrast, another factor model called *component factor analysis* is concerned with patterning all the variation in a set of variables, whether it may be a common or unique one.

Other factor models are image analysis, canonical analysis, and alpha analysis. Image analysis has the same purpose as common factor analysis, but more elegant mathematical properties. Canonical analysis defines common factors for a sample of cases that are the best estimates of those for the population; it enables tests of significance. Alpha analysis defines common factors for a sample of variables that are the best estimates of those in a universe of content. The important general factor models are geometric and algebraic model.

FACTOR ROTATION

The following are the characteristics of various factor rotation,

1. Character of Unrotated Factors

For the most popular factor analysis techniques (centroid and principal axes), the factor patterns define decreasing amounts of variation in the data. Each pattern may involve all or almost all the variables, and the variables may therefore have moderate or high loadings for several factor patterns. To uncover the first pattern, a factor is fitted to the data to account for the greatest regularity; each successive factor is fitted to best define the remaining regularity. *The result of this is that the first unrotated factor may be located between independent clusters of interrelated variables.* These clusters cannot be distinguished in terms of their loadings on the first factor, although they will have loadings different in sign on the second and subsequent factors.

The first unrotated factor delimits the most comprehensive classification, the widest net of linkages, or the greatest order in the data. For comparative political data, a first factor could be a "political institutions" pattern, and a second might define the democratic and totalitarian poles. For international relations, the first factor could be participation in international relations, and a second factor might reflect a polarization between cooperation and conflict. For variables measuring heat, the first factor could be temperature and a second might delineate the extremes of hot and cold. For physiological measurements on adults, the first factor could be size and a second might mirror a polarization between height and girth.

2. Character of Rotated Factors:

A scientist may rotate factors to see if a hypothesized cluster of relationships exists. This can be done by postulating the loadings of a hypothetical factor matrix and then rotating the factors to a best fit with this matrix. The truth of the hypothesis is tested by the difference between the fitted and hypothesized factor loadings.

A simple structure rotation has several characteristics of interest here:

- Each variable is identified with one or a small proportion of the factors. If the factors are viewed as explanations, causes, or underlying influences, this is equivalent to minimizing the number of agents or conditions needed to account for the variation of distinct groups of variables.
- The number of variables loading highly on a factor is minimized. This changes the unrotated factor patterns from being general to the largest number of variables to patterns involving separate groups of variables. The rotation attempts to define a small number of distinct clusters of interrelated phenomena. The moderate and large factor loadings are indicated by x and small loadings are left blank.
- A major ontological assumption underlying the use of simple structure is that, whenever possible, our model of reality should be simplified. If phenomena can be described equally well using simpler factors, then the principle of parsimony is that we should do so. Simple structure maximizes parsimony by shifting from general factors involving all the variables to group factors involving different sets of variables.
- A goal of research is to generalize factor results. The unrotated factor solution, however, depends on all the variables. Add or subtract a variable from the study and the results are altered. The unrotated solution should be adjusted, then, so that the factors will be invariant of the variables selected. An invariate factor solution will delineate the same clusters of relationships regardless of the extraneous variables included in the analysis.

One of the chief justifications for simple structure rotation is that it determines invariant factors. This enables a comparison of the factor results of different studies. Very seldom do different scientists study exactly the same variables. But when variables overlap between studies and each study employs simple structure rotation, tests can be made to see if the same patterns are consistently emerging.

3. Orthogonal Simple Structure Rotation

One important type of simple structure rotation is an orthogonal one. A second type is oblique simple structure. Factors rotated to orthogonal simple structure are usually reported simply as "orthogonal factors." Occasionally, the *varimax* or *quartimax* criteria for achieving the rotation are used to designate the factors.

Orthogonality is a restriction placed on the simple-structure search for the clusters of interdependent variables. The total set of factors is rotated as a rigid frame, with each factor immovably fixed to the origin at a right angle (orthogonal) to every other factor. This system of factors is rotated around the origin until the system is maximally aligned with the separate clusters of variables. If all the clusters are uncorrelated with each other, each orthogonal factor will be aligned with a distinct cluster. The more correlated the separate clusters are, however, the less clearly can orthogonal rotation discriminate them. Simple structure can then only be approximated, not achieved.

Whether or not uncorrelated clusters of relationship exist in the data, orthogonal rotation will still define uncorrelated patterns of relationships. These patterns may not completely overlap with the distinct clusters, but the delineation of these uncorrelated factors is useful. Results involving uncorrelated patterns are easier to communicate, and the loadings can be interpreted as correlations. Moreover, orthogonal factors are more amenable to subsequent mathematical manipulation and analysis.

4. Oblique Simple Structure Rotation

Whereas in orthogonal simple structure rotation the final factors are necessarily uncorrelated, in oblique rotation the factors are allowed to become correlated. In orthogonal rotation the whole factor structure is moved around the origin as a rigid frame (like the spokes of a wheel around the hub) to fit the configuration of clusters of interrelated variables. In oblique rotation to simple structure, however, the factors are rotated individually to fit each distinct cluster. The relationship between the resulting factors then reflects the relationship between the clusters.

Orthogonal rotation is a subset of oblique rotations. If the clusters of relationships are in fact uncorrelated, then oblique rotation will result in orthogonal factors. Therefore, the difference between orthogonal and oblique rotation is not in discriminating uncorrelated or correlated factors but in determining whether this distinction is empirical or imposed on the data by the model.

Controversy exists as to whether orthogonal or oblique rotation is the better scientific approach. Proponents of oblique rotation usually advocate it on two grounds:

- it generates additional information; there is a more precise definition of the boundaries of a cluster, and the central variables in a cluster can be identified by their high loadings;
- the correlations between the clusters are obtained, and these enable the researcher to gauge the degree to which his data approximate orthogonal factors.

Besides yielding more information, oblique rotation is justified on epistemological grounds. One justification is that the real world should not be treated asthough phenomena coagulate in unrelated clusters. As phenomena can be interrelated in clusters, so the clusters themselves can be related. Oblique rotation allows this reality to be reflected in the loadings of the factors and their correlations. A second justification is that correlations between the factors now allow the scientific search for uniformity to be carried to the second order.

TO SUM UP

- Factor analysis is a method for investigating whether a number of variables of interest are linearly related to a smaller number of unobservable factors.
- In the special vocabulary of factor analysis, the parameters of these linear functions are referred to as *loadings.*
- Under certain conditions, the theoretical variance of each variable and the covariance of each pair of variables can be expressed in terms of the loadings and the variance of the error terms.
- There exist an infinite number of sets of loadings yielding the same theoretical variances and covariances.
- Factor analysis usually proceeds in two stages. In the first, one set of loadings is calculated which yields theoretical variances and covariances that fit the observed ones as closely as possible according to a certain criterion. These loadings, however, may not agree with the prior expectations, or may not lend themselves to a reasonable interpretation. Thus, in the second stage, the first loadings are "rotated" in an effort to arrive at another set of loadings that fit equally well the observed variances and covariances, but are more consistent with prior expectations or more easily interpreted.
- A method widely used for determining a first set of loadings is the *principal component method*. This method seeks values of the loadings that bring the estimate of the total communality as close as possible to the total of the observed variances.
- When the variables are not measured in the same units, it is customary to *standardize* them prior to subjecting them to the principal component method so that all have mean equal to zero and variance equal to one.
- The *varimax* rotation method encourages the detection of factors each of which is related to few variables. It discourages the detection of factors influencing all variables.
- There is considerable subjectivity in determining the number of factors and the interpretation of these factors. There are several methods for

obtaining first and rotated factor solutions, and each such solution may give rise to a different interpretation.

INTERPRETATION OF FACTORS

One part of the output from a factor analysis is a matrix of factor loadings. A *factor loading* or *factor structure matrix* is a n by m matrix of correlations between the original variables and their factors, where n is the number of variables and m is the number of retained factors. When an oblique rotation method is performed, the output also includes a *factor pattern matrix*, which is a matrix of standardized regression coefficients for each of the original variables on the rotated factors. The meaning of the rotated factors is inferred from the variables significantly loaded on their factors. A decision needs to be made regarding what constitutes a significant loading. A rule of thumb frequently used is that factor loadings greater than .30 in absolute value are considered to be significant. This criterion is just a guideline and may need to be adjusted. As the sample size and the number of variables increase, the criterion may need to be adjusted slightly downward; it may need to be adjusted upward as the number of factors increases. The procedure described next outlines the steps of interpreting a factor matrix.

1. Identifying significant loadings: The analyst starts with the first variable (row) and examines the factor loadings horizontally from left to right, underlining them if they are significant. This process is repeated for all the other variables. You can instruct SAS to perform this step by using the FUZZ= option in the PROC FACTOR statement. For instance, FUZZ=.30 prints only the factor loadings greater than or equal to .30 in absolute value.

Ideally, we expect a single significant loading for each variable on only one factor: across each row there is only one underlined factor loading. It is not uncommon, however, to observe *split loadings*, a variable which has multiple significant loadings. On the other hand, if there are variables that fail to load significantly on any factor, then the analyst should critically evaluate these variables and consider deriving a new factor solution after eliminating them.

2. Naming of Factors: Once all significant loadings are identified, the analyst attempts to assign some meaning to the factors based on the patterns of the factor loadings. To do this, the analyst examines the significant loadings for each factor (column). In general, the larger the absolute size of the factor loading for a variable, the more important the variable is in interpreting the factor. The sign of the loadings also needs to be considered in labeling the factors. It may be important to reverse the scoring of the negatively worded items in Likert-type instruments to prevent ambiguity. That is, in Likert-type instruments some items are often negatively worded so that high scores on these items actually reflect low degrees of the attitude or construct being measured.

Remember that the factor loadings represent the correlation or linear association between a variable and the latent factor(s). Considering all the variables' loading on a factor, including the size and sign of the loading, the investigator makes a determination as to what the underlying factor may represent.

APPLICATION OF FACTOR ANALYSIS IN PSYCHOLOGY

Factor analysis is used to identify "factors" that explain a variety of results on different tests. For example, intelligence research found that people who get a high score on a test of verbal ability are also good on other tests that require verbal abilities. Researchers explained this by using factor analysis to isolate one factor, often called crystallized intelligence or verbal intelligence, which represents the degree to which someone is able to solve problems involving verbal skills.

Factor analysis in psychology is most often associated with intelligence research. However, it also has been used to find factors in a broad range of domains such as personality, attitudes, beliefs, etc. It is linked to psychometrics, as it can assess the validity of an instrument by finding if the instrument indeed measures the postulated factors.

Advantages

- Reduction of number of variables, by combining two or more variables into a single factor. For example, performance at running, ball throwing, batting, jumping and weight lifting could be combined into a single factor such as general athletic ability. Usually, in an item by people matrix, factors are selected by grouping related items. In the Q factor analysis technique, the matrix is transposed and factors are created by grouping related people: For example, liberals, libertarians, conservatives and socialists, could form separate groups.
- Identification of groups of inter-related variables, to see how they are related to each other. For example, Carroll used factor analysis to build his Three Stratum Theory. He found that a factor called "broad visual perception" relates to how good an individual is at visual tasks. He has also found a "broad auditory perception" factor, relating to auditory task capability. Furthermore, he also found a global factor, called "g" or general intelligence, which relates to both "broad visual perception" and "broad auditory perception". This means someone with a high "g" is likely to have both a high "visual perception" capability and a high "auditory perception" capability, and that "g" therefore explains a good part of why someone is good or bad in both of those domains.

Disadvantages

- "...each orientation is equally acceptable mathematically. But different factorial theories proved to differ as much in terms of the orientations of factorial axes for a given solution as in terms of anything else, so that model fitting did not prove to be useful in distinguishing among theories." (Sternberg, 1977). This means all rotations represent different underlying processes, but all rotations are equally valid outcomes of standard factor analysis optimization. Therefore, it is impossible to pick the proper rotation using factor analysis alone.
- Factor analysis can be only as good as the data allows. In psychology, where researchers often have to rely on less valid and reliable measures such as self-reports, this can be problematic.
- Interpreting factor analysis is based on using a "heuristic", which is a solution that is "convenient even if not absolutely true". More than one interpretation can be made of the same data factored the same way, and factor analysis cannot identify causality.

LIMITATIONS OF FACTOR ANALYSIS

- The utility of this technique largely depends to a large extent on the judgment of the researcher. He has to make number of decisions as to how the factor analysis will come out. Even with a given set of decisions, different results will emerge from different group of respondents, different mixes of data as also different ways of getting data. In other words, factor analysis is unable to give a unique solution of result.
- As any other method of analysis, a factor analysis will be of little use if the appropriate variable has not been measured, or if the measurements are inaccurate, or if the relationships in the data are non-linear.
- In the view of on-going limitations, the exploratory nature of factor analysis becomes clear. *Thurston* mentions the use of factor analysis should not be made where fundamental and fruitful concepts are already well formulated and tested. It may be used especially in those domains where basic and fruitful concepts are essentially lacking and where crucial experiments have been difficult to conceive.

Exploratory Factor Analysis

Learning Objectives

This chapter helps to understand the following

- Exploratory Factor analysis Using SPSS
- Procedure to Run Exploratory Factor analysis Using SPSS
- Results and Discussion of Exploratory Factor analysis Output

Exploratory Factor analysis Using SPSS

Factor analysis is performed most often with metric variables. The sample must have more observations than variables or minimum absolute sample size should be 50 observations or else maximize the number of observations per variable, with a desired ratio of 5 observations per variable. For testing the assumption of factor analysis, a well-built conceptual foundation needs to support the assumption that a structure does exist before the factor analysis is performed. The Bartlett's test of sphericity which is less than .05 indicates statistical significance to proceed further. For choosing the number of factors or factor model, 30 or above variable or communalities above .60 is required for most of the variables. Eigen value should be more than 1 for good model. The best method for data reduction is component method, for well specified theoretical built common factor method is appropriate.

As well to consider enough factors, total variance explained should be more than 60%. For accessing the factor loading .30 or .40 is sufficient but above .50 shows significance. For interpreting the results of factor analysis, take away the communalities below .50 for better results.

Test Procedure in SPSS

Following are the steps to analyse data using Factor Analysis in SPSS.

Step - 1: Click Analyze > Dimension Reduction > Factor analysis as shown in the figure -1, the Factor analysis dialogue box will appear (as given in the figure - 2).

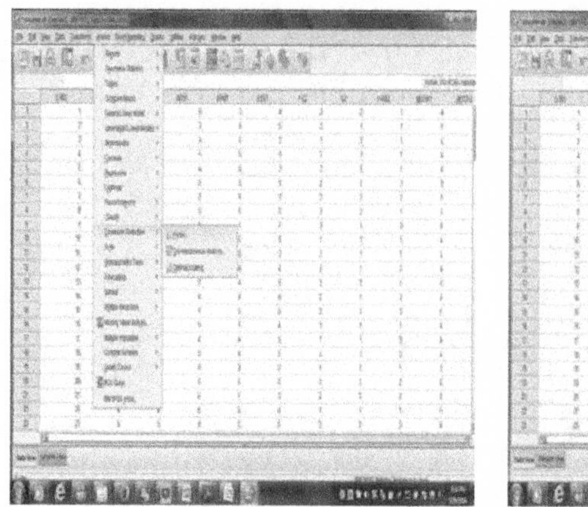

Figure – 1

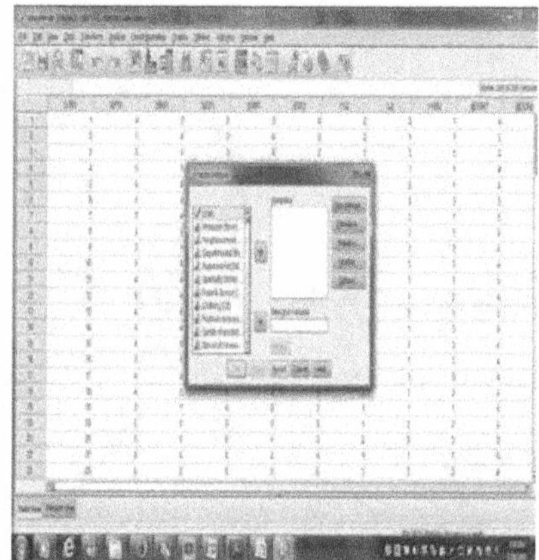

Figure – 2

Step - 2: From the factor analysis dialogue box, Select the required variable as given in figure -3 and transfer those using the arrow tab in variable box as given in figure - 4.

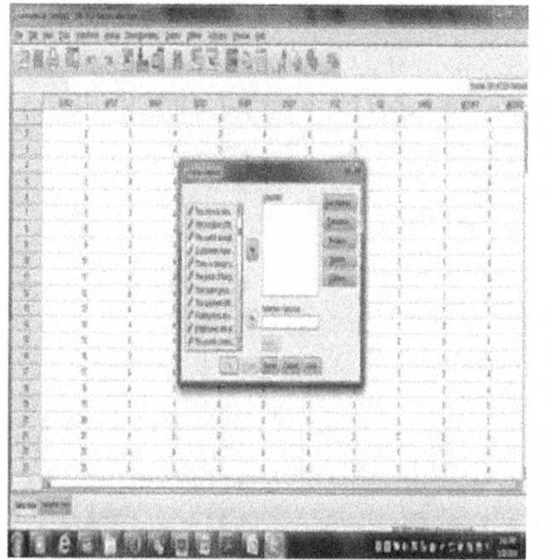

Figure – 3

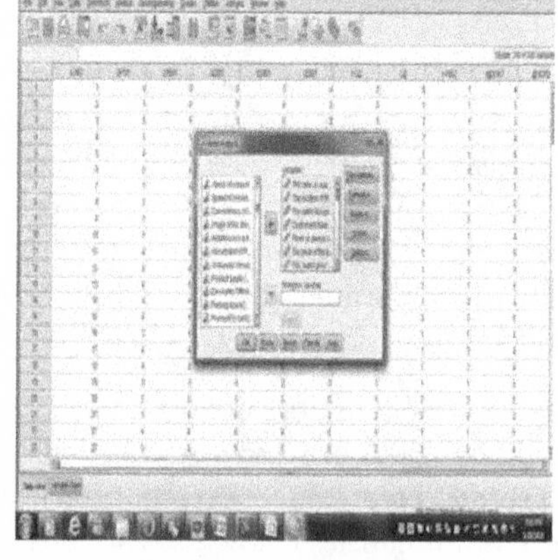

Figure – 4

Step - 3: In the factor analysis dialogue box, Click Descriptives tab, a dialogue box will appear as given in figure -5, Select Univariate Descriptives in statistics and in correlation matrix, select KMO and Bartlett's test of sphericity and click continue tab to return to factor analysis dialogue box.

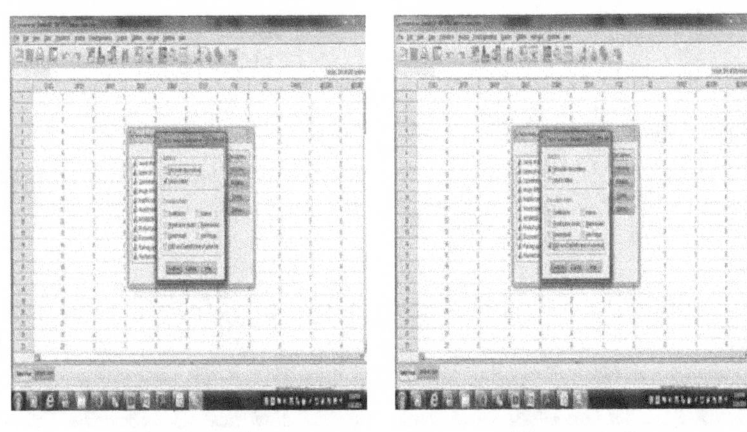

Figure – 5 **Figure – 6**

Step - 4: In factor analysis dialogue box, click extraction tab, a dialogue box will appear as given in figure -7, select default method principal components, select correlation matrix in analyze, in display select unrotated factor solution and screen plot, Give Eigen value greater than 1 as given in figure -8 then click continue tab to return to factor analysis dialogue box.

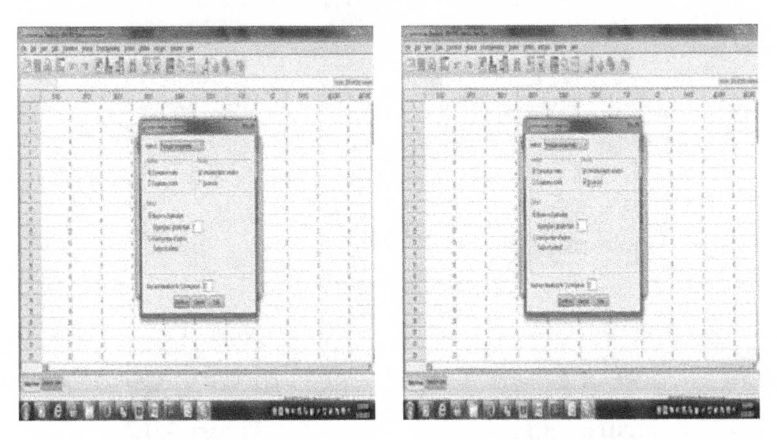

Figure – 7 **Figure – 8**

Step - 5: In factor analysis dialogue box, click Rotation tab, a dialogue box will appear as given in figure -9, select varimax method as given in figure 10. And click continue tab to return to factor analysis dialogue box.

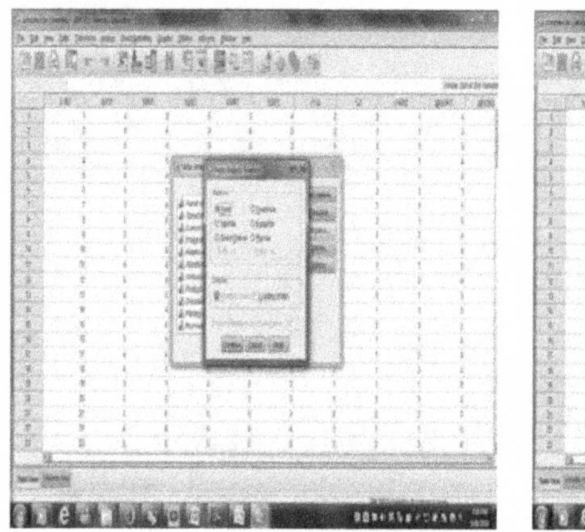

Figure – 9

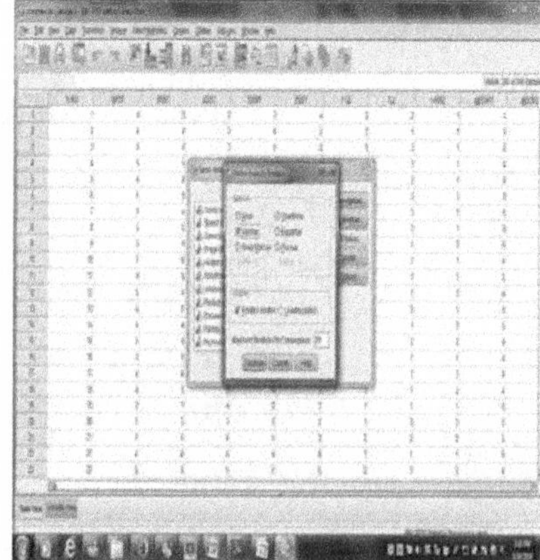

Figure – 10

Step - 6: In factor analysis dialogue box, click on factor scores tab, a dialogue box will appear as given in figure -11, select save as variables method regression method as given in figure -12. And click continue tab to return to factor analysis dialogue box.

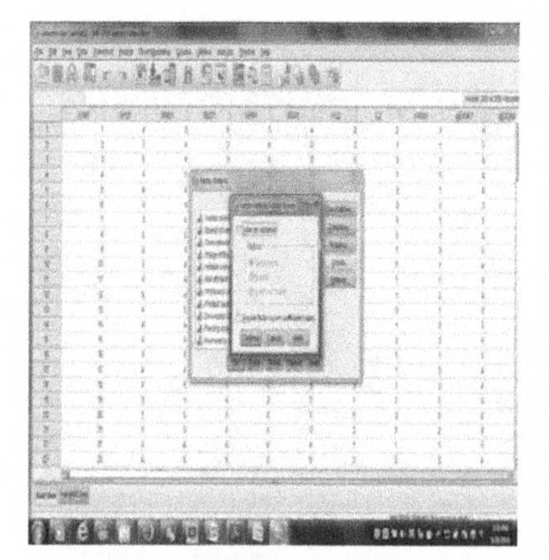

Figure – 11

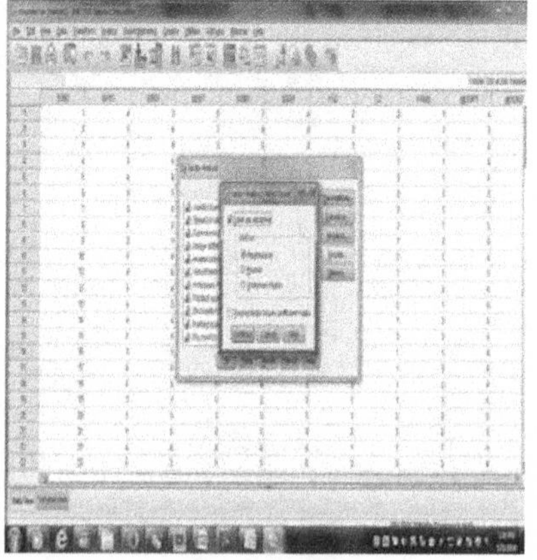

Figure – 12

Step - 7: In factor analysis dialogue box, click on options tab, a dialogue box will appear as given in figure -13, select exclude cases listwise and in coefficient display format select sorted by size and suppress small coefficients and in absolute value below .40 instead of .10 as given in figure -14. And click continue tab to return to factor analysis dialogue box.

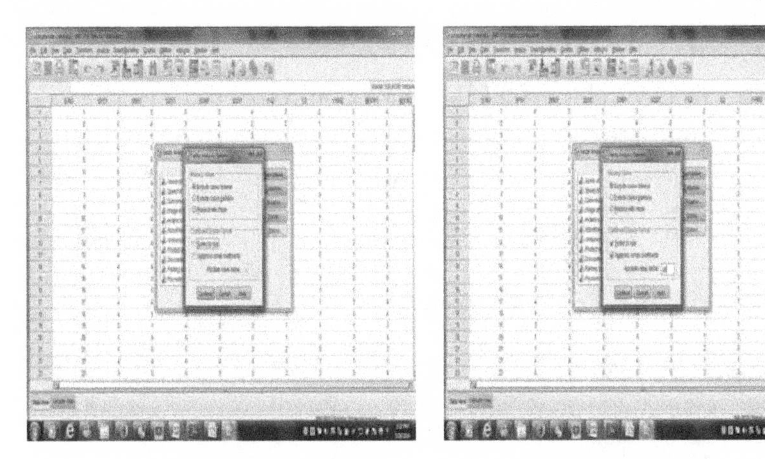

Figure – 13 **Figure – 14**

Step - 8: Then click ok tab in factor analysis dialogue box to generate output (figure-15).

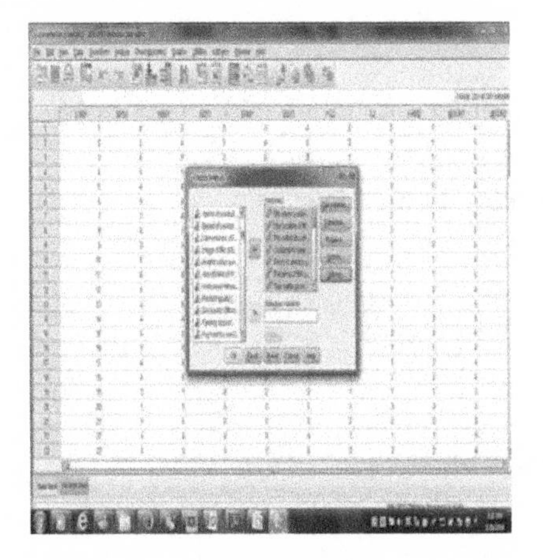

Figure – 15

SPSS Output of the Factor Analysis

Table 1 Descriptive Statistics

	Mean	Std. Deviation	Analysis N
The store is visually appealing, kept clean and run efficiently	3.74	1.155	516
The location of the store is convenient.	3.58	1.154	516
The outlet design helps customers to move around with easily and find products.	3.86	1.155	516
Customers have adequate parking space for their vehicles	3.74	1.143	516
There is always stock of products/brands desired by customers.	3.62	1.184	516
The price of the products are clearly indicated	3.79	1.181	516
This outlet gives appropriate and punctual information on its sales promotions	3.57	1.175	516
The cashiers bill products chosen by customers accurately	3.64	1.218	516
Waiting time at cash counter are short	3.64	1.205	516
Employees are always willing to help customers	3.47	1.234	516
The public contact staff (shelf stackers, cash registers, perishable section, information staff, and security personnel) is always polite to customers	3.38	1.298	516
Employees give individual attention in understanding specific requirement of customers	3.32	1.263	516

	Mean	Std. Deviation	Analysis N
Materials associated with the store service (such as shopping bags, catalogs, or statements) are visually appealing.	3.63	1.194	516
Employees in the store have the knowledge to answer customers' questions.	3.60	1.237	516
The behaviour of employees in the store persuade confidence in customers	3.53	1.201	516
Customers feel safe in their transactions with this store	3.81	1.226	516
Employees in this store give prompt service to customers.	3.67	1.182	516
The store gives customers individual attention	3.60	1.205	516
Employees of this store treat customers courteously on the telephone.	3.37	1.237	516
The store willingly handles returns and exchanges.	3.43	1.218	516
When a customer has problem, this store shows a sincere interest in solving it.	3.63	1.171	516
Employees of this store are able to handle customer complaints directly and immediately.	3.52	1.233	516
The store has operating hour convenient to all their customers	3.57	1.261	516
The store accept most credit cards	3.68	1.274	516
The fruits and vegetables that are sold fresh	3.50	1.227	516
The meat and fish product sold in this outlet are fresh	3.62	1.200	516

	Mean	Std. Deviation	Analysis N
The retailers own brand products are of high quality	3.34	1.226	516
The quality of other products that are sold in this outlet is good.	3.54	1.277	516
All well known brands of products are available in the store	3.61	1.276	516
A broad assortment of product and brands are offered	3.42	1.304	516
The store not selling expired and damaged products.	2.96	1.272	516
The store offers product with reasonable price.	2.76	1.313	516
The price of the product is higher compared to market price.	2.92	1.426	516
The product's pricing policy is not fair.	2.91	1.326	516
The product's pricing policy is ethical.	2.83	1.332	516
The product's pricing policy is acceptable.	2.76	1.404	516
The price is cheaper or same as other similar stores.	3.67	1.228	516

Table 2 KMO and Bartlett's Test

Kaiser-Meyer-Olkin Measure of Sampling Adequacy.		.958
Bartlett's Test of Sphericity	Approx. Chi-Square	13104.398
	df	666
	Sig.	.000

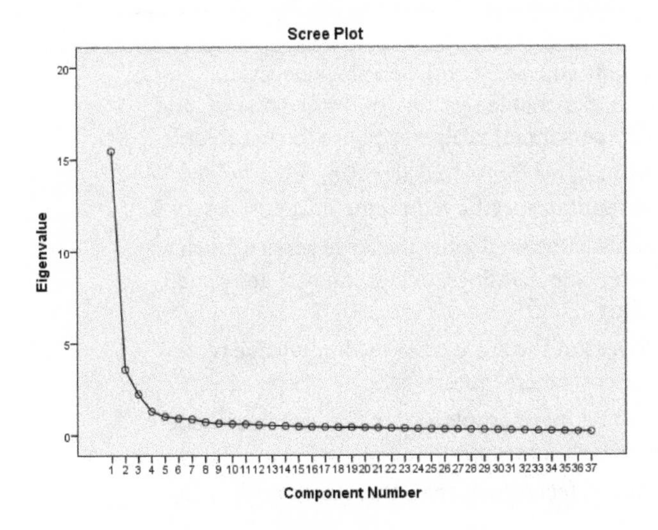

Figure – 16

Table - 3 Communalities

	Extraction
The store is visually appealing, kept clean and run efficiently	.604
The location of the store is convenient.	.553
The outlet design helps customers to move around with easily and find products.	.639
Customers have adequate parking space for their vehicles	.586
There is always stock of products/brands desired by customers.	.628
The price of the products are clearly indicated	.649
This outlet gives appropriate and punctual information on its sales promotions	.637
The cashiers bill products chosen by customers accurately	.677
Waiting time at cash counter are short	.614
Employees are always willing to help customers	.738

	Extraction
The public contact staff (shelf stackers, cash registers, perishable section, information staff, and security personnel) is always polite to customers	.629
Employees give individual attention in understanding specific requirement of customers	.633
Materials associated with the store service (such as shopping bags, catalogs, or statements) are visually appealing.	.651
Employees in the store have the knowledge to answer customers' questions.	.653
The behaviour of employees in the store persuade confidence in customers	.691
Customers feel safe in their transactions with this store	.640
Employees in this store give prompt service to customers.	.680
The store gives customers individual attention	.582
Employees of this store treat customers courteously on the telephone.	.624
The store willingly handles returns and exchanges.	.542
When a customer has problem, this store shows a sincere interest in solving it.	.645
Employees of this store are able to handle customer complaints directly and immediately.	.678
The store has operating hour convenient to all their customers	.623
The store accept most credit cards	.576
The fruits and vegetables that are sold fresh	.545
The meat and fish product sold in this outlet are fresh	.654
The retailers own brand products are of high quality	.650
The quality of other products that are sold in this outlet is good.	.626
All well known brands of products are available in the store	.640
A broad assortment of product and brands are offered	.680

	Extraction
The store is not selling expired and damaged products.	.559
The store offers product with reasonable price.	.716
The price of the product is higher compare to market price.	.801
The product's pricing policy is not fair.	.746
The product's pricing policy is ethical.	.748
The product's pricing policy is acceptable.	.689
The price is cheaper or same as other similar stores.	.426

Extraction Method: Principal Component Analysis.

Table - 4 Total Variance Explained

Component	Extraction Sums of Squared Loadings			Rotation Sums of Squared Loadings		
	Total	**% of Variance**	**Cumulative %**	**Total**	**% of Variance**	**Cumulative %**
1	15.473	41.819	41.819	7.979	21.566	21.566
2	3.591	9.707	51.526	5.728	15.480	37.046
3	2.242	6.060	57.586	4.583	12.387	49.433
4	1.319	3.565	61.151	3.494	9.443	58.876
5	1.028	2.779	63.929	1.870	5.054	63.929

Extraction Method: Principal Component Analysis.

Table - 5 Rotated Component Matrixa

	Component				
	1	**2**	**3**	**4**	**5**
Employees are always willing to help customers	.800				
The public contact staff (shelf stackers, cash registers, perishable section, information staff, and security personnel) is always polite to customers	.768				
The behaviour of employees in the store persuade confidence in customers	.745				

	Component				
	1	2	3	4	5
Employees give individual attention in understanding specific requirement of customers	.738				
There is always stock of products/ brands desired by customers.	.699				
Employees in the store have the knowledge to answer customers questions.	.698				
The cashiers bill products chosen by customers accurately	.696				
This outlet gives appropriate and punctual information on its sales promotions	.677				
Waiting time at cash counter are short	.651				
Materials associated with the store service (such as shopping bags, catalogs, or statements) are visually appealing.	.646	.441			
The outlet design helps customers to move around with easily and find products.	.623				
Customers have adequate parking space for their vehicles	.598				
The price of the products are clearly indicated	.562				
The location of the store is convenient.	.561				.402
Employees of this store are able to handle customer complaints directly and immediately.		.763			
The store has operating hour convenient to all their customers		.691			
Employees in this store give prompt service to customers.		.689			

	Component				
	1	**2**	**3**	**4**	**5**
When a customer has problem, this store shows a sincere interest in solving it.		.675			
Employees of this store treat customers courteously on the telephone.		.652			
Customers feel safe in their transactions with this store	.456	.617			
The store gives customers individual attention		.583			
The store accept most credit cards		.569			.414
The store willingly handles returns and exchanges.		.567			
The price is cheaper or same as other similar stores.		.454			
The price of the product is higher compare to market price.			.873		
The product's pricing policy is ethical.			.847		
The product's pricing policy is not fair.			.837		
The store offers product with reasonable price.			.828		
The product's pricing policy is acceptable.			.813		
The store not selling expired and damaged products.			.664		
A broad assortment of product and brands are offered				.724	
The quality of other products that are sold in this outlet is good.				.667	
The retailers own brand products are of high quality				.638	.410
The meat and fish product sold in this outlet are fresh				.620	

	Component				
	1	2	3	4	5
All well known brands of products are available in the store		.470		.601	
The fruits and vegetables that are sold fresh				.583	
The store is visually appealing, kept clean and run efficiently	.484				.553

Extraction Method: Principal Component Analysis.
Rotation Method: Varimax with Kaiser Normalization.

a. Rotation converged in 8 iterations.

Results & Discussion

Descriptive Statistics

Table-1 represents the descriptive statistics for the variables taken into the study. This table contains mean i.e. mean score of each item for 516 respondents/ customers and standard deviation i.e. degree of variability in scores for each item and the fourth column shows the number of observations i.e. sample size. Next to factor analysis the factor analysis give correlation matrix.

KMO and Bartlett's Test

Table -2 represents KMO measure of sampling adequacy and the Bartlett's test of sphericity for judging the appropriateness of a factor model. KMO statistics compares the magnitude of the observed correlation coefficient with the magnitude of the partial correlation coefficients. The high value of this statistics (0.5 to 1) indicates the appropriateness of the factor analysis, Kaiser has presented the ranges as follows i.e. statistics above .90 is marvelous, above .80 is meritorious, above .70 is middling, 0.6 mediocre and less than 0.5 is unacceptable. In this example, the computed KMO Statistics is .958, which indicates the value is excellent in this model.

Bartlett's test of sphericity tests the hypothesis whether the population correlation matrix is an identity matrix. The chi square statistics is 13104.398 with 666 degree of freedom. This value is significant at 0.01 level. Both the results, that is the KMO Statistics and Bartlett's test of sphericity, indicates the appropriate factor analysis model.

Scree Plot

One of the most popular guides for determining how many factors should be retailed in the factor analysis is the scree test. A scree plot is a plot of the eignvalues and also represents the optimum number of factors to be retained in the final solution. The objective of the scree plot is to visually isolate an elbow, which can be defined as the point where the Eigen values form a linear descending trend. (Figure-16)

Communalities

Table -3 represents communalities which describes the amount of variance a variable shares with all other variables taken into the study. Always the initial communality value is equal to 1. In this example the communalities ranges between .801 to .426.

Total Variance Explained

Table - 4 represents total variance explained which indicates the percentage of variance accounted for by each specific factor or component. The total variance explained presents initial Eigen value, extraction sums of squared loadings, rotation sum of squared loadings. In this example, the main focus of transformation is based on principal component method a set of inter related variables into a set of uncorrelated linear combination of these variables. This method is applied when the primary focus of the factor analysis is to determine the minimum number of factors that attributes maximum variance in the data. The total variance explained is 63.929% for 5 components.

Rotated Component Matrix

Table -5 represents rotated component matrix that is often referred as pattern matrix. The column in the table represents the factor loading for each variable, for the concerned factor, after rotation enhances the interpretability in terms of association of factor loadings with the concerned factor. In the given example, the loadings were sorted by size as well suppressed with absolute value above .40.

Confirmatory Factor Analysis

Learning Objectives

This chapter helps to understand the following

- Meaning of Confirmatory Factor Analysis
- Basic Elements of Confirmatory Factor Analysis
- Practical Example with Sample Data
- Steps to Run Confirmatory Factor Analysis
- Results and Discussion of Confirmatory Factor Analysis
- Modified Measurement Model of CFA
- Results and Discussion of Modified Measurement Model of CFA
- Higher Order Factor Analysis
- Results and Discussion of Higher Order Factor Analysis

Introduction

Confirmatory factor analysis is a measurement model. It is used to study the relationships between a set of observed variables and a set of continuous latent variables. The main advantage of CFA is that it allows for testing hypotheses about a particular factor structure. CFA is a unique case of the structural equation model (SEM), also known as the covariance structure (McDonald, 1978) or the linear structural relationship (LISREL) model (Joreskog & Sorbom, 2004). SEM consists of two components: a measurement model linking a set of observed variables to a usually smaller set of latent variables and a structural model linking the latent variables through a series of recursive and non-recursive relationships. Confirmatory factor analysis corresponds to the measurement model of SEM.

Confirmatory Factor Analysis (CFA) is useful when researchers have clear hypotheses about a scale – the number of factors or dimensions underlying its items, the links between specific items and specific factors, and the association between factors. Two rules for confirmatory factor models are given below.

The three –indicator rule

A model is identified if,

- Every factor has at least three indicators
- No manifest variable is indicator for more than one factor
- The error terms are not correlated

The two-indicator rule

- A confirmatory factor model with at least two indicators
- No manifest variable is indicator for more than one factor
- The error terms are not correlated
- The covariance matrix for the latent variables does not contain zeros.

Basic Elements of CFA-SEM Models

Exogenous constructs: latent, multi-item equivalent of independent variables that are not influenced by other variables in the model. They use a variate (linear combination) of measures to represent the construct, which acts as an independent variable in the model.

Endogenous constructs: latent, multi-item equivalent to dependent variables – they are affected by other variables in the theoretical model.

Unobserved variable: a hypothesized, latent construct (concept) that can only be approximated by observable or measurable indicator variables.

Observed variable: known as manifest or indicator variables, this type of data is collected from respondents through various data collection methods such as surveys, interviews or observations. These are measurable variables that are used to represent the latent constructs.

Examining Measurement Model Fit Indices

Good fit indicates that the hypothesized measurement model is consistent with observed data, providing support for that model. In contrast, poor fit indicates that the hypothesized measurement model is inconsistent with observed data, and it is interpreted as evidence against the adequacy of the model. Some fit indices emerge from other comparisons, but all are interpreted as support for or against the overall adequacy of the hypothesized model.

The most commonly-examined index is chi-square, which indicates the degree of mis-fit of the model. Small, non-significant chi-square values indicate little mis-fit, providing support for a hypothesized measurement model. In contrast, large significant chi-square values indicate large mis-fit, providing evidence against the hypothesized model. This "significant is bad" interpretation of chi-square differs from the typical perspective on inferential statistics, in which researchers generally hope for significant effects. Although chi-square is usually examined and reported in CFA, researchers and readers should recall that sample size affects chi-square. As with any inferential procedure, large samples produce large chi-square values, which produce statistical significance. This creates a paradox for CFA – large samples are required in order to obtain robust, reliable parameter estimates, but they increase the likelihood of significant chi-square values indicating inadequacy of a hypothesized model. For this reason (among others), researchers examine additional indices of model fit – most of which are not formal inferential statistics.

For example, researchers examine indices such as the Goodness of Fit Index (GFI), the Incremental fit Index (IFI), the Normed Fit Index (NFI), the Comparative Fit Index (CFI), the Non-normed Fit Index (NNFI, also known as the Tucker-Lewis Index or TLI), the Root Mean Square of Approximation (RMSEA), the Root Mean Square Residual (RMR), the Standardized Root Mean Square Residual (SRMR), and the Akaike Information Criterion (AIC), to name but a few. The fit indices have differing scales and norms for indicating model adequacy – for example, large values of the GFI (up to 1.0) indicate good fit, but small values of the RMR (down to 0) indicate good fit. Many sources provide guidance for interpreting the various fit indices (e.g., Hu & Bentler, 1995, 1999; Kline, 1998; Marsh, Balla, & McDonald, 1988).

Reference – fit indices

Measure	Threshold
Chi-square/df (cmin/df)	< 3 good; < 5 sometimes permissible
p-value for the model	> .05
CFI	> .95 great; > .90 traditional; > .80 sometimes permissible
GFI	> .95
AGFI	> .80
SRMR	< .09
RMSEA	< .05 good; .05 - .10 moderate; > .10 bad
PCLOSE	> .05

Source Hair, J., Black, W., Babin, B., and Anderson, R. (2010). *Multivariate data analysis* (7th ed.): Prentice-Hall, Inc. Upper Saddle River, NJ, USA.

PRACTICAL EXAMPLE WITH SAMPLE DATA

Consider a situation where a researcher is interested in studying factors that impact expectation of service quality in banks. After reviewing the relevant theory and literature, the researcher identified that the five factors tangibility, reliability, responsiveness, assurance and empathy have largest impact (Parasuraman et.al.1988). The measured variables for these five factors are evaluated using a 5- point, agree-disagree Likert scale.

The measured variables are represented by the following 22 items.

Tangibility

- Up-to-date equipment and instrument facilities of your bank (SQ_E1.1)
- Bank's physical facilities should be visually appealing (SQ_E1.2).
- Employees of your bank should be well dressed and appear neat (SQ_E1.3).
- The appearance of the physical facilities of the bank should be in keeping with the type of services provided (SQ_E1.4).

Reliability

- Dependability in handling customer service problems (SQ_E2.1).
- Providing services as promised (SQ_E2.2).
- Performing services right at the first time (SQ_E2.3).
- Bank should keep their records accurately (SQ_E2.4).
- Providing services at the promised time (SQ_E2.5).

Responsiveness

- Keeping customers informed about when services will be performed (SQ_E3.1).
- Bank should provide prompt service to the customers (SQ_E3.2).
- Employees of bank should always be willing to help customers (SQ_E3.3).
- Banker should reply to any query of the customers (SQ_E3.4).

Assurance

- When customers have problems, Bank should be sympathetic and reassuring (SQ_E4.1).

- Customers should feel safe in transactions with Banks (SQ_E4.2).
- Customers can trust employees of their Bank (SQ_E4.3).
- Employees of banks should be polite (SQ_E4.4).

Empathy

- Employees get adequate support from Bank to do their jobs well (SQ_E5.1).
- Bankers should give individual attention to the customers (SQ_E5.2).
- Bank should have operating hours convenient to all their customers (SQ_E5.3).
- Having the customer's best interest at heart (SQ_E5.4).
- Bankers should try to understand the needs of their customers (SQ_E5.5).

The following steps will explain the measurement model for the above five factors of service quality.

Step 1: **Start → Programs → Amos software → Amos Graphics**. The following screen will display. On the far left appear the different tools that can be used to create path diagrams. Just to the right of the toolbar buttons is a column that will display information about the model after estimates have been calculated. The remainder of the screen contains the area where the path diagram will be drawn.

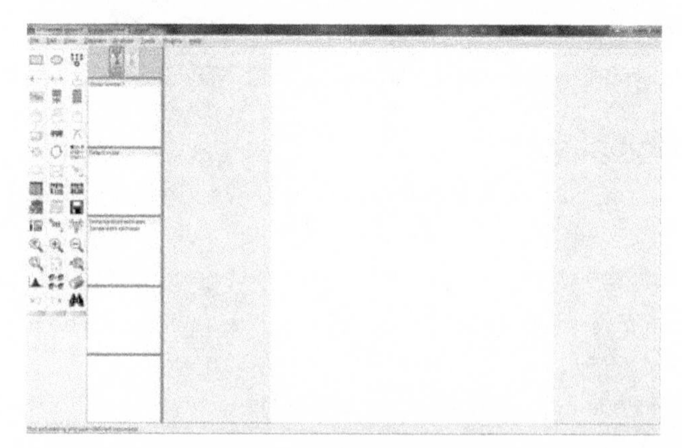

Step 2: To load the data, go to **File → Data Files**. The **Data Files** dialog box then opens. Click on **File Name** and navigate to the location where the data file is stored. By default, Amos looks for an SPSS file.

Step 3: Choose the SPSS data file for working in AMOS, click **Open** and then **OK**.

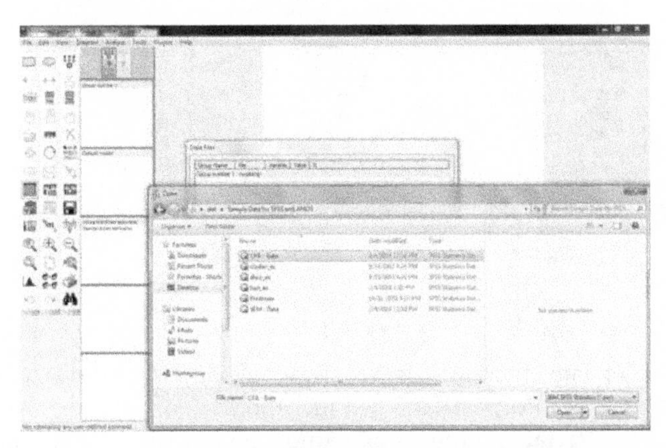

Step 4: Draw five latent variables with required number of indicators. Click the **Draw a latent variable or add an indicator to a latent variable** button, place the curser inside the oval, and click based on the indicator numbers. Amos adds the requested rectangles representing observed indicators along with ovals representing measurement error. The scales of the unique factors are automatically set by constraining the regression weights to equal one. The variables could then be named as described below.

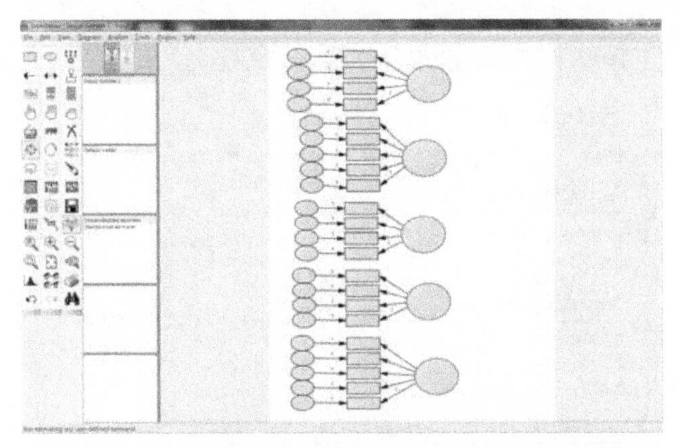

Step 5: Click **"List variables in data set"** and then drag the names of the measured variables into the rectangles as shown below.

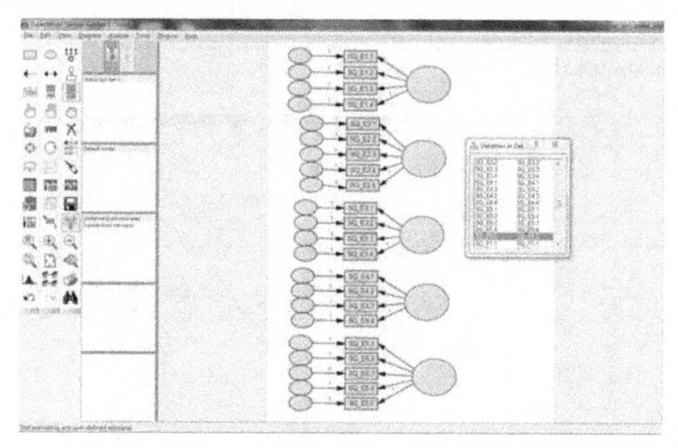

Step 6: To name the common latent variable right click inside it and choose **Object Properties**. Then click on the Text tab and enter the name in **Variable name** box. Amos applies the change immediately to the path diagram. It is also possible, if desired, to add a label describing the variable. To name the unique factors representing measurement error, follow the same process. Name these e1 through e22, yielding the following path diagram.

Step 7: Draw co-variances between all the latent variables using the two-headed arrow.

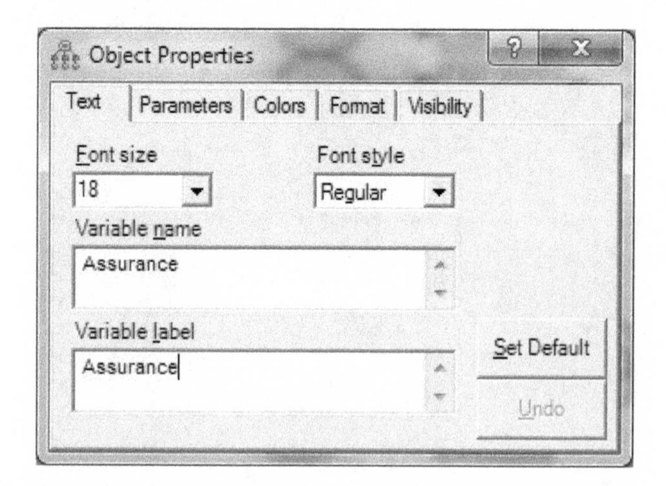

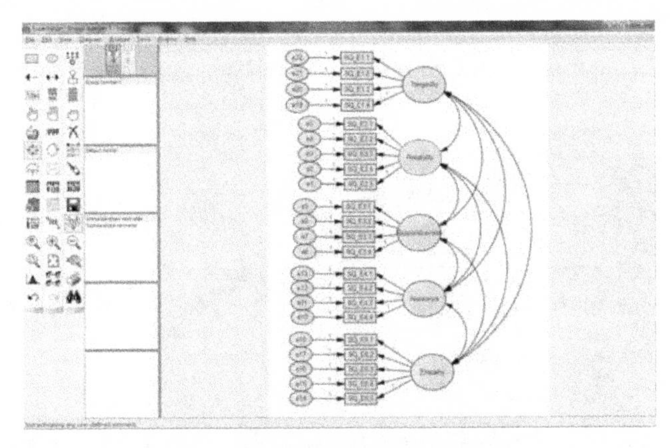

Step 8 : click " **Analysis properties**" and select the desired output.

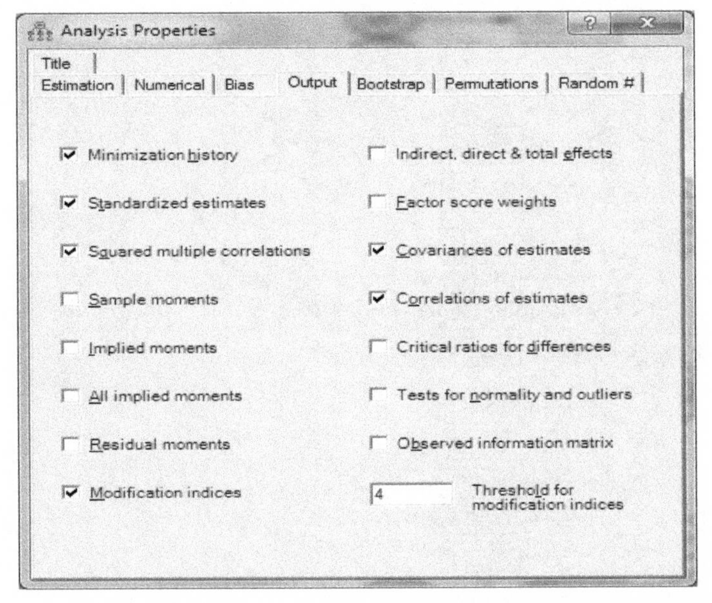

Step 9: To estimate this model, go to **Analyze → Calculate Estimates**. The standardized output can be viewed by clicking **View the output path diagram** button. By default the unstandardized estimates will display. To bring up the standardized estimates, click on the **Standardized estimates** option in the column between the tools and the drawing space.

The path diagram now displays the standardized regression weights (factor loadings) for the common factor and each of the indicators. The squared multiple correlation coefficients (R2), describing the amount of variance the common factor accounts for in the observed variables, are also displayed. Additionally, a X^2 (chi-square) statistic is listed in the column between the tools and the path diagram.

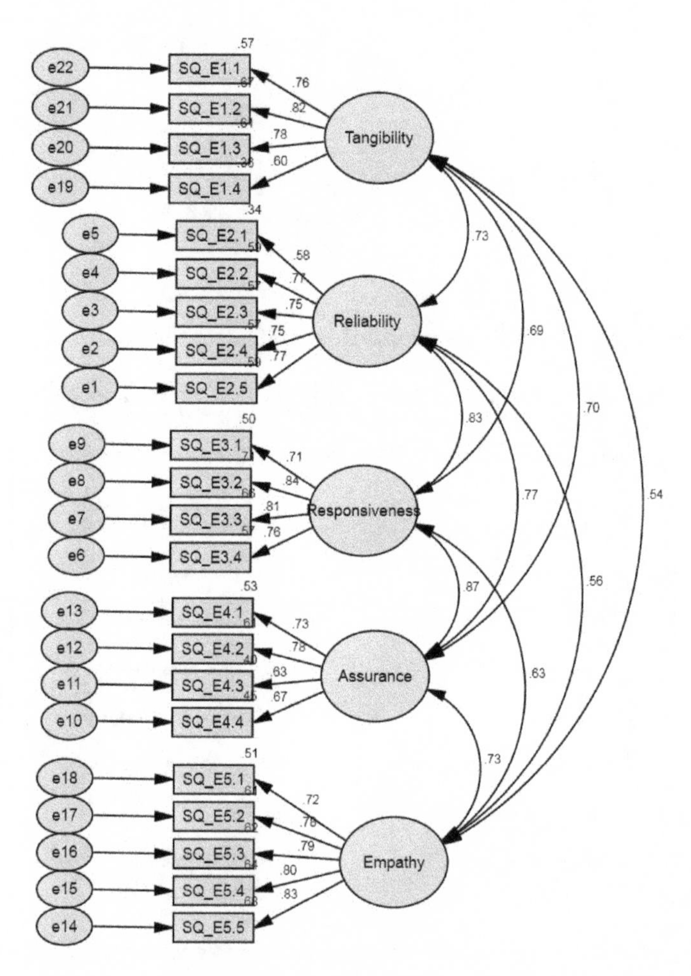

Look at the above figure with standardized estimates. It is possible to get more information about the model than what appears in the path diagram by going to **View → Text** output. This opens an output window giving information about the raw data, the model, estimation, model fit, and any additional information requested with the **Analysis Properties** box utilized earlier. Now consider only the parameter estimates, which are displayed in the output window as follows.

Result (Default model)

Minimum was achieved

Chi-square = 1077.902

Degrees of freedom = 199

Probability level = .000

Scalar Estimates (Group number 1 - Default model)

Maximum Likelihood Estimates

Regression Weights: (Group number 1 - Default model)

			Estimate	S.E.	C.R.	P	Label
SQ_E2.5	<---	Reliability	1.000				
SQ_E2.4	<---	Reliability	1.109	.072	15.496	***	par_1
SQ_E2.3	<---	Reliability	.983	.063	15.510	***	par_2
SQ_E2.2	<---	Reliability	.938	.059	15.805	***	par_3
SQ_E2.1	<---	Reliability	.813	.069	11.700	***	par_4
SQ_E3.4	<---	Responsiveness	1.000				
SQ_E3.3	<---	Responsiveness	1.059	.063	16.929	***	par_5
SQ_E3.2	<---	Responsiveness	1.091	.062	17.558	***	par_6
SQ_E3.1	<---	Responsiveness	.759	.052	14.527	***	par_7
SQ_E4.4	<---	Assurance	1.000				
SQ_E4.3	<---	Assurance	.913	.080	11.396	***	par_8
SQ_E4.2	<---	Assurance	1.111	.082	13.615	***	par_9
SQ_E4.1	<---	Assurance	1.094	.085	12.888	***	par_10
SQ_E5.5	<---	Empathy	1.000				
SQ_E5.4	<---	Empathy	.879	.048	18.426	***	par_11
SQ_E5.3	<---	Empathy	.878	.049	18.090	***	par_12
SQ_E5.2	<---	Empathy	.883	.049	17.901	***	par_13
SQ_E5.1	<---	Empathy	.868	.055	15.889	***	par_14
SQ_E1.4	<---	Tangibility	1.000				
SQ_E1.3	<---	Tangibility	1.298	.109	11.889	***	par_15
SQ_E1.2	<---	Tangibility	1.308	.107	12.169	***	par_16
SQ_E1.1	<---	Tangibility	1.376	.118	11.648	***	par_17

Standardized Regression Weights: (Group number 1 - Default model)

			Estimate
SQ_E2.5	<---	Reliability	.769
SQ_E2.4	<---	Reliability	.753
SQ_E2.3	<---	Reliability	.753
SQ_E2.2	<---	Reliability	.766
SQ_E2.1	<---	Reliability	.584
SQ_E3.4	<---	Responsiveness	.758
SQ_E3.3	<---	Responsiveness	.814
SQ_E3.2	<---	Responsiveness	.841
SQ_E3.1	<---	Responsiveness	.709
SQ_E4.4	<---	Assurance	.668
SQ_E4.3	<---	Assurance	.632
SQ_E4.2	<---	Assurance	.780
SQ_E4.1	<---	Assurance	.729
SQ_E5.5	<---	Empathy	.827
SQ_E5.4	<---	Empathy	.800
SQ_E5.3	<---	Empathy	.789
SQ_E5.2	<---	Empathy	.783
SQ_E5.1	<---	Empathy	.716
SQ_E1.4	<---	Tangibility	.598
SQ_E1.3	<---	Tangibility	.782
SQ_E1.2	<---	Tangibility	.816
SQ_E1.1	<---	Tangibility	.756

Correlations: (Group number 1 - Default model)

			Estimate
Reliability	<-->	Responsiveness	.830
Reliability	<-->	Assurance	.772
Reliability	<-->	Empathy	.561
Reliability	<-->	Tangibility	.727
Responsiveness	<-->	Assurance	.869
Responsiveness	<-->	Empathy	.626
Responsiveness	<-->	Tangibility	.694

			Estimate
Assurance	<-->	Empathy	.726
Assurance	<-->	Tangibility	.702
Empathy	<-->	Tangibility	.538

Covariances: (Group number 1 - Default model)

			M.I.	**Par Change**
e20	<-->	Responsiveness	15.631	.047
e19	<-->	Tangibility	18.023	-.051
e19	<-->	Reliability	23.409	.071
e17	<-->	e18	27.003	.091
e15	<-->	e18	25.839	-.085
e13	<-->	Responsiveness	15.635	.055
e13	<-->	e19	20.735	.095
e10	<-->	e13	21.115	-.098
e10	<-->	e11	53.369	.166
e9	<-->	Tangibility	32.126	-.058
e9	<-->	Reliability	73.924	.106
e7	<-->	Reliability	23.382	-.063
e6	<-->	Reliability	18.025	-.061
e6	<-->	e16	16.681	-.070
e6	<-->	e12	18.417	.078
e6	<-->	e9	16.575	-.070
e4	<-->	e15	19.877	-.065
e4	<-->	e12	19.085	-.069
e3	<-->	e19	21.293	.091
e3	<-->	e13	17.662	.081
e2	<-->	e17	23.687	-.091
e2	<-->	e16	20.021	.082
e1	<-->	Tangibility	19.374	-.049
e1	<-->	Responsiveness	16.696	.052
e1	<-->	e9	64.874	.130

Model Fit Summary

CMIN

Model	NPAR	CMIN	DF	P	CMIN/DF
Default model	54	1077.902	199	.000	**5.417**
Saturated model	253	.000	0		
Independence model	22	5904.582	231	.000	25.561

RMR, GFI

Model	RMR	GFI	AGFI	PGFI
Default model	**.044**	**.815**	**.765**	.641
Saturated model	.000	1.000		
Independence model	.301	.202	.126	.185

Baseline Comparisons

Model	NFI Delta1	RFI rho1	IFI Delta2	TLI rho2	CFI
Default model	**.817**	**.788**	**.846**	**.820**	**.845**
Saturated model	1.000		1.000		1.000
Independence model	.000	.000	.000	.000	.000

Parsimony-Adjusted Measures

Model	PRATIO	PNFI	PCFI
Default model	.861	.704	.728
Saturated model	.000	.000	.000
Independence model	1.000	.000	.000

NCP

Model	NCP	LO 90	HI 90
Default model	878.902	779.711	985.588
Saturated model	.000	.000	.000
Independence model	5673.582	5426.474	5927.053

FMIN

Model	FMIN	F0	LO 90	HI 90
Default model	2.597	2.118	1.879	2.375
Saturated model	.000	.000	.000	.000
Independence model	14.228	13.671	13.076	14.282

RMSEA

Model	RMSEA	LO 90	HI 90	PCLOSE
Default model	**.103**	**.097**	**.109**	**.000**
Independence model	.243	.238	.249	.000

AIC

Model	AIC	BCC	BIC	CAIC
Default model	1185.902	1192.238	1403.559	1457.559
Saturated model	506.000	535.689	1525.763	1778.763
Independence model	5948.582	5951.164	6037.257	6059.257

ECVI

Model	ECVI	LO 90	HI 90	MECVI
Default model	**2.858**	**2.619**	**3.115**	**2.873**
Saturated model	1.219	1.219	1.219	1.291
Independence model	14.334	13.738	14.945	14.340

HOELTER

Model	HOELTER .05	HOELTER .01
Default model	90	96
Independence model	19	20

RESULTS AND DISCUSSION

Under the Regression Weights heading the unstandardised loadings appear along with standard errors, a critical ratio, and p-values. All of the unconstrained estimates are significant. The Standardized Regression Weights can be interpreted as the correlation between the observed variable and the corresponding common factor. For this five-factor model the regression weights are all significant. The result modification indices make suggestions about loosening certain model parameters in order to improve the overall model fit. As long as any decisions made on the basis of modification indices are theoretically meaningful and do not result in an unidentified model they can be helpful in improving model specification.

The model fit results indicates that the CMIN/DF, 5.417 is not reasonable fit to the model. The Root Mean Square Error of Approximation (RMSEA) value of 0.103, CFI of 0.845, GFI of 0.815 and AGFI 0.765 also indicates that the model have to be improved further more to attain the good fit indices.

For the purpose of increasing model fit, the modification indices show the variables which are high co-variance and correlation with other variables in the diagram. From the model it is identified that the variables SQ_E1.4, SQ_E 2.5, SQ_E 3.4, SQ_E 4.3, SQ_E 5.1 are problematic indicators and these are deleted from the model. This improves the model fit for the diagram and the modified diagram presented as follows.

MODIFIED MEASUREMENT MODEL

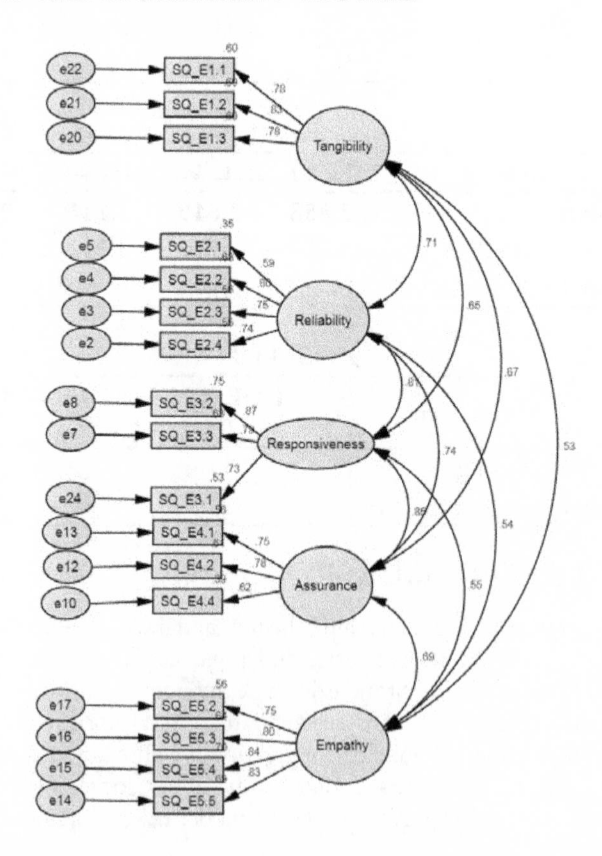

The results of the modified model based on the modification indices presented below.

Scalar Estimates (Group number 1 - Default model)

Maximum Likelihood Estimates

Regression Weights: (Group number 1 - Default model)

			Estimate	S.E.	C.R.	P	Label
SQ_E2.4	<---	Reliability	1.000				
SQ_E2.3	<---	Reliability	.896	.062	14.498	***	par_1
SQ_E2.2	<---	Reliability	.892	.058	15.342	***	par_2
SQ_E2.1	<---	Reliability	.757	.066	11.441	***	par_3
SQ_E3.3	<---	Responsiveness	1.333	.087	15.365	***	par_4
SQ_E3.2	<---	Responsiveness	1.451	.087	16.592	***	par_5
SQ_E4.4	<---	Assurance	1.000				
SQ_E4.2	<---	Assurance	1.196	.097	12.373	***	par_6
SQ_E4.1	<---	Assurance	1.201	.100	11.993	***	par_7
SQ_E5.5	<---	Empathy	1.000				
SQ_E5.4	<---	Empathy	.916	.047	19.279	***	par_8
SQ_E5.3	<---	Empathy	.888	.049	18.234	***	par_9
SQ_E5.2	<---	Empathy	.837	.050	16.611	***	par_10
SQ_E1.3	<---	Tangibility	1.000				
SQ_E1.2	<---	Tangibility	1.038	.063	16.433	***	par_11
SQ_E1.1	<---	Tangibility	1.098	.071	15.509	***	par_12
SQ_E3.1	<---	Responsiveness	1.000				

Standardized Regression Weights:(Group number 1-Default model)

			Estimate
SQ_E2.4	<---	Reliability	.742
SQ_E2.3	<---	Reliability	.750
SQ_E2.2	<---	Reliability	.796
SQ_E2.1	<---	Reliability	.594
SQ_E3.3	<---	Responsiveness	.794
SQ_E3.2	<---	Responsiveness	.868
SQ_E4.4	<---	Assurance	.624
SQ_E4.2	<---	Assurance	.784
SQ_E4.1	<---	Assurance	.747

			Estimate
SQ_E5.5	<---	Empathy	.830
SQ_E5.4	<---	Empathy	.836
SQ_E5.3	<---	Empathy	.800
SQ_E5.2	<---	Empathy	.745
SQ_E1.3	<---	Tangibility	.775
SQ_E1.2	<---	Tangibility	.833
SQ_E1.1	<---	Tangibility	.777
SQ_E3.1	<---	Responsiveness	.725

Covariances: (Group number 1 - Default model)

			Estimate	S.E.	C.R.	P	Label
Reliability	<-->	Responsiveness	.314	.034	9.314	***	par_13
Reliability	<-->	Assurance	.289	.035	8.204	***	par_14
Reliability	<-->	Empathy	.283	.036	7.777	***	par_15
Reliability	<-->	Tangibility	.309	.035	8.868	***	par_16
Responsiveness	<-->	Assurance	.250	.029	8.665	***	par_17
Responsiveness	<-->	Empathy	.218	.028	7.920	***	par_18
Responsiveness	<-->	Tangibility	.213	.025	8.427	***	par_19
Assurance	<-->	Empathy	.273	.033	8.276	***	par_20
Assurance	<-->	Tangibility	.222	.028	7.916	***	par_21
Empathy	<-->	Tangibility	.234	.030	7.760	***	par_22

Correlations: (Group number 1 - Default model)

			Estimate
Reliability	<-->	Responsiveness	.810
Reliability	<-->	Assurance	.739
Reliability	<-->	Empathy	.539
Reliability	<-->	Tangibility	.706
Responsiveness	<-->	Assurance	.848
Responsiveness	<-->	Empathy	.552
Responsiveness	<-->	Tangibility	.646
Assurance	<-->	Empathy	.686
Assurance	<-->	Tangibility	.668
Empathy	<-->	Tangibility	.527

Model Fit Summary

CMIN

Model	NPAR	CMIN	DF	P	CMIN/DF
Default model	44	437.890	109	.000	4.017
Saturated model	153	.000	0		
Independence model	17	4105.870	136	.000	30.190

RMR, GFI

Model	RMR	GFI	AGFI	PGFI
Default model	.033	.894	.851	.637
Saturated model	.000	1.000		
Independence model	.295	.249	.156	.222

Baseline Comparisons

Model	NFI Delta1	RFI rho1	IFI Delta2	TLI rho2	CFI
Default model	.893	.867	.918	.897	.917
Saturated model	1.000		1.000		1.000
Independence model	.000	.000	.000	.000	.000

Parsimony-Adjusted Measures

Model	PRATIO	PNFI	PCFI
Default model	.801	.716	.735
Saturated model	.000	.000	.000
Independence model	1.000	.000	.000

NCP

Model	NCP	LO 90	HI 90
Default model	328.890	268.215	397.127
Saturated model	.000	.000	.000
Independence model	3969.870	3764.391	4182.631

FMIN

Model	FMIN	F0	LO 90	HI 90
Default model	1.055	.793	.646	.957
Saturated model	.000	.000	.000	.000
Independence model	9.894	9.566	9.071	10.079

RMSEA

Model	RMSEA	LO 90	HI 90	PCLOSE
Default model	.085	.077	.094	.000
Independence model	.265	.258	.272	.000

AIC

Model	AIC	BCC	BIC	CAIC
Default model	525.890	529.880	703.240	747.240
Saturated model	306.000	319.874	922.695	1075.695
Independence model	4139.870	4141.412	4208.392	4225.392

ECVI

Model	ECVI	LO 90	HI 90	MECVI
Default model	1.267	1.121	1.432	1.277
Saturated model	.737	.737	.737	.771
Independence model	9.976	9.480	10.488	9.979

HOELTER

Model	HOELTER .05	HOELTER .01
Default model	128	139
Independence model	17	18

RESULTS AND DISCUSSION

CFA output includes many fit indices. Each SEM program (AMOS, LISREL, EQS etc.) includes a slightly different set, but all contain the key values such as the x^2 statistic, the CFI, and RMSEA. The values may appear in a different order or perhaps in a tabular format, but the researchers have to find enough information to evaluate the model's fit in any program.

The above result includes many fit statistics. The rule of thumb suggests that the model should obtain one absolute fit index and one incremental fit index, in addition to the x^2 results. The value for RMSEA, an absolute fit index is 0.085. This value provides additional support for model fit. The next absolute fit index CMIN/DF is 4.01. This measure is the chi-square value divided by the degrees of freedom. A number smaller than 2.0 is considered very good and between 2.0 and 5.0 is acceptable. Thus the normed chi-square suggests an acceptable fit for the model.

Moving to the measurement fit indices the CFA is the most widely used index. In the expected service quality CFA model CFI has a value of 0.917, which like the RMSEA, exceeds the CFI guidelines of greater than 0.90 for a model of this complexity and sample size. The other incremental fit indices also exceed suggested cutoff values. Although this model is not compared to other models, the parsimony index of AGFI has a value (), which reflects good model fit.

The CFA results suggest the expected service quality measurement model provides a reasonably good fit, and thus it is suitable to proceed to further examination of the model results.

HIGHER ORDER FACTOR ANALYSIS

The CFA model described in part-1 is a first order factor model. A first order factor model means that the covariances between measured items are explained with a single latent factor layer. For now, think of a layer as one level of latent constructs. Researchers increasingly are employing higher order factor analyses although this aspect of measurement theory is not new. Higher order CFAs most often test a second order factor structure that contains two layers of latent constructs. The second order latent factor(s) causes multiple first order latent factors, which in turn cause the measured variables (x). Theoretically this process can be extended to any number of multiple layers. Thus the term higher order factor analysis. Researchers seldom examine theories beyond a second order model.

Both theoretical and empirical considerations are associated with higher-order CFA. All CFA models must account for the relationships among constructs. In first order CFA model, these covariance terms are typically free (estimated) unless the researcher has a strong theoretical reason to hypothesize independent dimensions. Therefore, all of the factors are interrelated but without a specific causal construct. Higher order factors can be thought of as explicitly representing the causal constructs that impact the first order factors. An additional way to view a higher order factor is that it accounts for covariance between constructs just as first order factors account for covariance between observed variables. In other words, the first order factors now act as indicators of the second-order factor. All the considerations and rules of thumb apply to second order factors just as first order factors. The difference is that the researcher must consider the first order constructs as indicators of the second order construct.

Practical example

Using the example of first order confirmatory analysis, the following second order factor model is drawn.

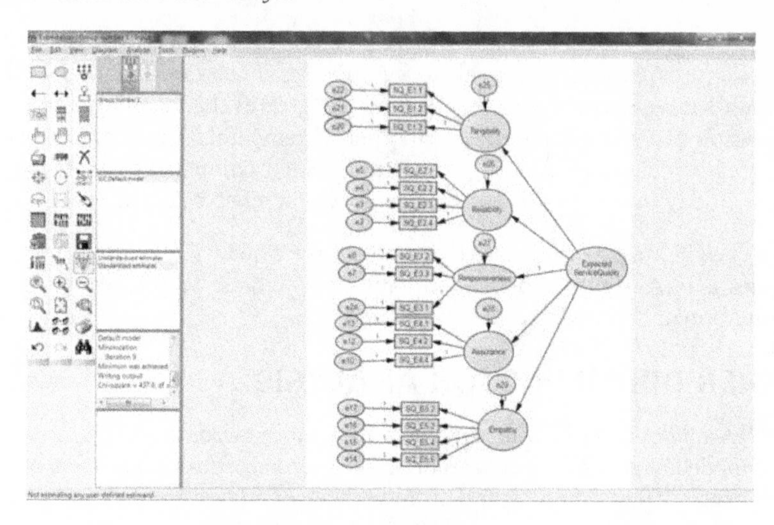

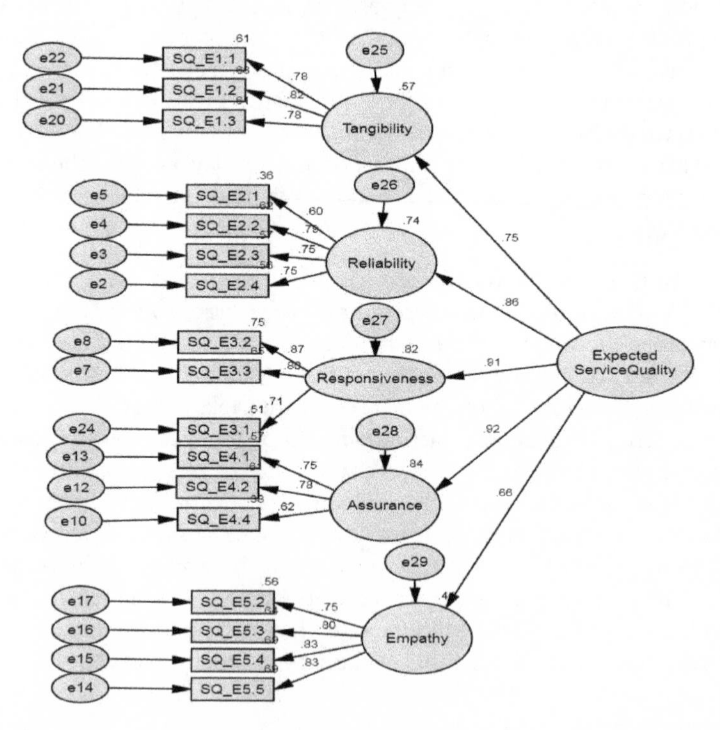

Note: First order factors (treated as endogeneous): Tangibility, Reliability, Responsiveness, Assurance and Empathy.

Second order factor (treated as exogeneous): Expected Service Quality.

The following are the results of second order factor analysis.

Scalar Estimates (Group number 1 - Default model)

Maximum Likelihood Estimates

Regression Weights: (Group number 1 - Default model)

			Esti mate	S.E.	C.R.	P	Label
Tangibility	<---	F1	.962	.087	10.998	***	par_13
Empathy	<---	F1	1.011	.096	10.536	***	par_14
Reliability	<---	F1	1.298	.111	11.681	***	par_15
Responsiveness	<---	F1	1.000				
Assurance	<---	F1	1.031	.099	10.392	***	par_16
SQ_E2.4	<---	Reliability	1.000				
SQ_E2.3	<---	Reliability	.894	.061	14.595	***	par_1
SQ_E2.2	<---	Reliability	.873	.058	15.167	***	par_2
SQ_E2.1	<---	Reliability	.756	.066	11.516	***	par_3
SQ_E3.3	<---	Responsiveness	1.375	.091	15.182	***	par_4
SQ_E3.2	<---	Responsiveness	1.478	.092	16.137	***	par_5
SQ_E4.4	<---	Assurance	1.000				
SQ_E4.2	<---	Assurance	1.204	.099	12.161	***	par_6
SQ_E4.1	<---	Assurance	1.219	.103	11.866	***	par_7
SQ_E5.5	<---	Empathy	1.000				
SQ_E5.4	<---	Empathy	.908	.048	19.101	***	par_8
SQ_E5.3	<---	Empathy	.890	.049	18.311	***	par_9
SQ_E5.2	<---	Empathy	.839	.050	16.687	***	par_10
SQ_E1.3	<---	Tangibility	1.000				
SQ_E1.2	<---	Tangibility	1.019	.062	16.330	***	par_11
SQ_E1.1	<---	Tangibility	1.098	.070	15.643	***	par_12
SQ_E3.1	<---	Responsiveness	1.000				

Standardized Regression Weights: (Group number 1 - Default model)

			Estimate
Tangibility	<---	F1	.754
Empathy	<---	F1	.665
Reliability	<---	F1	.863

			Estimate
Responsiveness	<---	F1	.906
Assurance	<---	F1	.918
SQ_E2.4	<---	Reliability	.747
SQ_E2.3	<---	Reliability	.754
SQ_E2.2	<---	Reliability	.785
SQ_E2.1	<---	Reliability	.597
SQ_E3.3	<---	Responsiveness	.804
SQ_E3.2	<---	Responsiveness	.868
SQ_E4.4	<---	Assurance	.619
SQ_E4.2	<---	Assurance	.782
SQ_E4.1	<---	Assurance	.752
SQ_E5.5	<---	Empathy	.832
SQ_E5.4	<---	Empathy	.830
SQ_E5.3	<---	Empathy	.803
SQ_E5.2	<---	Empathy	.748
SQ_E1.3	<---	Tangibility	.780
SQ_E1.2	<---	Tangibility	.823
SQ_E1.1	<---	Tangibility	.782
SQ_E3.1	<---	Responsiveness	.712

Variances: (Group number 1 - Default model)

	Estimate	S.E.	C.R.	P	Label
F1	.231	.032	7.210	***	par_17
e25	.162	.023	7.196	***	par_18
e26	.134	.025	5.350	***	par_19
e27	.050	.012	4.381	***	par_20
e28	.046	.014	3.258	.001	par_21
e29	.298	.034	8.832	***	par_22
e2	.414	.036	11.510	***	par_23
e3	.317	.028	11.378	***	par_24
e4	.248	.023	10.698	***	par_25
e5	.538	.041	13.155	***	par_26
e7	.290	.027	10.934	***	par_27
e8	.201	.023	8.596	***	par_28
e10	.470	.037	12.861	***	par_29

	Estimate	S.E.	C.R.	P	Label
e12	.268	.026	10.237	***	par_30
e13	.333	.030	11.034	***	par_31
e14	.239	.023	10.286	***	par_32
e15	.198	.019	10.322	***	par_33
e16	.233	.021	11.081	***	par_34
e17	.296	.024	12.104	***	par_35
e20	.241	.023	10.576	***	par_36
e21	.186	.020	9.238	***	par_37
e22	.289	.027	10.541	***	par_38
e24	.274	.022	12.516	***	par_39

Squared Multiple Correlations: (Group number 1 - Default model)

	Estimate
Tangibility	.568
Empathy	.442
Assurance	.842
Responsiveness	.821
Reliability	.744
SQ_E3.1	.507
SQ_E1.1	.611
SQ_E1.2	.677
SQ_E1.3	.609
SQ_E5.2	.560
SQ_E5.3	.645
SQ_E5.4	.690
SQ_E5.5	.692
SQ_E4.1	.565
SQ_E4.2	.612
SQ_E4.4	.383
SQ_E3.2	.754
SQ_E3.3	.647
SQ_E2.1	.357
SQ_E2.2	.616
SQ_E2.3	.569
SQ_E2.4	.558

Model Fit Summary

CMIN

Model	NPAR	CMIN	DF	P	CMIN/DF
Default model	39	467.362	114	.000	4.100
Saturated model	153	.000	0		
Independence model	17	4105.870	136	.000	30.190

RMR, GFI

Model	RMR	GFI	AGFI	PGFI
Default model	.035	.887	.848	.661
Saturated model	.000	1.000		
Independence model	.295	.249	.156	.222

Baseline Comparisons

Model	NFI Delta1	RFI rho1	IFI Delta2	TLI rho2	CFI
Default model	.886	.864	.911	.894	.911
Saturated model	1.000		1.000		1.000
Independence model	.000	.000	.000	.000	.000

Parsimony-Adjusted Measures

Model	PRATIO	PNFI	PCFI
Default model	.838	.743	.764
Saturated model	.000	.000	.000
Independence model	1.000	.000	.000

NCP

Model	NCP	LO 90	HI 90
Default model	353.362	290.459	423.823
Saturated model	.000	.000	.000
Independence model	3969.870	3764.391	4182.631

FMIN

Model	FMIN	F0	LO 90	HI 90
Default model	1.126	.851	.700	1.021
Saturated model	.000	.000	.000	.000
Independence model	9.894	9.566	9.071	10.079

RMSEA

Model	RMSEA	LO 90	HI 90	PCLOSE
Default model	.086	.078	.095	.000
Independence model	.265	.258	.272	.000

AIC

Model	AIC	BCC	BIC	CAIC
Default model	545.362	548.899	702.559	741.559
Saturated model	306.000	319.874	922.695	1075.695
Independence model	4139.870	4141.412	4208.392	4225.392

ECVI

Model	ECVI	LO 90	HI 90	MECVI
Default model	1.314	1.163	1.484	1.323
Saturated model	.737	.737	.737	.771
Independence model	9.976	9.480	10.488	9.979

HOELTER

Model	HOELTER .05	HOELTER .01
Default model	125	136
Independence model	17	18

RESULTS AND DISCUSSION

The goodness-of-fit statistics related to the second order factor model are presented in the above tables. It is observed that the CMIN/DF value is 4.10 indicates moderate fit to the model. Thus, it is more reasonable and appropriate to base decisions on other indices of fit. Primary among these in the AMOS

Output are the CFI and RMSEA values. Furthermore, the ECVI is also of interest. In reviewing these fit indices, the hypothesized model is relatively well fitting as indicated by a CFI of .911 and a RMSEA value of .086, which is well within the recommended range of acceptability (< .05 to .10). In addition, the ECVI for this initially hypothesized model is 1.134, indicating a reasonable fit to the model.

Chapter - 7

Cluster Analysis

Learning Objectives

This chapter helps to understand the following

- Meaning of Cluster Analysis
- Objectives of Cluster Analysis
- Area of Application of Cluster Techniques
- Advantage and Disadvantage of Cluster Analysis
- Requirements for Cluster Analysis
- Difference Between Cluster Analysis with Factor analysis and Discriminant Analysis
- Types of Cluster Analysis
- Steps for conducting Cluster Analysis
- Clustering Procedure
- Select A Clustering Algorithm
- Difference between Hierarchical and Non -Hierarchical Cluster Analysis
- Deciding on the Numbers of Cluster

Introduction

The term cluster analysis does not have any particular statistical technique or model, as do discriminant analysis, factor analysis, and regression has. Cluster analysis is a group of multivariate data mining techniques and its primary purpose is to identify homogenous group data objects based on information found in the data that defines the objects, characteristics and their relationships they possess. Cluster analysis is also known as Q analysis, typology construction, classification analysis, numerical taxonomy, and exploratory data analysis.

Clustering is defined as "the process of organizing objects into groups whose members are similar in some way".

A cluster is a group of objects which are "similar" between them and are "dissimilar" to the objects belonging to other clusters. It is a group of relative variables or homogeneous observation. The essence of all clustering approaches is the classification of data as suggested by "natural" groupings of the data themselves.

What is Cluster Analysis?

Cluster analysis is a collection of statistical methods subdividing a sample into homogeneous classes to produce an operational classification. (E.g. people, things, events, brands, companies) It is used to organize pragmatic data into meaningful structures, taxonomies, groups or clusters. It deals with finding a structure in a collection of unlabeled data.

Cluster analysis is a technique used for combining group observations into homogenous classes –clusters and observations should be

1. HOMOGENOUS **within clusters**: Each group has impact with respect to certain characteristics i.e., observations in the same cluster should be similar (not dissimilar)
2. HETEROGENOUS **between clusters**: Each group should be different from other groups with respect to the characteristics. i.e., observations from different clusters should be quite distinct (dissimilar)

Clusters should exhibit high internal homogeneity and high external heterogeneity. We are interested in determining groups of observations internally characterized by a high level of structure. The goal is that the objects within a group be similar (or related) to one another with respect to variables or attributes and different from (or unrelated to) the objects in others groups.

1. Cluster Analysis is a way of grouping objects of data based on the resemblance of responses to several variables.
2. Cluster analysis is also used to divide the observations or variables into homogeneous and distinct groups.

Clustering occurs in almost every aspect of our day-to-day life. Take, for example, In supermarket similar types of items are always displayed in the same or nearby location - vegetables, meat, soda, paper products, etc. In the field of marketing, clusters of consumer segments are often required for effective marketing strategies.

OBJECTIVES

Following are the main objectives of cluster analysis,

- Identifying natural groups within the data.(i.e. Taxonomy description)
- The ability to analyze groups of similar observations instead of all individual observations.
- Cluster analysis embraces a variety of techniques, the main objective of which is to group observations or variables into similar and dissimilar clusters.
- Relationship identification – the simplified structure from cluster analysis reveals relationship between variables.
- Theoretical, conceptual and practical considerations must be observed when selecting clustering variables for cluster analysis.

CLUSTER ANALYSIS - AREA OF APPLICATION

Clustering techniques have been applied to a wide variety of fields.

1. **Biology** : clustering to find groups of genes, kingdom, class, order, and family that have similar function
2. **Stock market** : Group stocks with similar price fluctuations
3. **Business:** Clustering can be used to segment customers into a small number of groups of additional analysis and marketing actives like purchasing behavior, database of customer etc. Company similarity based on various company financial metrics.
4. **Psychology:** clustering has been used to identify different types of depression.
5. **Medicine**: cluster analysis can also be used to detect patterns in the spatial, clustering diseases, cures for diseases or symptoms of diseases can lead to very useful taxonomies.
6. **Information Retrieval:** Group related documents for browsing; web pages grouped into data of similar access patterns.
7. **Insurance:** identifying groups of insurance policy holders ; identifying frauds;
8. **Archeology:** clustering of stone tools, funeral objects etc.,

ADVANTAGES OF CLUSTER ANALYSIS

1. Good for quick overview of data in groups.
2. Good for the nearest neighbours, ordination better for the closer relationships.
3. Cluster analysis reveals associations and structure in data and the results of cluster analysis may contribute to formal classification scheme or taxonomy.
4. Cluster analysis is relatively simple, and can use a variety of input data.
5. It has flexibility to meet user needs. So, it is used and acceptance in various fields of research and major statistical drivers are implemented.
6. Its "transparent" norm in peer grouping restricts the analyst bias and suspicion of discrimination.

DISADVANTAGE OF CLUSTER ANALYSIS

1. It is "descriptive, a theoretical and there is no inferential in cluster analysis"
2. It has high influence on the interpretation of analyst and it's difficult to control.
3. The different methods of clustering usually give very different results and it's difficult to validate.
4. Cluster analysis methods are not clearly established; therefore, most of the guidelines for using cluster analysis are rules of thumb.
5. There are no completely satisfactory methods for determining the appropriate number of clusters and the analysis is not stable.
6. The cluster solution is not generalizable because it is totally dependent upon the variables used as a basis for the similarity measure.

REQUIREMENTS FOR CLUSTER ANALYSIS

The main requirements that a clustering algorithm should satisfy are:

- **Scalability** - clustering technique that deals with large sets of data must be scalable both in term of speed and space.
- **Attributes** – it deals with different types of attributes such as interval, binary data etc.
- **Discovering clusters with arbitrary shape** – clustering algorithm should be capable of distinguish cluster of random shape. It should not be restricted to only distance measures.

- **Determination input parameter** - Minimal requirements for domain knowledge to determine input parameters.
- **Ability to deal with noise** – it is an effective means of detecting and dealing with noise and outliers. It can be detected by applying a test and finding the data that belongs to a given clusters, so that noise can be eliminated.
- **Insensitivity to order of input records** – clustering algorithm depends on the order of the input. If the order of data points processed changes, then the resulting clusters may change.
- **High dimensionality** – the clustering procedure should be able to handle low – dimensional data as well as should have ability to handle high dimensional spaces properly.
- **Interpretability** – clustering result should highly influence on the interpretability and usability of data.

DIFFERENCE BETWEEN CLUSTER ANALYSIS AND FACTOR ANALYSIS

Both Factor and Cluster analysis are data reduction techniques. Both of them are unsupervised learning methods in which the outcome is not determined by response from some objective. These two forms of analysis are heavily used in the natural and behavior sciences. Both cluster analysis and factor analysis allow the user to group parts of the data into "clusters" or onto "factors," depending on the type of analysis. While cluster analysis and factor analysis seem similar on the surface, they differ in many ways, including in their overall objectives and applications.

CLUSTER ANLAYSIS	FACTOR ANALYSIS
Cluster analysis is commonly used to classify a group of respondents.	Factor analysis is a data reduction technique.
"Cluster analysis or clustering is the task of assigning a set of objects into groups (called clusters) so that the objects in the same cluster are more similar (in some sense or another) to each other than to those in other clusters."	"Factor analysis is a statistical method used to describe variability among observed, correlated variables in terms of a potentially lower number of unobserved variables called factors."

CLUSTER ANLAYSIS	FACTOR ANALYSIS
Cluster analysis is a grouping of observations.	Factor analysis is a grouping of variables.
Cluster analysis reduces the number of observations or cases by grouping them into a smaller set of clusters.	Whereas Factor analysis reduces the number of variables by grouping them into a smaller set of factors
cluster analysis is a form of categorization	whereas factor analysis is a form of simplification
Cluster works on (ordinal, nominal, interval and ratio) data scaling.	Whereas factor works on likert scale (interval).

DIFFERENCE BETWEEN CLUSTER ANALYSIS AND DISCRIMINANT ANALYSIS

Following are the difference between cluster analysis and discriminant analysis

CLUSTER ANALYSIS	DISCRIMINANT ANALYSIS
Cluster analysis helps to identify objects that are similar to one another, based on some specified criterion	Discriminant analysis helps to identify the independent variables that discriminate a nominally scaled dependent variable.
Cluster analysis classifies unknown groups	While discriminate function analysis classifies known groups.
Cluster analysis is a grouping of observations	It is a linear combination of independent variables indicates the discriminating function showing the large difference that exists in the two group means.
Cluster analysis allows many choices about the nature of the algorithm for combining groups. Each choice may result in a different grouping structure.	On the other hand, the procedure for doing a discriminate function analysis is well established. There are few options, other than type of output, that need to be specified when doing a discriminate function analysis

CLUSTER ANALYSIS	DISCRIMINANT ANALYSIS
Clustering a form of unsupervised learning	Discriminant analysis is a form of supervised learning
Cluster works on to derive segments.	In discriminate anlaysis independent variables measured on an interval or ratio scale.

DIFFERENT TYPES OF CLUSTER ANALYSIS

The term cluster does not have a detailed definition. However, several operative definitions of a cluster are commonly used to visually illustrate the differences among these types of cluster.

1. **Well-Separated Cluster**: A cluster is a set of object, such that any object in a cluster is closer (or more similar) to every other object in the cluster than to any object not in the cluster. Sometimes a threshold is used to specify that all the points in a cluster must be sufficiently close (or similar) to one another.

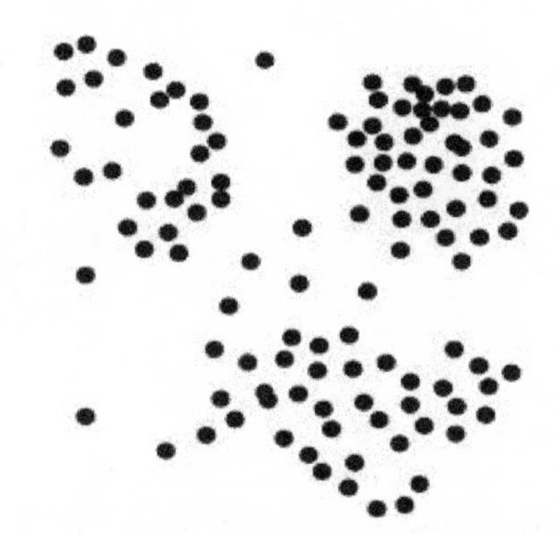

Three well-separated clusters of 2 dimensional points

However, in many sets of data, a point on the edge of a cluster may be closer (or more similar) to some objects in another cluster than to objects in its own cluster. Consequently, many clustering algorithms use the following criterion. Well separated clusters do not need to be globular, but can have any shape.

2. **Center-based Cluster**: A cluster is a set of objects such that an object in a cluster is closer (more similar) to the "center" of a cluster, than to the center of anyother cluster. The center of a cluster is often a centroid,

i.e., the average of all the points in the cluster. When a centroid is not meaningful it is often a medoid, the most "representative" point of a cluster. Center based cluster is also known as prototype based cluster

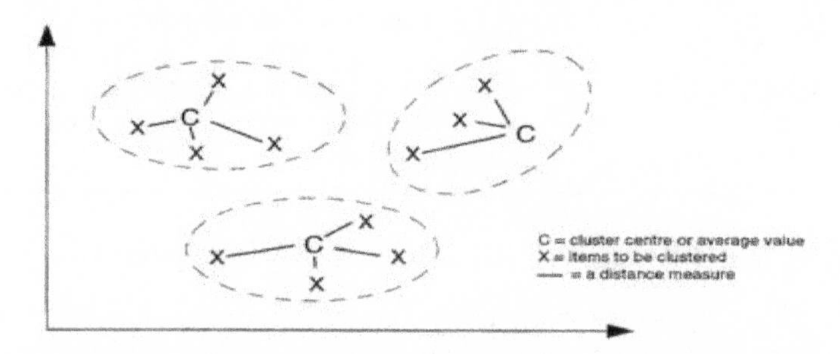

3. **Contiguous Cluster (Nearest neighbor or Transitive Clustering)**: A group of objects that are connected to one another, but that have no connection to objects outside the group is known has connected component. A cluster is a set of points such that a point in a cluster is closer (or more similar) to one or more other points in the cluster than to any point not in the cluster.

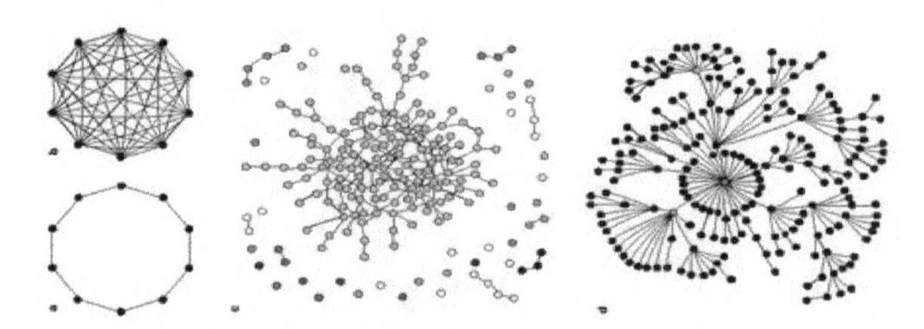

4. **Density-based**: A cluster is a dense region of points, which is separated by low-density regions, from other regions of high density. This definition is more often used when the clusters are irregular or intertwined, and when noise and outliers are present.

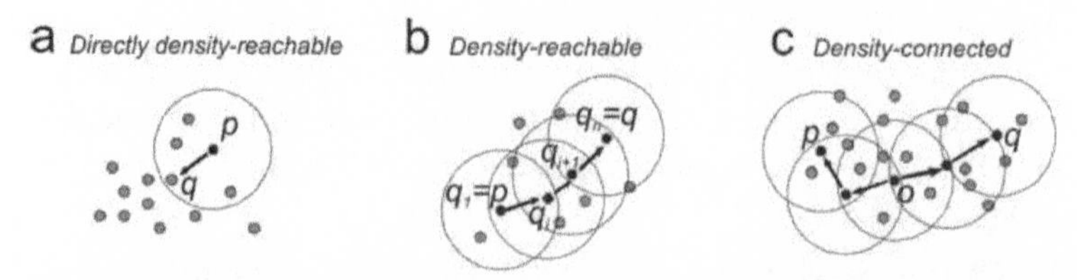

5. **Shared property Cluster (Similarity-based cluster or conceptual cluster)**: we can define a cluster as a set of objects that share some

property with respect to particular concept. A cluster is a set of objects that are "similar", and objects in other clusters are not "similar." A variation on this is to define a cluster as a set of points that together create a region with a uniform local property, e.g., density or shape.

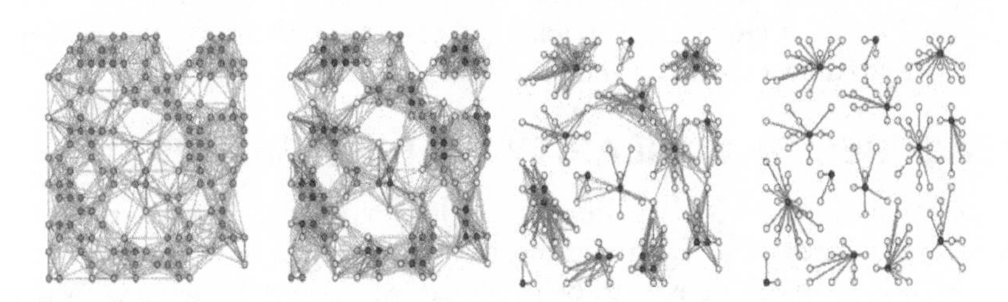

STEPS FOR CONDUCTING CLUSTER ANALYSIS

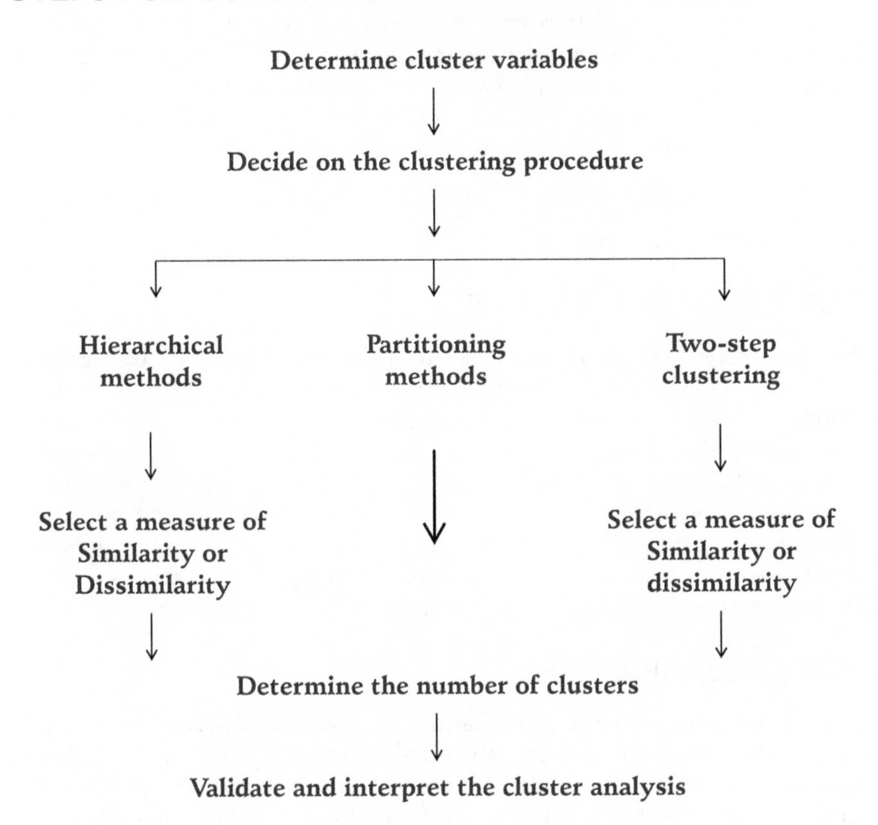

To create the clustering process, we have to select appropriate variables based on theoretical, conceptual and practical consideration and it should be observed for clustering data analysis. In certain circumstances, clustering the variables is important rather than the observation. Clustering variables should be homogeneous to evade use of different scaling values among the variables. There are several types of clustering variables and they are classified under similarity measures (proximity measure) and difference measures (distance measure).

MEASURES OF SIMILARITY OR DISSIMILARITY

Cluster analysis identifies the objects that are similar and group them into clusters. Similarity is a pragmatic measure of proximity, correspondence, or resemblance, between objects to be clustered. Many techniques use an index of similarity or proximity between each pair of observations into separate groups. Proximity measures can be extracted directly from individuals object. A convenient measure of proximity is the distance between two observations. Dissimilarity is measured by estimating the distance between the pairs of objects. Object with smaller distances between one another are more similar whereas objects with larger distances are more dissimilar. Two important elements for describing similarity between variables are as follows

- Correlation measures
- Distance measures

CORRELATION MEASURE

We can use the correlation co-efficient to measure the similarity between the pair of variables. It tells us whether as one variable changes the other variable changes by a similar extent. Correlation co-efficient is a standardized measure (not influenced by different scales) and it has the advantage that it is unaffected by dispersion difference through variables. But, there is a problem with using a correlation coefficient to compare the similarity between the variable and it ignores distance between two objects. There are several techniques to group the objects based on their similarity coefficients. Utmost these methods work in hierarchical technique to cluster the variables.

DISTANCE MEASURE

Distance measures uses the dissimilarities or distances between objects while creating the clusters. It represents the variable with the larger variance. Cluster analysis is based on dissimilarity between observations and groups. Distance measure can be constructed on a single dimension or multiple dimensions. We can use distance measure when variables are measured on common scale or calculate distance based on standardized value when variables are not on the similar scale. Distance can be measured in various methods as mentioned below,

Types of Distance Measures

- Euclidean distance.
- Squared (or absolute) Euclidean
- City-block (Manhattan) distance.
- Chebychev distance.
- Mahalanobis distance (D2).

Euclidean distance

Euclidean (straight line) distance is the most commonly used to measure the distance. Euclidean can be measured with a "ruler". It is a symmetrical distance in the multidimensional space. The distance between two vectors is the square root of the sum of the squared differences in the variables value. It is most commonly used to measure similarity between objects and also used in analyzing ratio and interval scale data. It is applicable for variables that are uncorrelated and have equal variances.

Squared Euclidean distance

The squared Euclidean distance is the most used method than Euclidean distance. It is the sum of squared distance and it provides information about the absolute difference between the variables. Squared Euclidean distance is the proposed measure for the ward's and centroid methods of clustering. Euclidean (and squared Euclidean) distances frequently work out from raw data, and not from standardized data.

City-block distance

City block distance uses the sum of the variables absolute (unsquared) difference. In this method the weight of single large difference is reduced (since they are not squared).It is also called as manhattan metric.

Chebychev distance

Chebychev distance is the maximum of the absolute difference in the clustering variables values. This measure is applicable, when two objects are different on any one of the dimension. It is most commonly used in analyzing ordinal data or metric.

Mahalanobis distance

Mahalanobis distance computes the data with adjustment made for correlation between variables equally. When variables are highly inter-correlated, Mahalanobis distance is most appropriate method.

CLUSTERING PROCEDURE

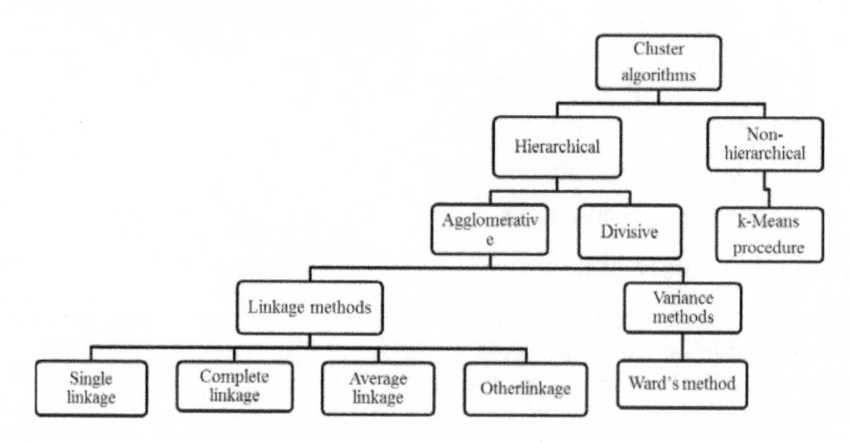

SELECT A CLUSTERING ALGORITHM

After selecting the distance or similarity measure, we need to decide which clustering algorithm to apply. The main purpose of algorithm is to merge or split the objects to form a cluster. There are several agglomerative measures and they can be distinguished by the way they state the similarity or distance from a newly shaped cluster to a certain object, or to other clusters in the solution. There are several different types of cluster model. The two most commonly used are as follows,

- Hierarchical methods
- Partitioning methods (k-means), and
- Two-step clustering (it is a combination of the first two methods).

These methods follow a different approach in grouping the most similar (distinct) objects into a cluster and to determining each object's cluster association. In simple words, where an object in certain cluster should be as similar as possible to all the other objects in the similar cluster, where an object in a definite cluster should be as similar as possible to all the other objects in the same cluster, it should also be as distinct as possible from objects in different clusters.

HIERARCHICAL METHOD

There are various methods to form cluster. Hierarchical clustering is one of the most straight forward methods. It usually starts with an individual objects and gradually group objects together to form single (large) cluster. These techniques are characterized by the construction of a hierarchy or tree-like structure. A tree like diagram that records the sequences of merges or splits of cluster and it is also known as a **dendrogram.** In application, hierarchical clustering is useful in determining if objects are similar rather than mutually exclusive. It can be either agglomerative or divisive.

HIERARCHY CLUSTERING- The Algorithm

A set of n items to be clustered into g clusters is given by hierarchical clustering algorithms which involves a successive practice.

Step 1: Initially, each object constitutes an individual cluster (n). The distance or similarity is calculated between all objects in clusters.

Step 2: Find the two most similar clusters and merge them sequentially to form a single new cluster.

Step 3: Find distances (similarities) between the new cluster and each of the old clusters.

Step 4: Repeat step 2 and 3 until all cases eventually form into single cluster.

Step 3 can be done in various ways, which differentiate single linkage from complete linkage and average linkage clustering.

Hierarchical clustering follows one of two approaches,

- Agglomerative methods (built upwards)
- Divisive methods (break downwards).

AGGLOMERATIVE METHOD

Agglomerative hierarchical clustering starts with each case/object as a cluster and in each step grouping the objects/cases to form clusters (i.e., similar objects are merged to form a cluster), until there is only one large cluster. Once a cluster is formed, it cannot be split and it can only be combined with other clusters.

DIVISIVE METHOD

Divisive hierarchical clustering starts with one large cluster and go on splitting into smaller/individual cluster items that are most dissimilar. Divisive method has a tendency to form new clusters rather than joining the objects to existing clusters.

Divisive algorithm needs much more calculating power so in practical only agglomerative methods are used.

SINGLE LINKAGE (NEAREST NEIGHBORS IN SPSS/ WIN)

Hierarchical clustering algorithm in which similarity is defined as the minimum distance between two clusters is determined by the distance of the two closest objects (nearest neighbors) in the different clusters. It is based on short term distance between objects. It groups the objects together to form clusters, and the resulting clusters epitomize long "chains." Single linkage clustering also called the connectedness or minimum method.

Single-Linkage Clustering: The Algorithm

1. Each case begins as a cluster.
2. Find the two most similar cases/clusters (e.g. **A** & **B**) by looking at the similarity coefficients between pairs of cases (e.g. the correlations or Euclidean distances). The cases/clusters with the highest similarity are merged to form the nucleus of a larger cluster.
3. The next case/cluster (C) to be merged with this larger cluster with the highest similarity coefficient to either A or B.
4. The next case to be merged is the one with the highest similarity to A, B or C, and so on.

COMPLETE LINKAGE (FURTHEST NEIGHBOR IN SPSS/ WIN)

It has opposite approach to single linkage method, which uses the maximum distance between two clusters, in which similairty is determined by the distance of the two closest objects in the different clusters. It is based on the largest distance between objects. Complete linkage clustering is also called the diameter or maximum method.

COMPLETE LINKAGE: The Algorithm

This method is the logical opposite to simple linkage. To begin with the procedure is the same as simple linkage.

Step 1: Finding the two objects with the highest similarity (in terms of their correlation or average Euclidean distance). These two cases (**A** & **B**) form the nucleus of the cluster.

Step 2: Cluster that has the highest similarity score to both **A** and **B**. The case (**C**) with the highest similarity to both **A** and **B** is added to the cluster.

Step 3: The next object to be added to the cluster is the one with the highest similarity to **A**, **B** and **C**. This method reduces dissimilarity within a cluster because it is based on overall similarity to members of the cluster (rather than similarity to a single member of a cluster).

AVERAGE LINKAGE

The distance between two clusters is defined as the average distance between all pairs of the two clusters members. It is based on the average distance between objects. This approach tends to combine clusters with small variance. This method is also very efficient when the objects form natural distinct "clumps," however, it performs equally well with elongated, "chain" type clusters.

AVERAGE LINKAGE: The Algorithm

This method is another variation on simple linkage. Again, we begin by

Step 1: Finding the two most similar cases (based on their correlation or average Euclidean distance). These two cases (**A** & **B**) form the nucleus of the cluster.

Step 2: At this stage the average similarity within the cluster is calculated. To determine which case (**C**) is added to the cluster we compare the similarity of each remaining cases to the average similarity of the cluster.

Step 3: The next case to be added to the cluster is the one with the highest similarity to the average similarity value for the cluster. Once this third case has been added, the average similarity within the cluster is re-calculated.

Step 4: The next case (**D**) to be added to the cluster is the one most similar to this new value of the average similarity.

CENTROID LINKAGE

In this method, centroid of each cluster is computed first. The distance between two clusters is determined as the difference between centroids. The *centroid* of a cluster is the average point in the multidimensional space defined by the dimensions. It is based on the distance between cluster centroids. Here the distance between clusters is that between their centroids (mean vectors).

WARD'S METHOD

This method is different from all other methods because it uses an analysis of variance approach to evaluate the distances between clusters. Ward's method applies squared Euclidean Distance as the distance or similarity measure. In short, this method attempts to minimize the Sum of Squares (SS) of any two (hypothetical) clusters that can be formed at each step. Ward's method, which is also known as the incremental sum of squares method, uses the within cluster and between cluster distances. In general, this method is regarded as very efficient; however, it tends to create clusters of small size.

WARD'S METHOD: The Algorithm

1. Initially the entire set of observations is considered as one set. The group is split based on the one variable which makes the greatest contribution to within-group sum of squares.
2. Group centroids are re-computed and subject distances to all group centroids are computed. The subject that would best improve the objective function is re-assigned.
3. This process is repeated until a finite number of transfers are performed, no further improvement in within-groups sum of squares is found, or a local optimum is reached.
4. The group with the largest within-groups sum of squares is selected for splitting. Steps 2 and 3 are then repeated until the desired number of clusters is identified.

FEATURES

Procedure	Proximity measure	Remark
Single linkage	distance or similarity	tendency to form chains
Complete linkage	distance or similarity	tendency to smaller groups of same size
Average linkage	distance or similarity	"between" single and complete linkage
Other linkage	only distance	No remark
Ward's method	only distance	tendency to groups of same size

Advantages of hierarchical cluster

1. Does not require information about the number of clusters in advance.
2. Easy to implement and gives best result in some cases.
3. Computes a complete hierarchy of clusters

Disadvantages of hierarchical cluster

1. Once a conclusion is made to combine two clusters, it cannot be undone.
2. Difficulty handling different sized clusters and convex shapes.
3. No objective function is directly reduced.
4. Sometimes it is difficult to identify the correct number of clusters by the dendogram.

NONHIERARCHICAL CLUSTER

A second most popular method for forming clusters is non-hierarchical clustering technique. They are designed to group object, rather than variables, into a collection of *K* clusters. Each cluster has a seed point and all objects within a prescribed distance are included in that cluster. In non-hierarchical methods object can leave the assigned cluster and can join another one cluster. The number of clusters K is specified in advance or determined as part of the clustering procedure. One of the more popular nonhierarchical procedures is K-means method discussed below.

K - MEANS CLUSTER

K -Means is one of most important non-hierarchical clustering methods. It is designed to cluster the objects and not variables. K-means clustering is used when you already have hypotheses relating to the number of clusters in cases or variables. It splits objects into k mutually exclusive clusters so that object within a cluster as similar as possible and objects between clusters as different as possible. Each cluster is then observed by its mean, or center point.

k- Means cluster algorithm

K-mean algorithm is implemented in four steps:

Step 1: Starts with splitting of the data into k number of clusters.

Step 2: Assign all points to the cluster with the nearest centroid.

Step 3: Compute the centroid of each cluster after all points are assigned to clusters.

Step 4: Reiterate step 2 and 3 until the centroids don't change.

Evaluating K-means Clusters

Usually, we use the **sum of squared error (SSE)** to compute the error of k-means.

- We calculate the error of each data point, i.e., its Euclidean distance to the closest centroid, and then compute the total sum of the squared errors.
- When we have two different clusters produced by two different k-means, we prefer the one with smallest SSE. Error can be reduced easily by increasing the k number of clusters, but a cluster with smaller K can have a lower SSE than a cluster with higher K.

Advantages of k – Mean cluster

1. K- Means is simple and can be used for a variety of data types.
2. K- Mean's algorithm splits the n samples into k clusters to minimize the sum of squared distance between cluster centers.

Disadvantages of k – Means cluster

1. Different size and density of cluster cannot be handled under this method.
2. K – Means cluster is constrained to data and concept start from center (centroid) and it does not start with specified means.
3. It is not suitable for all type of data and results depend on the value of k.

DIFFERENCE BETWEEN HIERARCHICAL AND NON- HIERARCHICAL CLUSTERINGS

HIERARCHICAL CLUSTER	NON-HIERARCHICAL CLUSTER
A set of nested clusters organized as a hierarchical tree.	Non-hierarchical techniques create a one-level (un nested) partitioning of the data objects into non-overlapping clusters.ie each data object is exactly in one cluster.
Hierarchical clustering as a process of merging two clusters or splitting one cluster into two.	This technique is based on the idea that a center point can represent a cluster.
Doesn't require knowledge about the number of clusters	Need to specify the number of clusters (arbitrary).
Hierarchical clustering is applied when there is no prior knowledge of clusters.	K-means clustering is used when you already have hypotheses relating to the number of clusters in cases or variables
Hierarchical Method generally requires only the proximity matrix (similarity) among the objects	Whereas non-hierarchical expect the data in the form of a pattern (shape) matrix
Hierarchical techniques are popularly used in the field of biology, social and behavior science because of the need of contract taxonomies.	Non-hierarchical techniques are frequently used in engineering application where single partition is important.
It is very slow and rigid.	It is very faster and more reliable.

DECIDING ON THE NUMBER OF CLUSTER

One of the major delinquent with cluster analysis is identifying the number of clusters to be retained from the data. As the merging process continues increasingly dissimilar cluster must be merged. Hierarchical method provides very less assistance over making decision. Deciding upon the optimum number of clusters is largely subjective, although looking at a dendrogram. Clusters variables are exclusively interpreted in terms of variables included in them. Clusters should also contain at least four elements. Once we drop to three or two elements it ceases to be meaningful.

This decision should be guided by theoretical and practicality of the results, along with use of the inter-cluster distances at successive steps. It is based on certain criterion such as between-groups sum of squares or prospect; this can be plotted against the number k of clusters in a scree graph. Also, the likelihood can be used in model selection criteria such as AIC (Akaike's Information Criterion) or BIC (Bayesian Information Criterion) to estimate k.

K - Mean Cluster Analysis

Learning Objectives

This chapter helps to understand the following

- K Mean Cluster Analysis Using SPSS
- Procedure to Run K Mean Cluster Analysis Using SPSS
- Results and Discussion of K Mean Cluster Analysis Using SPSS

Introduction

This procedure attempts to identify relatively homogeneous groups of cases based on selected characteristics, using an algorithm that can handle large numbers of cases. However, the algorithm necessitates specifying the number of clusters. You can specify initial cluster centers if you know this information. It is possible to select one of two methods for classifying cases, either updating cluster centers iteratively or classifying only. There is an option to save cluster membership, distance information, and final cluster centers. Optionally, specify a variable whose values are used to label casewise output. There is possibility request analysis of variance *F* statistics. While these statistics are opportunistic (the procedure tries to form groups that do differ), the relative size of the statistics provides information about each variable's contribution to the separation of the groups.

Example: What are some identifiable groups of television shows that attract similar audiences within each group? With *k*-means cluster analysis, cluster television shows (cases) into *k* homogeneous groups based on viewer characteristics. This process can be used to identify segments for marketing. Or cluster cities (cases) into homogeneous groups so that comparable cities can be selected to test various marketing strategies.

Statistics: Complete solution: initial cluster centers, ANOVA table. Each case: cluster information, distance from cluster center.

Data: Variables should be quantitative at the interval or ratio level. If the variables are binary or counts, use the Hierarchical Cluster Analysis procedure.

Case and initial cluster center order: The default algorithm for choosing initial cluster centers is not invariant to case ordering. The Use running means option in the Iterate dialog box makes the resulting solution potentially dependent on case order, regardless of how initial cluster centers are chosen. Use the methods, to obtain several different solutions with cases sorted in different random orders to verify the stability of a given solution. Specifying initial cluster centers and not using the Use running means option will avoid issues related to case order. However, ordering of the initial cluster centers may affect the solution if there are tied distances from cases to cluster centers. To assess the stability of a given solution, it helps to compare results from analyses with different permutations of the initial center values.

Assumptions: Distances are computed using simple Euclidean distance. If needed use another distance or similarity measure, use the Hierarchical Cluster Analysis procedure. Scaling of variables is an important consideration. If the variables are measured on different scales (for example, one variable is expressed in dollars and another variable is expressed in years), the results may be misleading. In such cases, consider standardizing variables before performing the *k*-means cluster analysis (this task can be done in the Descriptives procedure).

The procedure assumes that appropriate number of clusters that includes all relevant variables have been selected. The results may be misleading while choosing an inappropriate number of clusters or omitting important variables.

To Obtain a K-Means Cluster Analysis

From the menus choose:

Analyze > Classify > K-Means Cluster...

- Select the variables to be used in the cluster analysis.
- Specify the number of clusters. (The number of clusters must be at least 2 and must not be greater than the number of cases in the data file.)
- Select either Iterate and classify or Classify only.
- Optionally, select an identification variable to label cases.

K-Means Cluster Analysis Efficiency

The *k*-means cluster analysis command is efficient primarily because it does not compute the distances between all pairs of cases, as do many clustering algorithms, including the algorithm that is used by the hierarchical clustering command.

For maximum efficiency, take a sample of cases and select the Iterate and classify method to determine cluster centers. Select Write final as. Then restore the entire data file and select Classify only as the method and select Read initial from to classify the entire file using the centers that are estimated from the sample. Datasets are available for subsequent use in the same session but are not saved as files unless explicitly saved prior to the end of the session. Dataset names must conform to variable-naming rules.

K-Means Cluster Analysis Iterate

Note: These options are available to select the Iterate and classify method from the K-Means Cluster Analysis dialog box.

Maximum Iterations: Limits the number of iterations in the *k*-means algorithm. Iteration stops after this much iteration even if the convergence criterion is not satisfied. This number must be between 1 and 999.

To reproduce the algorithm used by the Quick Cluster command prior to version 5.0, set Maximum Iterations to 1.

Convergence Criterion: Determines when iteration ceases. It represents a proportion of the minimum distance between initial cluster centers, so it must be greater than 0 but not greater than 1. If the criterion equals 0.02, for example, iteration ceases when a complete iteration does not move any of the cluster centers by a distance of more than 2% of the smallest distance between any initial cluster centers.

Use running means: Allows to apply for cluster centers to be updated after each case is assigned. If the select option is ignored, new cluster centers are calculated after all cases have been assigned.

K-Means Cluster Analysis Save

Used to save information about the solution as new variables to be used in subsequent analyses:

Cluster membership: Creates a new variable indicating the final cluster membership of each case. Values of the new variable range from 1 to the number of clusters.

Distance from cluster center: Creates a new variable indicating the Euclidean distance between each case and its classification center.

K-Means Cluster Analysis Options

Statistics: To select the following statistics: initial cluster centers, ANOVA table, and cluster information for each case.

- **Initial cluster centers:** First estimate the variable means for each of the clusters. By default, a number of well-spaced cases equal to the number of clusters is selected from the data. Initial cluster centers are used for a first round of classification and are then updated.
- **ANOVA table:** Displays an analysis-of-variance table which includes univariate F tests for each clustering variable. The F tests are only descriptive and the resulting probabilities should not be interpreted. The ANOVA table is not displayed if all cases are assigned to a single cluster.
- **Cluster information for each case:** Displays for each case the final cluster assignment and the Euclidean distance between the case and the cluster center used to classify the case. Also displays Euclidean distance between final cluster centers.

Missing Values: Available options are Exclude cases listwise or Exclude cases pairwise.

- **Exclude cases listwise:** Excludes cases with missing values for any clustering variable from the analysis.
- **Exclude cases pairwise:** Assigns cases to clusters based on distances that are computed from all variables with nonmissing values.

QUICK CLUSTER Command Additional Features

The K-Means Cluster procedure uses QUICK CLUSTER command syntax. The command syntax language also allows to:

- Accept the first *k* cases as initial cluster centers, thereby avoiding the data pass that is normally used to estimate them.
- Specify initial cluster centers directly as a part of the command syntax.
- Specify names for saved variables.

See the *Command Syntax Reference* for complete syntax information.

SPSS

Test Procedure in SPSS

Following are the steps to analyse data using Correspondence Analysis in SPSS.

Step -1: Click Analyze > Classify > K-Means Cluster as shown in the figure -1, the K-Means Cluster Analysis dialogue box will appear (as given in the figure - 2).

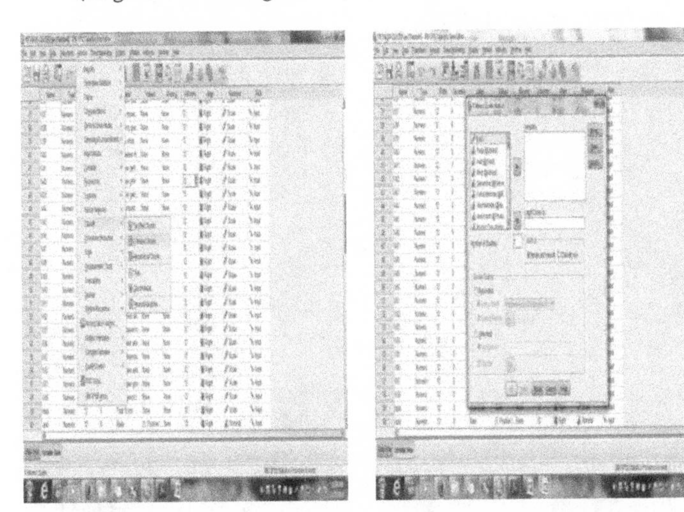

Figure - 1 **Figure - 2**

Step -2: in K-Means Cluster analysis dialogue box, Transfer the variables into variable box by using the arrow tab or by dragging the variable as given in figure -3. Give number of clusters as 5 followed by methods, click iterate and classify.

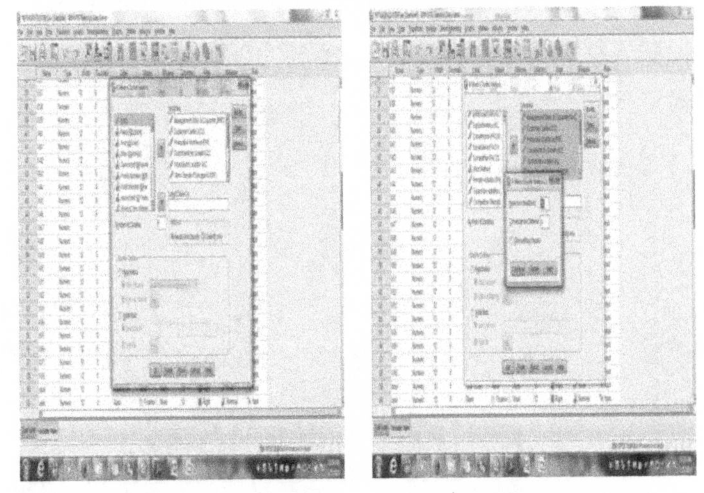

Figure - 3 **Figure - 4**

Step - 3: In K-Means Cluster analysis box, click iterate tab, a dialogue will appear. Give maximum iteration as 10 and convergence criteria as 0 and click continue as given in figure -4 to return to K-Means cluster dialogue box.

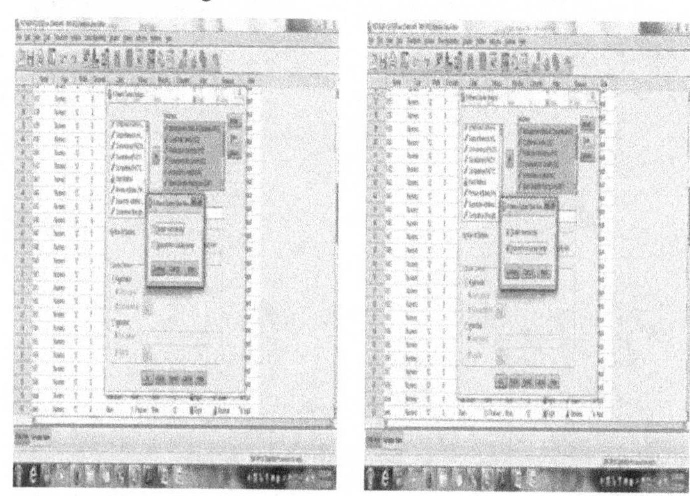

Figure - 5 **Figure - 6**

Step - 4: In K-Means Cluster dialogue box, click save tab, a dialogue box will appear as given in figure-5. Select cluster membership and distance from cluster centre checkbox as given in figure -6 and click Continue to return to K-means cluster analysis dialogue box.

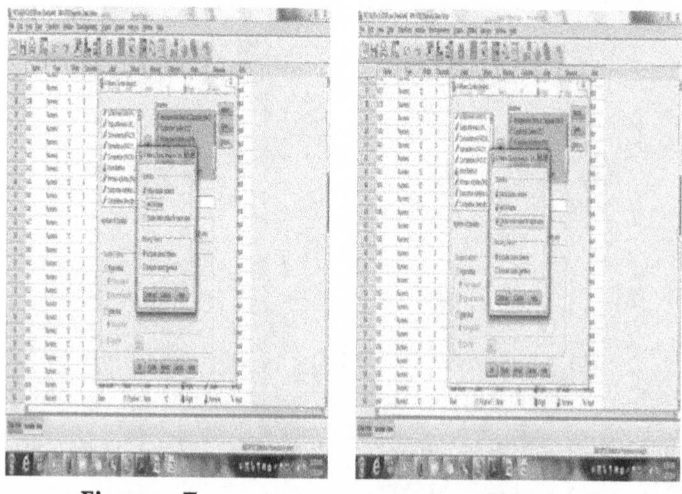

Figure - 7 **Figure - 8**

Step - 5: Click option tab, a dialogue will appear as given in figure -7. Select Initial cluster centres, Anova Table, Cluster Information for each case checkbox in statistics and exclude cases listwise in missing values as given in figure 8 and click continue to return to K Means Cluster analysis dialogue box.

***Step - 6*:** Click ok to generate output.

SPSS Output of the K- Means Cluster Analysis

Table 1 Initial Cluster Centers

	Cluster				
	1	2	3	4	5
Management Skills & Capability	3.63	.88	2.50	2.75	4.38
Customer Centric	4.71	1.00	1.57	3.71	5.00
Productive Workforce	4.33	2.33	3.00	2.17	4.33
Constraints for Growth	5.00	5.00	3.00	2.50	1.00
Accessible Location	5.00	5.00	3.00	2.75	5.00
Store Specific Planogram	3.25	2.00	2.00	3.50	5.00
Adaptiveness	5.00	2.67	1.67	3.33	5.00
Employees Skill	1.00	1.00	5.00	1.00	5.00
Supportiveness of Local Authorities	4.25	1.00	5.00	2.00	5.00
Convenience	3.67	3.67	3.67	2.67	3.67
Surveillance	1.50	3.00	3.00	2.00	3.00
Competition	5.00	5.00	3.00	1.00	5.00

Table 2 Iteration History[a]

Iteration	Change in Cluster Centers				
	1	2	3	4	5
1	4.095	2.835	3.732	2.102	3.314
2	.525	.799	.807	.549	.744
3	.619	.133	2.477	2.729	1.616
4	.190	.022	.582	.323	.121
5	.005	.004	.043	.003	.035
6	.000	.001	.001	2.572E-005	.000

Iteration	Change in Cluster Centers				
	1	2	3	4	5
7	2.761E-006	.000	9.848E-006	2.296E-007	5.424E-006
8	6.735E-008	1.712E-005	1.492E-007	2.050E-009	6.780E-008
9	1.643E-009	2.853E-006	2.261E-009	1.831E-011	8.475E-010
10	4.007E-011	4.755E-007	3.425E-011	1.633E-013	1.059E-011

a. Iterations stopped because the maximum number of iterations was performed. Iterations failed to converge. The maximum absolute coordinate change for any center is 2.481E-007. The current iteration is 10. The minimum distance between initial centers is 6.206.

Table -3 Cluster Membership

Case No.	Cluster	Distance
1	5	2.721
2	5	2.721
3	1	2.445
4	1	3.279
5	1	2.395
6	1	3.108
7	4	2.118
8	1	1.566
9	3	2.016
10	4	3.089
11	1	2.719
12	4	3.141
13	5	2.299
14	4	1.664
15	3	3.015
16	5	2.099
17	5	3.282
18	3	2.411
19	3	2.648
20	3	1.890
21	3	1.980
22	5	2.401
23	5	2.401
24	3	2.549
25	3	2.364
26	3	1.636
27	5	2.528
28	3	1.752
29	5	2.376
30	3	1.935
31	5	2.261
32	5	2.251
33	5	2.376
34	3	1.996
35	3	3.097
36	5	1.995
37	5	1.962
38	5	2.146
39	5	2.146

Case No.	Cluster	Distance
40	4	2.426
41	4	3.105
42	4	2.363
43	4	2.773
44	3	3.272
45	3	3.272
46	1	2.945
47	1	2.945
48	4	3.419
49	4	3.419
50	5	1.523
51	4	3.714
52	3	2.884
53	5	2.024
54	1	3.596
55	3	1.375
56	3	2.567
57	3	1.767
58	4	3.427
59	5	2.537
60	1	2.945
61	4	2.205
62	4	2.296
63	4	2.296
64	4	2.296
65	3	1.533
66	5	2.139
67	3	1.356
68	4	1.891
69	4	2.280
70	5	3.712
71	3	2.953
72	4	2.976
73	4	1.876
74	1	1.389
75	4	2.476
76	4	2.430
77	4	2.807
78	3	3.948
79	4	1.909
80	4	1.909
81	4	1.909
82	4	2.511
83	4	2.511
84	4	2.511
85	3	2.705
86	1	3.042
87	5	1.878
88	5	1.878
89	5	2.923
90	5	1.751
91	5	1.772
92	5	1.800
93	1	2.607
94	3	3.100
95	5	2.139
96	4	2.680
97	3	2.789
98	3	2.550
99	4	1.791
100	3	3.969
101	4	3.389
102	4	3.389
103	4	3.389
104	4	3.021
105	4	3.021

Case No.	Cluster	Distance	Case No.	Cluster	Distance
106	1	1.637	139	1	3.108
107	1	1.637	140	3	2.392
108	5	1.261	141	3	2.067
109	5	1.261	142	4	1.955
110	5	1.261	143	3	2.057
111	4	2.778	144	3	2.818
112	1	3.480	145	5	1.856
113	4	3.150	146	5	2.916
114	4	2.256	147	3	1.678
115	1	3.065	148	3	1.690
116	4	3.666	149	3	1.517
117	5	1.913	150	4	1.391
118	4	2.251	151	3	2.151
119	1	3.131	152	5	2.119
120	5	2.969	153	4	3.105
121	3	2.998	154	5	3.131
122	2	2.824	155	5	1.499
123	1	2.510	156	5	1.656
124	4	1.699	157	3	2.369
125	3	1.947	158	1	2.413
126	1	2.859	159	5	1.689
127	4	2.635	160	4	4.467
128	3	2.523	161	4	4.467
129	1	2.074	162	5	2.329
130	4	3.604	163	5	2.329
131	5	2.985	164	3	2.049
132	5	1.499	165	4	2.440
133	4	2.353	166	4	1.003
134	2	1.425	167	4	1.134
135	2	1.425	168	5	2.985
136	2	1.425	169	5	2.985
137	4	2.458	170	1	2.615
138	5	1.453	171	5	4.424

Case No.	Cluster	Distance	Case No.	Cluster	Distance
172	4	2.680	205	3	1.889
173	4	2.680	206	3	2.578
174	4	1.801	207	5	2.447
175	5	2.930	208	4	3.104
176	2	3.556	209	3	1.940
177	4	2.313	210	3	2.781
178	5	3.425	211	5	2.146
179	4	2.704	212	4	2.426
180	3	2.403	213	4	3.105
181	4	3.604	214	4	3.974
182	4	3.604	215	4	2.426
183	1	2.967	216	4	2.363
184	5	3.187	217	5	1.878
185	4	3.000	218	3	2.614
186	5	2.985	219	4	2.841
187	4	1.225	220	3	2.432
188	4	1.646	221	3	2.306
189	1	2.761	222	3	2.306
190	5	1.499	223	4	1.973
191	1	2.451	224	5	2.985
192	1	2.451	225	5	2.985
193	1	2.451	226	4	1.932
194	1	3.582	227	1	1.637
195	3	1.917	228	4	3.604
196	4	3.270	229	1	2.550
197	1	2.469	230	5	2.985
198	5	2.177	231	4	1.225
199	4	1.250	232	4	1.225
200	5	2.593	233	5	2.721
201	1	2.184	234	4	3.757
202	5	1.836	235	4	2.068
203	4	3.070	236	4	2.068
204	3	2.228	237	5	2.758

Case No.	Cluster	Distance
238	4	3.418
239	1	3.211
240	5	2.950
241	1	2.445
242	5	1.882
243	4	1.910
244	4	1.889
245	4	3.041
246	5	1.954
247	3	2.482
248	1	3.279
249	5	2.370
250	4	2.676
251	1	3.299
252	1	1.789
253	4	1.843
254	5	1.970
255	4	4.613
256	3	2.016
257	4	2.118
258	4	2.118
259	5	2.762
260	5	2.762
261	4	3.089
262	4	3.682
263	5	3.490
264	4	1.402
265	4	1.466
266	4	1.300
267	1	2.352
268	3	2.083
269	1	3.386

Case No.	Cluster	Distance
270	1	2.517
271	1	2.517
272	1	2.517
273	4	1.515
274	4	1.686
275	4	3.225
276	3	1.688
277	5	1.856
278	5	1.856
279	3	1.728
280	3	1.338
281	3	1.611
282	3	1.387
283	3	1.752
284	5	2.257
285	3	2.061
286	3	1.622
287	5	2.176
288	5	2.176
289	5	2.393
290	5	2.393
291	3	2.358
292	3	2.095
293	3	2.288
294	3	1.572
295	4	4.378
296	4	3.044
297	5	2.899
298	3	2.130
299	3	2.130
300	4	3.419

Table 4 Final Cluster Centers

	Cluster				
	1	2	3	4	5
Management Skills & Capability	2.47	.88	3.86	3.38	3.68
Customer Centric	3.13	1.60	4.67	3.45	4.44
Productive Workforce	2.72	2.30	3.11	3.52	3.37
Constraints for Growth	2.72	5.00	1.91	2.27	1.89
Accessible Location	3.09	4.05	4.37	3.49	4.21
Store Specific Planogram	3.31	3.10	4.78	3.65	4.49
Adaptiveness	3.72	4.40	4.83	3.74	4.49
Employees Skill	1.80	1.00	2.54	4.08	2.95
Supportiveness of Local Authorities	3.13	1.50	4.30	3.40	4.06
Convenience	3.28	2.87	3.63	3.21	3.60
Surveillance	3.05	1.80	3.36	2.86	3.18
Competition	3.34	2.80	2.19	3.16	4.71

Table 5 Distances between Final Cluster Centers

Cluster	1	2	3	4	5
1		4.123	3.684	2.720	3.423
2	4.123		6.724	5.847	6.720
3	3.684	6.724		3.129	2.647
4	2.720	5.847	3.129		2.722
5	3.423	6.720	2.647	2.722	

Table 6 ANOVA

	Cluster		Error		F	Sig.
	Mean Square	df	Mean Square	df		
Management Skills & Capability	22.504	4	.364	295	61.812	.000
Customer Centric	34.220	4	.493	295	69.369	.000
Productive Workforce	6.677	4	.546	295	12.235	.000
Constraints for Growth	16.276	4	.780	295	20.871	.000
Accessible Location	16.932	4	.463	295	36.548	.000

	Cluster		Error		F	Sig.
	Mean Square	df	Mean Square	df		
Store Specific Planogram	24.239	4	.535	295	45.284	.000
Adaptiveness	16.585	4	.464	295	35.766	.000
Employees Skill	54.276	4	.753	295	72.127	.000
Supportiveness of Local Authorities	20.681	4	.618	295	33.492	.000
Convenience	3.190	4	.296	295	10.784	.000
Surveillance	4.792	4	.547	295	8.753	.000
Competition	61.704	4	.702	295	87.887	.000

The F tests should be used only for descriptive purposes because the clusters have been chosen to maximize the differences among cases in different clusters. The observed significance levels are not corrected for this and thus cannot be interpreted as tests of the hypothesis that the cluster means are equal.

Table -7 Number of Cases in each Cluster

Cluster	1	44.000
	2	5.000
	3	69.000
	4	102.000
	5	80.000
Valid		300.000
Missing		.000

Results & Discussion

Initial Cluster Centers

The first step in k-means clustering is to discover the cluster centers.

Iteration History

Table -2 represents Iteration history which deals with modification of the default appeared to be unnecessary here, as the convergence threshold was achieved by the tenth iteration. One concern of researchers in performing a k-means cluster analysis is making sure that the number of cases assigned to each cluster in the final solution is acceptable.

Cluster Membership

Table -3 represents cluster membership, and this has been saved in dataset for further analysis

Final Cluster Centers

This table – 4 shows the heart of the results and is used to characterize the clusters that there are relatively extreme values of the seed points that have been substantially moderated.

Distance between Final Cluster Centers

Table -5 reveals that Cluster results are good when all non-zero numbers are relatively large.

ANOVA

The ANOVA summary table - 6 is important that the analysis needs to be treated as strictly exploratory. Here, there are statistically significant group differences in all of the clustering variables. The primary use of this table to researchers is to recognize when a clustering variable does not yield a statistically significant effect; this would indicate that the variable is not effective for its chosen purpose and might suggest to researchers that they may wish to repeat the analysis swapping it out for another variable if one is reasonable and available.

Number of Cases in Each Cluster

Table – 7 presents data for the number of units in each cluster as well as their total number and missing units. There are 5 clusters in the study; this table represents no. of respondents in each cluster. Cluster 1 consists of 44 respondents, Cluster 2 consists of 5 respondents, cluster 3 consists of 69 respondents, cluster 4 consists of 102 respondents and cluster 5 consists of 80 respondents.

Hierarchical Cluster Analysis

Learning Objectives

This chapter helps to understand the following

- Hierarchical Cluster Analysis Using SPSS
- Procedure to Run Hierarchical Cluster Analysis
- Results and Discussion of Hierarchical Cluster Analysis Output
- Procedure to run the results of Hierarchical Cluster Analysis with ANOVA

This procedure attempts to identify relatively homogeneous groups of cases (or variables) based on selected characteristics, using an algorithm that starts with each case (or variable) in a separate cluster and combines clusters until only one is left. This procedure attempts to analyze raw variables or to choose from a variety of standardizing transformations. Distance or similarity measures are generated by the Proximities procedure. Statistics are displayed at each stage to help to select the best solution.

Example: Are there identifiable groups of television shows that attract similar audiences within each group? With hierarchical cluster analysis, cluster television shows (cases) into homogeneous groups based on viewer characteristics. This can be used to identify retailer characteristics. Or to cluster cities (cases) into homogeneous groups so that comparable cities can be selected to test various marketing strategies.

Statistics: Agglomeration schedule, distance (or similarity) matrix, and cluster membership for a single solution or a range of solutions. Plots: dendrograms and icicle plots.

Data: The variables can be quantitative, binary, or count data. Scaling of variables is an important issue—differences in scaling may affect the cluster solution(s). If the variables have large differences in scaling (for example, one variable is measured in dollars and the other is measured in years), It should consider standardizing them (this can be done automatically by the Hierarchical Cluster Analysis procedure).

Case order: If tied distances or similarities exist in the input data or occur among updated clusters during joining, the resulting cluster solution may depend on the order of cases in the file. To obtain several different solutions with cases sorted in different random orders to verify the stability of a given solution.

Assumptions: The distance or similarity measures used should be appropriate for the data analyzed (see the Proximities procedure for more information on choices of distance and similarity measures). Also, it should include all relevant variables in the analysis. Omission of influential variables can result in a misleading solution. Because hierarchical cluster analysis is an exploratory method, results should be treated as tentative until they are confirmed with an independent sample.

To Obtain a Hierarchical Cluster Analysis

From the menus choose:

Analyze > Classify > Hierarchical Cluster...

Hierarchical Cluster Analysis dialog box

For clustering cases, select at least one numeric variable. For clustering variables, select at least three numeric variables. Optionally, select an identification variable to label cases.

Hierarchical Cluster Analysis Method

Cluster Method: Available alternatives are between-groups linkage, within-groups linkage, nearest neighbor, furthest neighbor, centroid clustering, median clustering, and Ward's method.

Measure: Allows to specify the distance or similarity measure to be used in clustering. Select the type of data and the appropriate distance or similarity measure:

- **Interval:** Available alternatives are Euclidean distance, squared Euclidean distance, cosine, Pearson correlation, Chebychev, block, Minkowski, and customized.
- **Counts:** Available alternatives are chi-square measure and phi-square measure.
- **Binary:** Available alternatives are Euclidean distance, squared Euclidean distance, size difference, pattern difference, variance, dispersion, shape, simple matching, phi 4-point correlation, lambda, Anderberg's *D*, dice, Hamann, Jaccard, Kulczynski 1, Kulczynski 2, Lance and Williams, Ochiai, Rogers and Tanimoto, Russel and Rao, Sokal and Sneath 1,Sokal and Sneath 2, Sokal and Sneath 3, Sokal and Sneath 4, Sokal and Sneath 5, Yule's *Y*, and Yule's *Q*.

Transform Values: Allows to standardize data values for either cases or values before computing proximities (not available for binary data). Available standardization methods are z scores, range −1 to 1, range 0 to 1, maximum magnitude of 1, mean of 1, and standard deviation of 1.

Transform Measures: Allows to transform the values generated by the distance measure. They are applied after the distance measure has been computed. Available alternatives are absolute values, change sign, and rescale to 0–1 range.

Hierarchical Cluster Analysis Statistics

Hierarchical Cluster Analysis Statistics dialog box

Agglomeration schedule: Displays the cases or clusters combined at each stage, the distances between the cases or clusters being combined, and the last cluster level at which a case (or variable) joined the cluster. **Proximity matrix:** Gives the distances or similarities between items.

Cluster Membership: Displays the cluster to which each case is assigned at one or more stages in the combination of clusters. Available options are single solution and range of solutions.

Hierarchical Cluster Analysis Plots

Dendrogram: Displays a **dendrogram**. Dendrograms can be used to assess the cohesiveness of the clusters formed and can provide information about the appropriate number of clusters to keep.

Icicle: Displays an **icicle plot**, including all clusters or a specified range of clusters. Icicle plots display information about how cases are combined into clusters at each iteration of the analysis.

Orientation allows to select a vertical or horizontal plot.

Hierarchical Cluster Analysis Save New Variables

Cluster Membership: Allows to save cluster memberships for a single solution or a range of solutions. Saved variables can then be used in subsequent analyses to explore other differences between groups.

SPSS

Test Procedure in SPSS

Following are the steps to analyse data using Hierarchical cluster analysis in SPSS.

Model 1 – Between-groups linkage

Step - 1: Click Analyze > Classify > Hierarchical cluster as shown in the figure -1, the Hierarchical cluster dialogue box will appear (as given in the figure - 2).

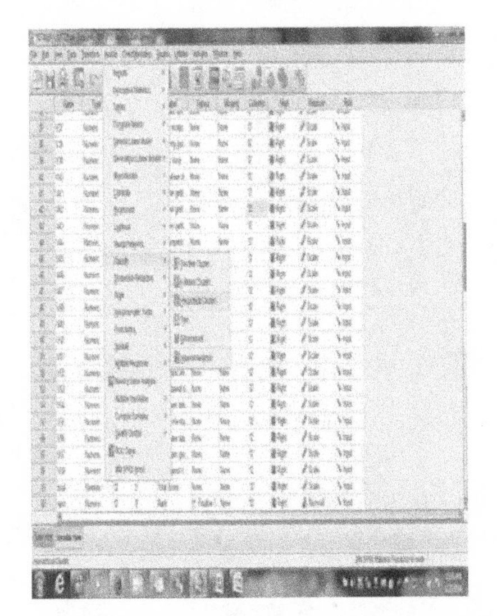

Figure - 1

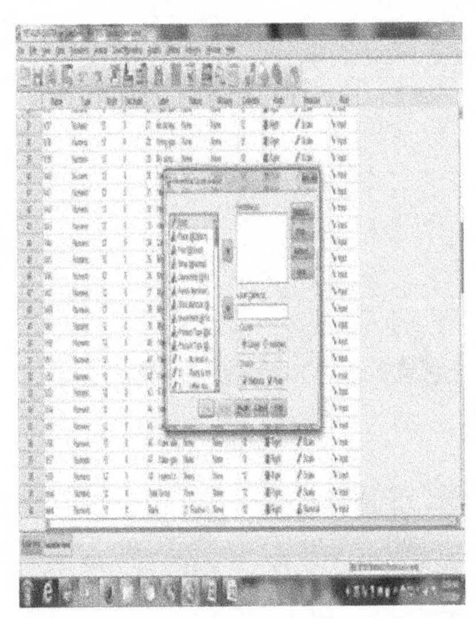

Figure - 2

Step - 2: Transfer the dependent variable such as Management Skills & Capability, Customer Centric, Productive Workforce, Constraints for Growth, Accessible Location, Store Specific Planogram, Adaptiveness, Employees Skill, Supportiveness of Local Authorities, Convenience, Surveillance and Competition into the variable by drag-and-dropping the variables into their respective boxes or by using the SPSS Right Arrow Button. Select the cluster and display according to the requirement. In this example, cases were chosen in cluster and statistics and plots were chosen in display section. The result is shown below in figure -3

Step - 3: In Hierarchical cluster dialogue box, click statistics button the sub dialogue box will appear (The default image given in figure -4).

Select Agglomeration schedule check box and proximity matrix. As well in cluster membership, select single solution and give 5 in number of clusters (depends upon requirement of the working data) and click on continue to return to hierarchical cluster dialogue box. The result is shown in figure -5.

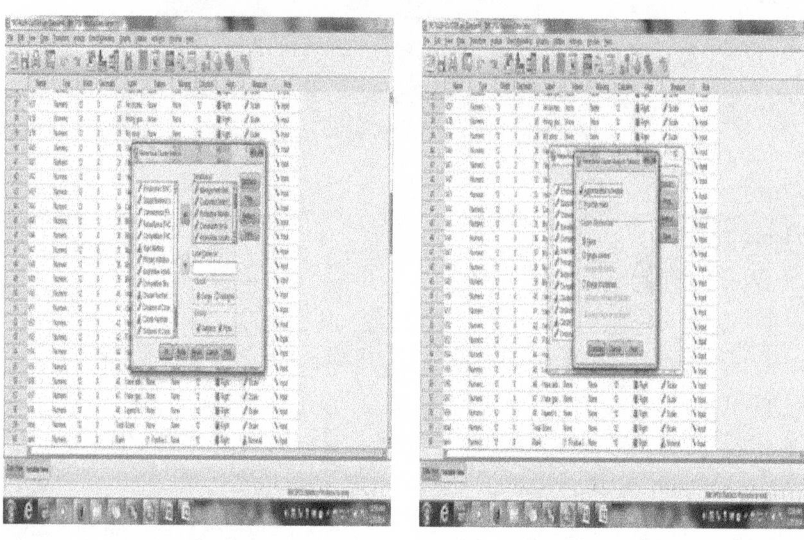

Figure - 3 **Figure - 4**

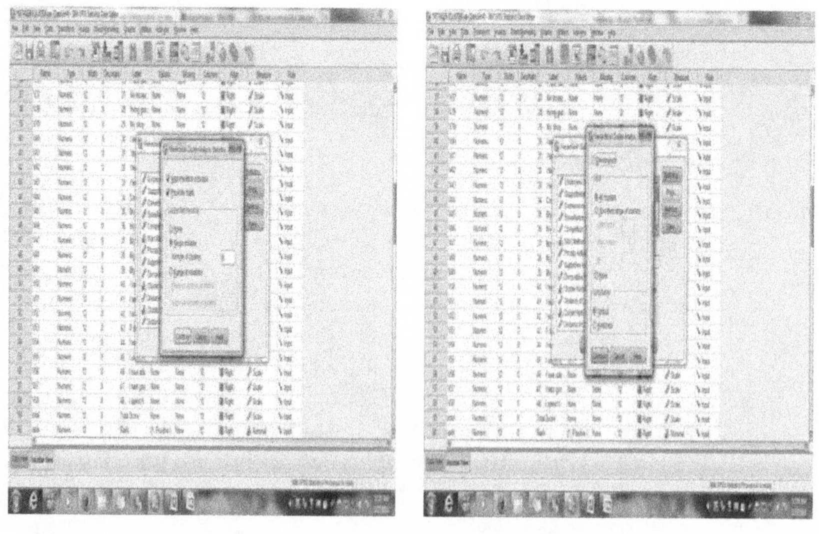

Figure - 5 **Figure - 6**

Step - 4: In Hierarchical cluster dialogue box, click plots button where the sub dialogue box will appear (the default image given in figure -6). Select Dendrogram as given in figure -7 and click on continue to return to hierarchical cluster dialogue box.

Step - 5: In Hierarchical cluster dialogue box, click methods button; then the sub dialogue box will appear (the default image given in figure -8). Select – Wards Method in cluster method and in measures, choose interval – Squared Euclidean Distance (depends on data type) and click on continue to return to hierarchical cluster dialogue box, choose cases in cluster as given in the figure.

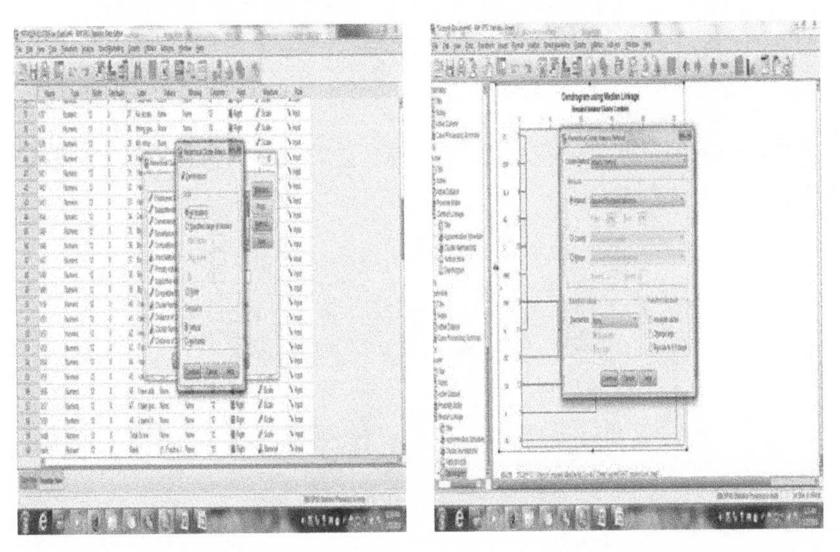

Figure - 7 **Figure - 8**

Step - 6: In hierarchical cluster dialogue box, click save button the sub dialogue box will appear (the default image given in figure -9). In cluster membership select single solution - Number of clusters, give 5 and click on continue to return to hierarchical cluster dialogue box

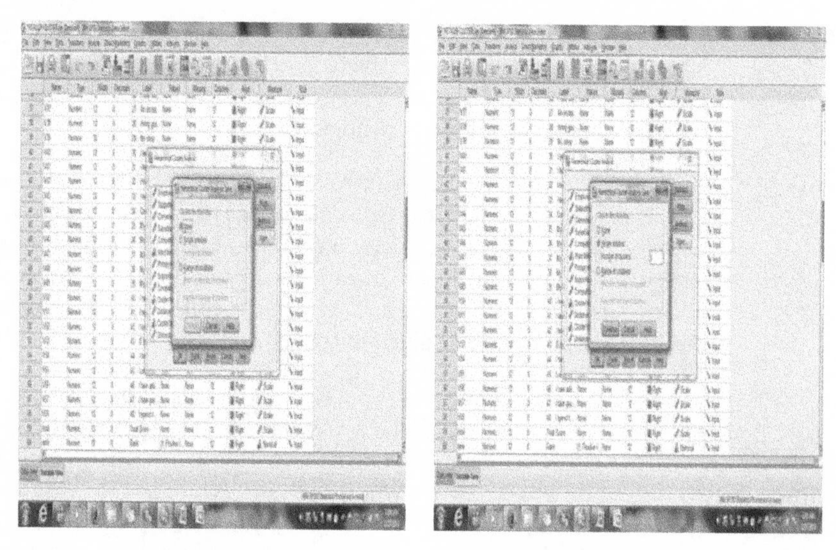

Figure - 9 Figure - 10

Step - 7: Click the Ok button to generate the output.

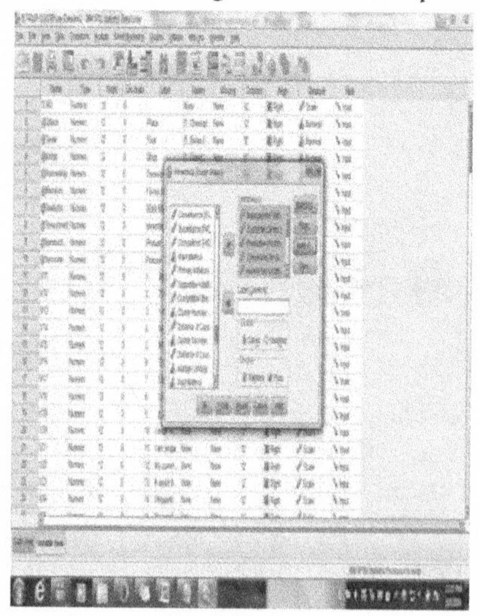

Figure - 11

SPSS Output of the Hierarchical cluster Analysis

Table 1 Proximity Matrix

Case	Squared Euclidean Distance											
	1	2	3	4		5	295	296	297	298	299	300
1	0	0	21.04	31.671	. . .	16.971	22.55	39.85	13.835	19.091	10.545	30.816
2	0	0	21.04	31.671	. . .	16.971	22.55	39.85	13.835	19.091	10.545	30.816
3	21.04	21.04	0	17.039	. . .	6.442	27.802	14.01	10.173	26.67	22.225	44.794
4	31.671	31.671	17.039	0	. . .	20.876	26.905	12.105	16.862	26.867	33.453	42.981
5	16.971	16.971	6.442	20.876	. . .	0	24.91	17.111	9.632	18.8	13.518	39.289
6	41.411	41.411	14.965	12.354	. . .	18.057	43.535	12.409	19.253	35.645	32.635	55.621
7	16.502	16.502	29.902	31.064	. . .	18.854	15.09	39.167	11.007	7.575	9.669	23.196
.	.	.	.	.	. . .	.	.	.	.	.	.	.
.	.	.	.	.	. . .	.	.	.	.	.	.	.
.	.	.	.	.	. . .	.	.	.	.	.	.	.
.	.	.	.	.	. . .	.	.	.	.	.	.	.
.	.	.	.	.	. . .	.	.	.	.	.	.	.
.	.	.	.	.	. . .	.	.	.	.	.	.	.
.	.	.	.	.	. . .	.	.	.	.	.	.	.
294	20.503	20.503	31.398	42.113	. . .	23.171	32.088	41.77	13.592	17.575	12.556	50.007
295	20.673	20.673	36.016	30.386	. . .	36.306	18.71	42.78	29.769	26.203	32.32	24.229

Case	Squared Euclidean Distance											
296	25.571	25.571	16.957	22.835	. . .	9.283	15.625	19.917	9.559	16.802	12.389	24.598
297	14.732	14.732	28.127	41.259	. . .	25.317	27.151	45.374	19.655	35.929	7.701	46.278
298	18.805	18.805	20.958	26.698	. . .	15.17	15.564	30.356	7.324	10.29	9.887	30.228
299	18.805	18.805	20.958	26.698	. . .	15.17	15.564	30.356	7.324	10.29	9.887	30.228
300	31.967	31.967	37.238	34.696	. . .	28.705	17.566	52.495	20.242	18.457	23.454	34.891

This is a dissimilarity matrix

(to fit into the page, the table was edited)

Table 2 Agglomeration Schedule

Stage	Cluster Combined		Coefficients	Stage Cluster First Appears		Next Stage
	Cluster 1	Cluster 2		Cluster 1	Cluster 2	
1	49	300	0.000	0	0	62
2	298	299	0.000	0	0	179
3	289	290	0.000	0	0	113
4	287	288	0.000	0	0	88
5	277	278	0.000	0	0	6
6	145	277	0.000	0	5	200
7	271	272	0.000	0	0	8
8	270	271	0.000	0	7	121
9	10	261	0.000	0	0	254
10	259	260	0.000	0	0	155
.	.	.	.	.	.	.
.	.	.	.	.	.	.
.	.	.	.	.	.	.
.	.	.	.	.	.	.
.	.	.	.	.	.	.
.	.	.	.	.	.	.
.	.	.	.	.	.	.
.	.	.	.	.	.	.
.	.	.	.	.	.	.
.	.	.	.	.	.	.
.	.	.	.	.	.	.
.	.	.	.	.	.	.
.	.	.	.	.	.	.
.	.	.	.	.	.	.
.	.	.	.	.	.	.
.	.	.	.	.	.	.
290	3	104	1568.893	283	289	294
291	16	37	1629.014	269	282	296
292	1	17	1690.552	274	284	296
293	9	44	1765.235	275	281	295

Stage	Cluster Combined		Coefficients	Stage Cluster First Appears		Next Stage
	Cluster 1	Cluster 2		Cluster 1	Cluster 2	
294	3	8	1864.572	290	288	299
295	7	9	1964.606	287	293	297
296	1	16	2105.256	292	291	298
297	7	18	2320.963	295	285	298
298	1	7	2581.050	296	297	299
299	1	3	3063.664	298	294	0

(to fit into the page, the table was edited)

Table 3 Cluster Membership

Case	5 Clusters
1	1
2	1
3	2
4	2
5	2
6	2
7	3
8	2
9	3
10	3
.	..
.	.
.	.
.	.
.	
287	4
288	4
289	4
290	4
291	5

Case	**5 Clusters**
292	5
293	5
294	5
295	3
296	4
297	4
298	3
299	3
300	1

(to fit into the page, the table was edited)

Figure - 12 Vertical Icicle plot

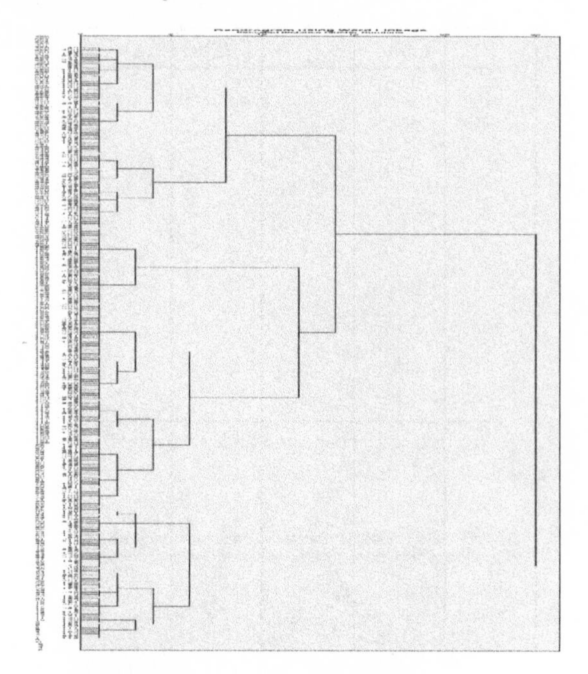

Figure - 13 Dendogram

Results & Discussion

Proximity Matrix

Table -1 represents proximity matrix, and is the matrix of proximity between subjects. The values in the table represent dissimilarities between each pair of items. The distance measure is used to measure the dissimilarities based on squared Euclidean distance. In this example, the dissimilarity of the larger values indicates items that are very different to fit into the page. The table is edited, this gives some initial figure about clustering of object.

Agglomeration Schedule

Table -2 represents agglomeration schedule, which gives the information of how the cluster was grouped into each stage in this analysis. The cluster analysis starts with 300 clusters at stage 1, where case 11 and case 300 are combined (being minimum squared Euclidean distance). The coefficient column in the table indicates the squared Euclidean distance between two clusters joined at each stage. After coefficient column the stage cluster first appears as shown in the column. At the initial stage of analysis, single cases existed, so it is indicated by zero. The last column indicates the subsequent stage at which the cases combined in this cluster, which will be combined to another cases i.e. named as next stage. Likewise it goes on for all the stages i.e. upto 299 stages. Interpretation of the other part of agglomeration schedule can be done on the same lines.

Cluster Membership

Table -3 represents cluster membership. Normally the spss has presented two to five cluster solutions.

Vertical Icicle Plot

Figure – 12 represents vertical icicle plot that displays some important information graphically.

Dendrogram

Figure -13 represents dendrogram i.e. a relative similarity between subjects considered for cluster analysis. The dendrogram also helps to determine number of clusters in the cluster analysis.

After clustering it is important to find the meaning of the concern clusters in term of finding some natural or compelling structure in data. For this cluster Centroids tool has to be used, which can be obtained through discriminant analysis. (Continued in discriminant analysis chapter).

SPSS Output of the Hierarchical cluster Analysis

Following are the steps to analyse data using one-way ANOVA in SPSS for confirming the Hierarchical cluster analysis.

Step - 1: Click Analyze > Compare means > one-way Anova as shown in the figure -1, the one –way dialogue box will appear (as given in the figure - 2).

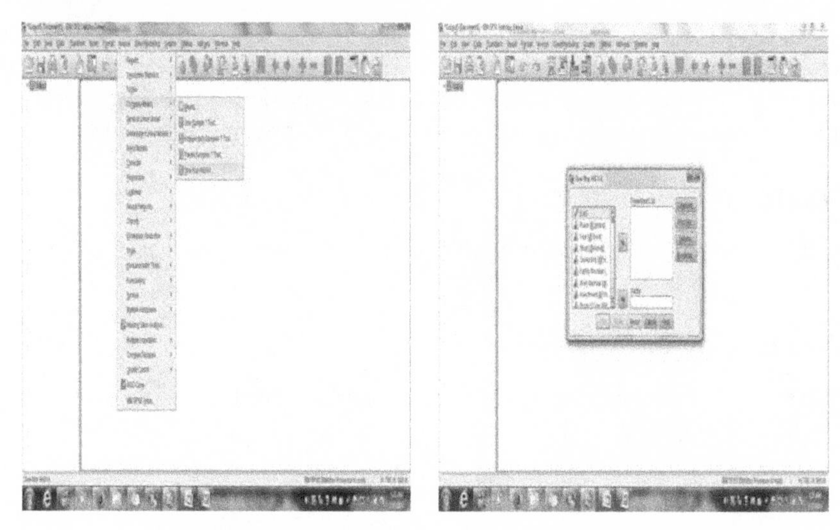

Figure - 1 **Figure- 2**

Step - 2: Transfer the dependent variable such as Management Skills & Capability, Customer Centric, Productive Workforce, Constraints for Growth, Accessible Location, Store Specific Planogram, Adaptiveness, Employees Skill, Supportiveness of Local Authorities, Convenience, Surveillance and Competition into the variable by drag-and-dropping the variables into their respective boxes or by using the SPSS Right Arrow Button. And move wards method group into the factor box as given in the figure

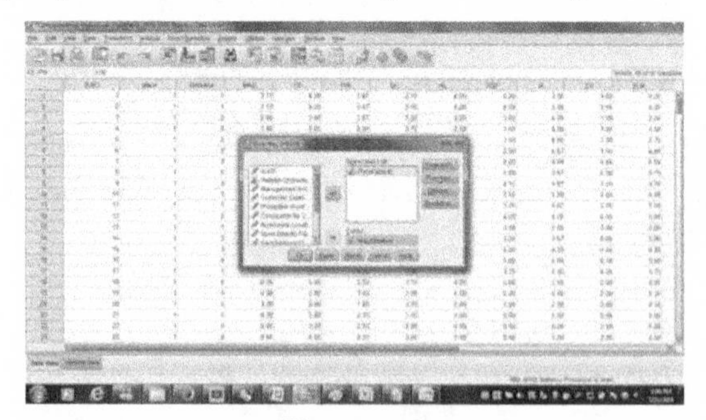

Figure - 3

Step -3: In one-way anova dialogue box, Click on option Button, the one-way anova: option dialogue box will appear (as given in figure -4)

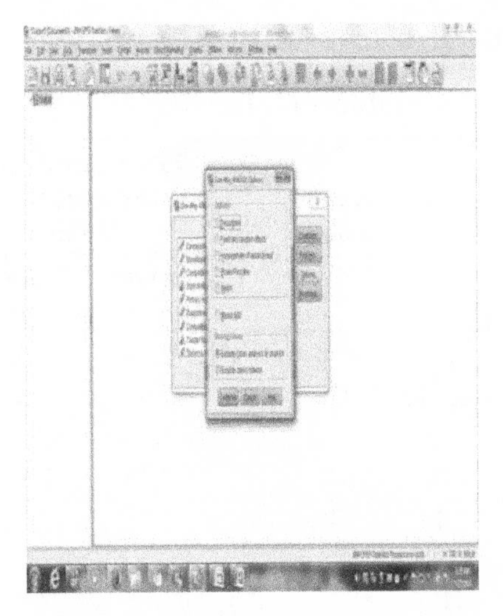

Figure - 4

Step - 3: In one-way anova: option dialogue box, select descriptive checklist box and click on continue button to return to one-way anova dialogue box

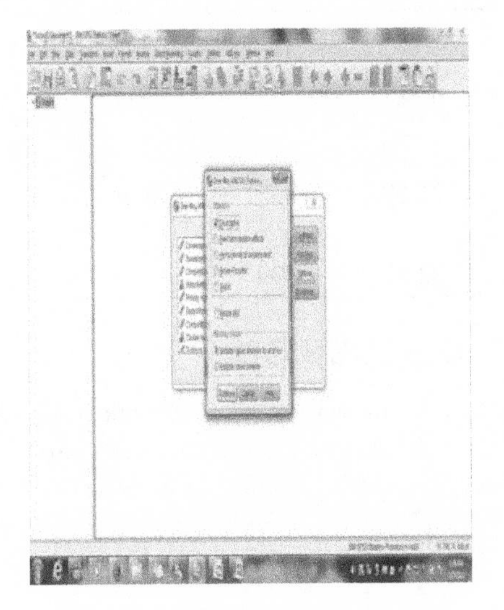

Figure - 5

Step - 9: Click the ok button to generate the output

SPSS Output of the one-way Analysis

Table - 1 Descriptives

Place	N	Mean	Std. Deviation	Std. Error	95% Confidence Interval for Mean		Minimum	Maximum
					Lower Bound	Upper Bound		
Innovator	54	2.37	1.033	.141	2.09	2.65	1	4
Diehard Pessimist	67	2.33	1.021	.125	2.08	2.58	1	4
Planner	103	2.68	1.122	.111	2.46	2.90	1	4
Active experimenter	38	2.45	1.245	.202	2.04	2.86	1	4
Blind Optimist	38	2.37	1.324	.215	1.93	2.80	1	4
Total	300	2.48	1.132	.065	2.35	2.61	1	4

Table - 2 ANOVA

Place	Sum of Squares	df	Mean Square	F	Sig.
Between Groups	6.804	4	1.701	1.334	.257
Within Groups	376.033	295	1.275		
Total	382.837	299			

RESULTS AND DISCUSSION

The descriptive results and one way ANOVA results are shown above. From the above, it is inferred that there is no significant mean variation between retailer characteristics with respect to geographical location (F=1.334, sig. value <.257).

Chapter - 8

Discriminant Analysis

Learning Objectives

This chapter helps to understand the following

- Meaning of Discriminant Analysis
- Objectives, Purpose, Usage of Discriminant Analysis
- Assumption of Discriminant Analysis
- Steps for Using Discriminant Analysis
- Methods to derive Discriminant Analysis
- Advantage and Disadvantage of Discriminant Analysis
- Discriminant Analysis Using SPSS
- Procedure to Run Discriminant Analysis using SPSS
- Results and Discussion of Discriminant Analysis Output

Introduction

Multiple discriminant analysis is a statistical technique which is used to reduce the difference between variables in order to classify the variables into set of broad group. Multiple discriminant analysis is also known as discriminant factor analysis and canonical discriminant analysis. MDA is an extension of discriminant analysis. MDA shares its techniques and ideas with multiple analyses of variance. The purpose of MDA is to classify cases into three or more categories or group by using continuous or duplicate variables as predictors. MDA is not directly used for classification.

OBJECTIVES

- MDA evaluates the difference between average score for a priori groups (known or assumed variable) on a set of variables.
- To determine which independent variables account for most of difference between groups.
- To classify the observation into groups.
- Interpreting the predictive equation to understand the better relationship that exists among variable.
- To evaluate accuracy of classification.

PURPOSE

Discriminant function analysis performs the task which is similar to multiple linear regressions by predicting an outcome. The purpose of using discriminant analysis is to predict outcome of variables, example, to predict outcomes like migrant and non migrant status, making profit or not, renting or paying mortgage for houses satisfied vs dissatisfied among employees etc. The procedure begins set of observation where both group membership and the value of internal variables are known. The end result of the procedure is a model that allows group membership prediction only when interval variables are known. The other purpose of discriminant function analysis is to understand the data set.

WHEN IT IS USED

- When the dependent variables is categorical with the predictors interval variable [Y axis (dependent variable) is an interval variable] such as income, age, year etc and although dummy variables can also be used as a predictor in multiple regression. The logistic regression interval variable can be at any level of measurement.
- When there are more than two dependent variables categories unlike logistic regression, which is similar to dichotomous dependent variable.

ASSUMPTION

- The random sample should be taken as observation.
- Each predictor variable is normally distributed.
- Each of the allocation for dependent categories in the initial classification is correctly classified.
- There must be at least two groups or categories. Each case should belong to same group so that the groups are mutually exclusive.

- Each group should be well defined and differentiate from other groups.
- Before collecting the data the groups or categories should be defined.
- The group size of the dependent variable should not be grossly different and it should be at least five times of the number of independent variable.

INTERPRETATION

- Discriminant function can be interpreted by determining which groups from it best separate.
- Correlation between a discriminant function and original dependent variable reveals what conceptual variable the discriminant function represents.
- The discriminant function determines which of the original dependent variables have contributed to overall significance.

STEPS FOR USING DISCRIMINANT ANALYSIS

Formulate problem

The problem has to be formulated by identifying the objectives, criterion and dependent variables. The criterion variables should consist of two or more exclusive and collectively exhaustive categories. The dependent variables should be ratio scale or interval. The predictor variable should be chosen on theoretical model.

Research design issue

While selecting the dependent and independent variables the researcher should specify it.

Sample division

Sample should be divided into two groups namely estimation or analysis sample and validation sample. The analysis sample is used for estimating discriminant function. The validation sample is for validating discriminant function.

METHOD TO DERIVE DISCRIMINANT FUNCTION

There are two computation methods used to derive discriminant function which is **direct and stepwise method**.

Direct method involve in estimating discriminant function so that all predictions are included simultaneously. Each dependent variable is included in direct method.

In Stepwise discriminant analysis the independent variables are entered one at a time based on the ability to discriminate among the group.

TEST OF SIGNIFICANCE

- For two groups the null hypothesis is mean of two groups on discriminant analysis, the centroids are equal.
- The mean discriminant analysis score for each group is called centroids
- Wilk's lambda is used to test significance difference between groups.
- Wilk's lambda is between 0 to 1.

FACTORS WHICH INFLUENCE CLASSIFICATION RESULTS FROM DISCRIMINANT ANALYSIS

In applied business research the multiple discriminant analysis has gained wide acceptance. The few applied research using discriminant analysis for methodological and statistical problems. The problems are caused due to the lack of in depth knowledge on statistical properties of discriminant analysis. The problems are associated with data typically employed in business studies and also due to lack of complete answer by statistician to the problem which associated with discriminant analysis.

The lack of understanding by researcher of the statistical consideration associated with classification discriminant analysis is employed. The factor or items (two groups) that may directly influence the reported classification results will be investigated.

The samples are variables are employed in first group. They are multivariate normality, number and independence of predictor variables, sample size, missing values, and initial classification.

The factor encompasses items that are under the direct control of researcher, it comes under second group. The factors are error rates, equal vs unequal dispersion, priori probabilities, cost of misclassification and discriminant space.

ADVANTAGES

- Discriminant analysis is multiple dependent variables.
- It reduces the error rates.

- Each discriminant function measures unique and different, so it is easier to interpret between groups.
- Discriminant analysis offers opportunity to classify as ungrouped on dependent variable.

DISADVANTAGES

- Interpretation of discriminant function is mystically like identifying factors in a factor analysis.
- Discriminant analysis is extremely sensitive to outliers.
- The relationship between variables is assumed to be linear in all groups.
- Each discriminant function is assumed to show approximately equal variances in each group.

DISCRIMINANT ANALYSIS

Discriminant analysis builds a predictive model for group membership. The model is composed of a discriminant function (or, for more than two groups, a set of discriminant functions) based on linear combinations of the predictor variables that provide the best discrimination between the groups. The functions are generated from a sample of cases for which group membership is known; the functions can then be applied to new cases that have measurements for the predictor variables but have unknown group membership.

DISCRIMINANT ANALYSIS USING SPSS

Discriminant analysis is used to predict membership in two or more mutually exclusive groups. There are three methods to classify group memberships:

- Maximum likelihood method: assign case to group k if the probability of membership is greater in group k than any other group
- Fisher (linear) classification functions: assign a membership to group k if its score on the function for group k is greater than any other function scores
- Distance function: assign membership to group k if its distance to the centroid of the group is minimum

PURPOSE OF DISCRIMINANT ANALYSIS

Discriminant analysis is used for

- Prediction

- Understand group differences
- Discriminant analysis characterize the relationship between a set of IVs with a categorical DV with relatively few categories
    - It creates a linear combination of the IVs that best characterizes the differences among the groups
    - Predictive discriminant analysis focus on creating a rule to predict group membership
    - Descriptive DA studies the relationship between the DV and the IVs.

ASSUMPTIONS OF DISCRIMINANT ANALYSIS

- Independent Variables are either dichotomous or measurement
- Normality
- Homogeneity of variances

Note: The grouping variable can have more than two values. The codes for the grouping variable must be integers, however, and need to specify their minimum and maximum values. Cases with values outside of these bounds are excluded from the analysis.

Statistics: For each variable: means, standard deviations, Univariate ANOVA. For each analysis: Box's *M*, within-groups correlation matrix, within-groups covariance matrix, separate-groups covariance matrix, total covariance matrix. For each canonical discriminant function: Eigen value, percentage of variance, canonical correlation, Wilks' lambda, chi-square. For each step: prior probabilities, Fisher's function coefficients, unstandardized function coefficients, Wilks' lambda for each canonical function.

Data: The grouping variable must have a limited number of distinct categories, coded as integers. Independent variables that are nominal must be recoded to dummy or contrast variables.

Assumptions: Cases should be independent. Predictor variables should have a multivariate normal distribution, and within-group variance-covariance matrices should be equal across groups. Group membership is assumed to be mutually exclusive (that is, no case belongs to more than one group) and collectively exhaustive (that is, all cases are members of a group). The procedure is most effective when group membership is a truly categorical variable; if group membership is based on values of a continuous variable (for example, high IQ versus low IQ), consider using linear regression to take advantage of the richer information that is offered by the continuous variable itself.

To Obtain a Discriminant Analysis

Analyze > Classify > Discriminant...

- Select an integer-valued grouping variable and click Define Range to specify the categories of interest.
- Select the independent, or predictor, variables. (If the grouping variable does not have integer values, Automatic Recode on the Transform menu will create a variable that does.)
- Select the method for entering the independent variables.
  - **Enter independents together:** Simultaneously enters all independent variables that satisfy tolerance criteria.
  - **Use stepwise method:** Uses stepwise analysis to control variable entry and removal.
- Optionally, select cases with a selection variable.

Discriminant Analysis Define Range

Specify the minimum and maximum value of the grouping variable for the analysis. Cases with values outside of this range are not used in the discriminant analysis but are classified into one of the existing groups based on the results of the analysis. The minimum and maximum values must be integers.

Discriminant Analysis Select Cases

Discriminant Analysis Set Value dialog box

To select cases for analysis:

- In the Discriminant Analysis dialog box, choose a selection variable.
- Click Value to enter an integer as the selection value.

Only cases with the specified value for the selection variable are used to derive the discriminant functions. Statistics and classification results are generated for both selected and unselected cases. This process provides a mechanism for classifying new cases based on previously existing data or for partitioning the data into training and testing subsets to perform validation on the model generated.

Discriminant Analysis Statistics

Descriptives: Available options are means (including standard deviations), univariate ANOVAs, and Box's M test.

- **Means:** Displays total and group means, as well as standard deviations for the independent variables.
- **Univariate ANOVAs:** Performs a one-way analysis-of-variance test for equality of group means for each independent variable.
- **Box's M:** A test for the equality of the group covariance matrices. For sufficiently large samples, a nonsignificant p value means there is insufficient evidence that the matrices differ.

The test is sensitive to departures from multivariate normality.

Function Coefficients: Available options are Fisher's classification coefficients and unstandardized coefficients.

- **Fisher's:** Displays Fisher's classification function coefficients that can be used directly for classification. A separate set of classification function coefficients is obtained for each group, and a case is assigned to the group for which it has the largest discriminant score (classification function value).
- **Unstandardized:** Displays the unstandardized discriminant function coefficients.

Matrices: Available matrices of coefficients for independent variables are within-groups correlation matrix, within-groups covariance matrix, separate-groups covariance matrix, and total covariance matrix.

- **Within-groups correlation:** Displays a pooled within-groups correlation matrix that is obtained by averaging the separate covariance matrices for all groups before computing the correlations.
- **Within-groups covariance:** Displays a pooled within-groups covariance matrix, which may differ from the total covariance matrix. The matrix is obtained by averaging the separate covariance matrices for all groups.
- **Separate-groups covariance:** Displays separate covariance matrices for each group.
- **Total covariance:** Displays a covariance matrix from all cases as if they were from a single sample.

Discriminant Analysis Stepwise Method

Method: Select the statistic to be used for entering or removing new variables. Available alternatives are Wilks' lambda, unexplained variance, Mahalanobis distance, smallest *F* ratio, and Rao's *V*. With Rao's *V*, specify the minimum increase in *V* for a variable to enter.

- **Wilks' lambda:** A variable selection method for stepwise discriminant analysis that chooses variables for entry into the equation on the basis of how much they lower Wilks' lambda. At each step, the variable that minimizes the overall Wilks' lambda is entered.
- **Unexplained variance:** At each step, the variable that minimizes the sum of the unexplained variation between groups is entered.
- **Mahalanobis distance:** A measure of how much a case's values on the independent variables differ from the average of all cases. A large Mahalanobis distance identifies a case as having extreme values on one or more of the independent variables.
- **Smallest F ratio:** A method of variable selection in stepwise analysis based on maximizing an F ratio computed from the Mahalanobis distance between groups.
- **Rao's V:** A measure of the differences between group means. Also called the Lawley-Hotelling trace. At each step, the variable that maximizes the increase in Rao's V is entered. After selecting this option, enter the minimum value a variable must have to enter the analysis.

Criteria: Available alternatives are Use F value and Use probability of F. Enter values for entering and removing variables.

- **Use F value:** A variable is entered into the model if its F value is greater than the Entry value and is removed if the F value is less than the Removal value. Entry must be greater than Removal, and both values must be positive. To enter more variables into the model, lower the Entry value. To remove more variables from the model, increase the Removal value.
- **Use probability of F:** A variable is entered into the model if the significance level of its F value is less than the Entry value and is removed if the significance level is greater than the Removal value. Entry must be less than Removal, and both values must be positive. To enter more variables into the model, increase the Entry value. To remove more variables from the model, lower the Removal value.

Display: Summary of steps displays statistics for all variables after each step; F for pairwise distances displays a matrix of pairwise *F* ratios for each pair of groups.

Discriminant Analysis Classification

Prior Probabilities: This option determines whether the classification coefficients are adjusted for a priori knowledge of group membership.

- **All groups equal:** Equal prior probabilities are assumed for all groups; this has no effect on the coefficients.
- **Compute from group sizes:** The observed group sizes in sample determine the prior probabilities of group membership. For example, if 50% of the observations included in the analysis fall into the first group, 25% in the second, and 25% in the third, the classification coefficients are adjusted to increase the likelihood of membership in the first group relative to the other two.

Display: Available display options are casewise results, summary table, and leave-one-out classification.

- **Casewise results:** Codes for actual group, predicted group, posterior probabilities, and discriminant scores are displayed for each case.
- **Summary table:** The number of cases correctly and incorrectly assigned to each of the groups based on the discriminant analysis. Sometimes called the "Confusion Matrix."
- **Leave-one-out classification:** Each case in the analysis is classified by the functions derived from all cases other than that case. It is also known as the "U-method."

Replace missing values with mean: Select this option to substitute the mean of an independent variable for a missing value during the classification phase only.

Use Covariance Matrix: Choose to classify cases using a within-groups covariance matrix or a separate-groups covariance matrix.

- **Within-groups:** The pooled within-groups covariance matrix is used to classify cases.
- **Separate-groups:** Separate-groups covariance matrices are used for classification. Because classification is based on the discriminant functions (not based on the original variables), this option is not always equivalent to quadratic discrimination.

Plots: Available plot options are combined-groups, separate-groups, and territorial map.

- **Combined-groups:** Creates an all-groups scatterplot of the first two discriminant function values. If there is only one function, a histogram is displayed instead.

- **Separate-groups:** Creates separate-group scatterplots of the first two discriminant function values. If there is only one function, histograms are displayed instead.
- **Territorial map:** A plot of the boundaries used to classify cases into groups based on function values. The numbers correspond to groups into which cases are classified. The mean for each group is indicated by an asterisk within its boundaries. The map is not displayed if there is only one discriminant function.

Discriminant Analysis Save

It will add new variables to active data file. Available options are predicted group membership (a single variable), discriminant scores (one variable for each discriminant function in the solution), and probabilities of group membership given the discriminant scores (one variable for each group).

Discriminant Analysis

Points to remember

In discriminant analysis design, the dependent variable must be non metric which representing groups of objects that is predictable to differ on the independent variables. In case of converting metric variables to a non metric scale for use as the dependent variable, consider using extreme groups to maximize the group differences. The Independent variables of discriminant analysis must identify differences between at least two groups. The sample size must be large enough to meet at least one more observation per group than the number of independent variables, but striving for at least 20 cases per group, to maximize the number of observations per variable a minimum ratio of five observations per independent variable required.

SPSS

Test Procedure in SPSS

Following are the steps to analyse data using Discriminant Analysis in SPSS.

Step - 1: Click Analyze > Classify > Discriminant as shown in the figure -1, the discriminant analysis dialogue box will appear (as given in the figure - 2).

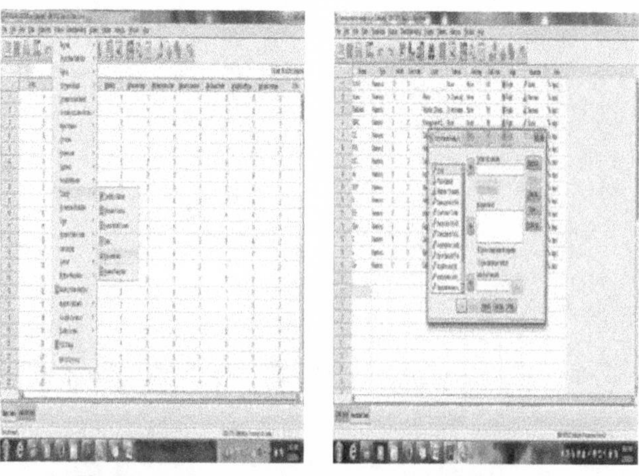

Figure - 1 **Figure - 2**

Enter grouping variables. **Click button – Define range**. Then define range lowest and lowest code value for grouping variables. Select the independent variables There are several methods for discriminant analysis. **"Enter independents together"** check box which is standard selected.

Step - 2: In discriminant analysis dialogue box, transfer the grouping variable in the box as given in figure -3 and then define its range by clicking the tab below grouping variable. A dialogue box will appear as given in figure -4. In this example, retailer characteristics is set as a grouping variable and define its range minimum "1" and maximum "5" (depends upon the group chosen) and click continue as given in figure -5.

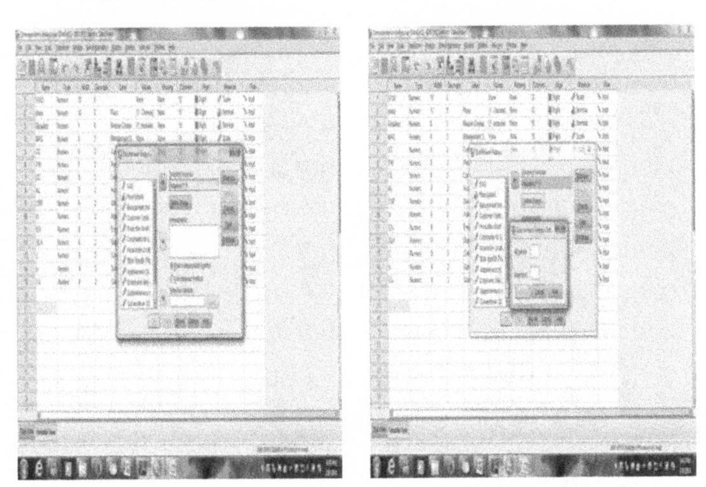

Figure - 3 **Figure - 4**

Step - 3: In discriminant analysis dialogue box, Transfer independent variable in the independent box. In this example, the factor loading are summated and transferred in the independent box. Click independent together as given in figure -6.

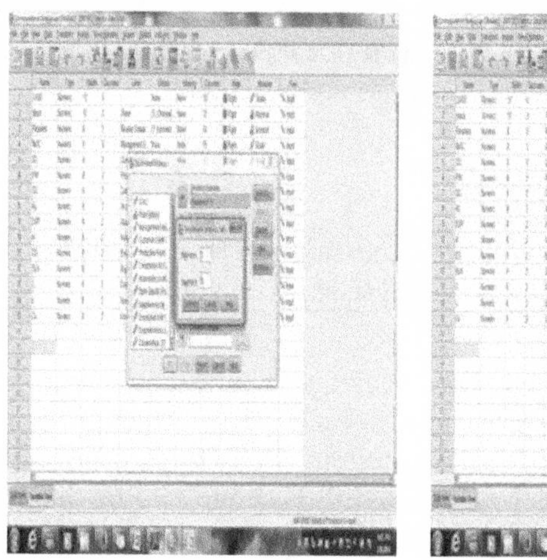

Figure - 5

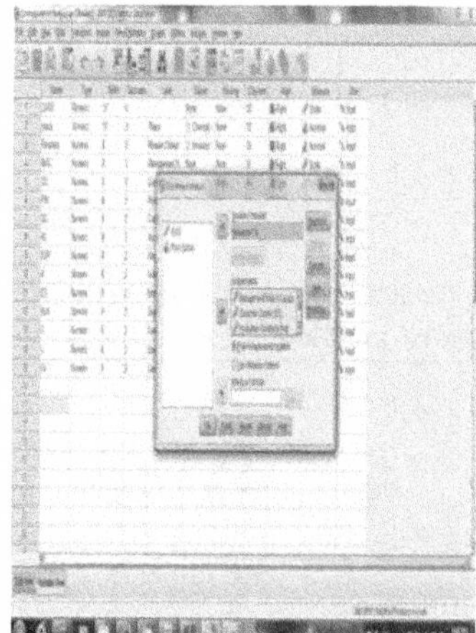

Figure - 6

Step - 4: In discriminant analysis dialogue box, Transfer independent variable in the independent box. In this example, the factor loading are summated and transferred in the independent box. Clicks enter independent together as given in figure -6.

Step - 5: In discriminant analysis dialogue box, click statistics tab, discriminant analysis: statistics dialogue box will appear as given in figure -7. In descriptives, click Means, Univariate ANOVAs, Box's M checkbox; in function coefficients, click unstandardized checkbox and in Matrices, click within-groups correlation checkbox as given in figure -8 and click continue to return to Discriminant analysis dialogue box.

Click button – Statistics. It indicates those statistics that are desire in discriminant analysis. It includes

- Mean generates: The mean and standard deviation for each variables and each groups.
- Univariate ANOVA is used to compare the mean values for group and used to verify significance difference between mean.
- Box'M is used to test the equality of group covariance matrices.
- Unstandardised function coefficient is based on raw score for discriminant variables.

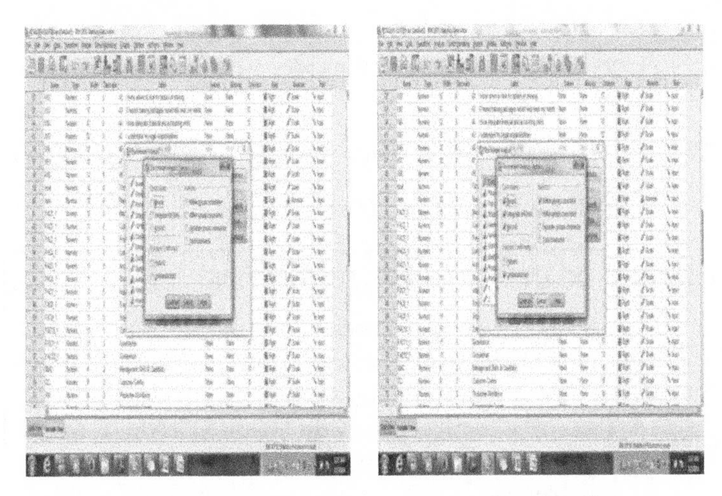

Figure - 7 Figure - 8

Step - 6: In discriminant analysis dialogue box, click classify tab, a dialogue box will appear as given in figure -9, click compute from group sizes in prior probabilities, followed by in display click summary table, leave-one-out classification check box, in use covariance matrix click within-groups, in plots, click combined-groups, separate groups, territorial map check box and click continue to return to discriminant analysis box.

Then **click button – classify**. Many classification options can be selected such as plots, priori probabilities.

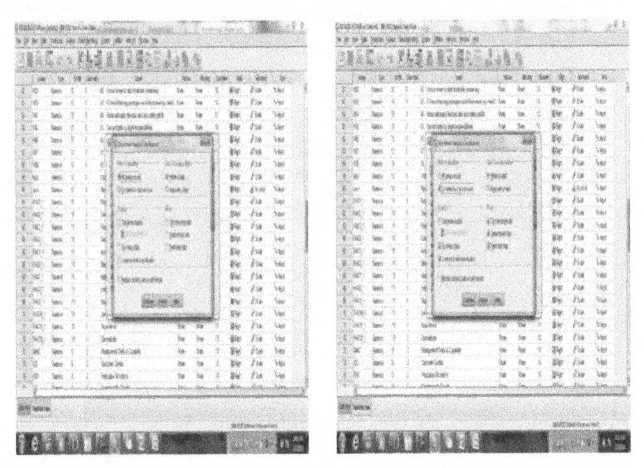

Figure - 9 Figure - 10

If needed Click on **save** tab in Discriminant Analysis Dialogue Box to save new variable. It predicts group membership, discriminant scores and probabilities of group member.

Step - 7: In discriminant analysis dialogue box, Click ok to generate output.

SPSS Output of the Discriminant Analysis

OUTPUT OF DISCRIMINANT ANALYSIS

'Enter independent together' are selected as discriminant analysis. The descriptive univariate, Anova's Box'm and unstandardised function coefficient are requested.

- **Test of equality of group mean** test the result of univariate Anova carried out for each independent variable.
- **Box test of equality of covariance matrix.**
- **Wilk's lambda** is the ratio of within group's sum of square of total sum of squares. It is the proportion of total variance in discriminant scores which are not explained by difference among groups. A lambda 1.00 occurs when desired group means are equal. A small lambda occurs when within-groups variability is small when compared to the total variability. Small lambda indicates the group means appear to differ. The associated significance value denotes whether the difference in significance.
- **Canonical discriminant function coefficient** indicates the independent variables unstandardisded scores.
- **Function of group centroid** indicates average discriminant score in two groups.
- **Classification results** is the summary of classified data. The leave-one-out classification is one of the cross validation method, results of which are also presented in classification result table.

Table 1 Group Statistics

	Ward Method	Mean	Std. Deviation	Valid N (listwise)	
				Unweighted	Weighted
Innovator	Management Skills & Capability	3.3171	.63073	54	54.000
	Customer Centric	3.5926	.78561	54	54.000
	Productive Workforce	3.2438	.59849	54	54.000
	Constraints for Growth	1.9306	.56095	54	54.000
	Accessible Location	3.8981	.59383	54	54.000
	Store Specific Planogram	4.0880	.52566	54	54.000
	Adaptiveness	4.0309	.87898	54	54.000
	Employees Skill	3.4352	.96175	54	54.000
	Supportiveness of Local Authorities	3.1389	.82939	54	54.000
	Convenience	3.3148	.68960	54	54.000
	Surveillance	3.2407	.58874	54	54.000
	Competition	4.5556	.50157	54	54.000

Ward Method		Mean	Std. Deviation	Valid N (listwise)	
				Unweighted	Weighted
Diehard Pessimist	Management Skills & Capability	2.6119	.77274	67	67.000
	Customer Centric	2.8827	.90108	67	67.000
	Productive Workforce	2.9005	.71836	67	67.000
	Constraints for Growth	3.3507	1.03083	67	67.000
	Accessible Location	3.1530	.74360	67	67.000
	Store Specific Planogram	3.1903	.95850	67	67.000
	Adaptiveness	3.7164	.83132	67	67.000
	Employees Skill	2.5149	1.32850	67	67.000
	Supportiveness of Local Authorities	2.9963	1.00707	67	67.000
	Convenience	3.1990	.49257	67	67.000
	Surveillance	2.7313	.69809	67	67.000
	Competition	3.1940	1.14467	67	67.000

	Ward Method	Mean	Std. Deviation	Valid N (listwise)	
				Unweighted	Weighted
Planner	Management Skills & Capability	3.4939	.65781	103	103.000
	Customer Centric	4.0402	.61116	103	103.000
	Productive Workforce	3.6974	.63561	103	103.000
	Constraints for Growth	1.8083	.64188	103	103.000
	Accessible Location	3.8058	.64162	103	103.000
	Store Specific Planogram	4.0485	.79139	103	103.000
	Adaptiveness	4.0615	.72423	103	103.000
	Employees Skill	3.5874	1.13420	103	103.000
	Supportiveness of Local Authorities	3.9029	.65701	103	103.000
	Convenience	3.4337	.56173	103	103.000
	Surveillance	3.0874	.63565	103	103.000
	Competition	2.6699	.86759	103	103.000

Ward Method		Mean	Std. Deviation	Valid N (listwise)	
				Unweighted	Weighted
Active experimenter	Management Skills & Capability	3.9507	.62638	38	38.000
	Customer Centric	4.8496	.33205	38	38.000
	Productive Workforce	3.5614	.85285	38	38.000
	Constraints for Growth	1.6974	.91378	38	38.000
	Accessible Location	4.3224	.96721	38	38.000
	Store Specific Planogram	4.7961	.56907	38	38.000
	Adaptiveness	4.8596	.28613	38	38.000
	Employees Skill	2.7500	1.15519	38	38.000
	Supportiveness of Local Authorities	4.5855	.49414	38	38.000
	Convenience	3.5789	.51805	38	38.000
	Surveillance	3.0526	.92114	38	38.000
	Competition	4.9211	.27328	38	38.000

Ward Method		Mean	Std. Deviation	Valid N (listwise)	
				Unweighted	Weighted
Blind Optimist	Management Skills & Capability	4.0461	.46537	38	38.000
	Customer Centric	4.9511	.18869	38	38.000
	Productive Workforce	2.3553	.31282	38	38.000
	Constraints for Growth	2.0921	.67639	38	38.000
	Accessible Location	4.5197	.62163	38	38.000
	Store Specific Planogram	4.9803	.08970	38	38.000
	Adaptiveness	5.0000	.00000	38	38.000
	Employees Skill	2.1842	.33734	38	38.000
	Supportiveness of Local Authorities	4.4013	.38791	38	38.000
	Convenience	3.7368	.46601	38	38.000
	Surveillance	3.3684	1.10089	38	38.000
	Competition	2.3421	.62715	38	38.000

Ward Method		Mean	Std. Deviation	Valid N (listwise)	
				Unweighted	Weighted
Total	Management Skills & Capability	3.3929	.81257	300	300.000
	Customer Centric	3.9190	.97185	300	300.000
	Productive Workforce	3.2506	.79230	300	300.000
	Constraints for Growth	2.1967	.99354	300	300.000
	Accessible Location	3.8325	.82680	300	300.000
	Store Specific Planogram	4.0767	.92324	300	300.000
	Adaptiveness	4.1989	.82424	300	300.000
	Employees Skill	3.0367	1.21184	300	300.000
	Supportiveness of Local Authorities	3.7125	.94123	300	300.000
	Convenience	3.4167	.57840	300	300.000
	Surveillance	3.0667	.77733	300	300.000
	Competition	3.3700	1.23214	300	300.000

Table 2 Tests of Equality of Group Means

	Wilks' Lambda	F	df1	df2	Sig.
Management Skills & Capability	.644	40.747	4	295	.000
Customer Centric	.460	86.712	4	295	.000
Productive Workforce	.665	37.181	4	295	.000
Constraints for Growth	.599	49.466	4	295	.000
Accessible Location	.715	29.436	4	295	.000
Store Specific Planogram	.594	50.363	4	295	.000
Adaptiveness	.704	30.947	4	295	.000
Employees Skill	.798	18.693	4	295	.000
Supportiveness of Local Authorities	.612	46.821	4	295	.000
Convenience	.913	6.990	4	295	.000
Surveillance	.930	5.568	4	295	.000
Competition	.427	98.900	4	295	.000

Table 3 Pooled Within-Groups Matrices

		Management Skills & Capability	Customer Centric	Productive Workforce	Constraints for Growth	Accessible Location	Store Specific Planogram	Adaptiveness	Employees Skill	Supportiveness of Local Authorities	Convenience	Surveillance	Competition
Correlation	Management Skills & Capability	1.000	.176	.193	-.001	-.104	.109	.002	.336	.117	-.061	-.005	-.045
	Customer Centric	.176	1.000	.057	-.010	.016	.214	.021	-.156	.199	.099	.102	-.011
	Productive Workforce	.193	.057	1.000	-.089	.056	.100	.054	.297	.092	.070	-.066	-.035
	Constraints for Growth	-.001	-.010	-.089	1.000	.130	.017	.105	-.060	-.040	-.088	.073	.047
	Accessible Location	-.104	.016	.056	.130	1.000	.116	.105	.016	-.015	.255	.149	.098
	Store Specific Planogram	.109	.214	.100	.017	.116	1.000	.154	-.145	.253	.002	.144	-.129
	Adaptiveness	.002	.021	.054	.105	.105	.154	1.000	-.127	-.054	.051	-.006	-.079
	Employees Skill	.336	-.156	.297	-.060	.016	-.145	-.127	1.000	.000	-.076	-.086	.115
	Supportiveness of Local Authorities	.117	.199	.092	-.040	-.015	.253	-.054	.000	1.000	.005	.143	-.023
	Convenience	-.061	.099	.070	-.088	.255	.002	.051	-.076	.005	1.000	.071	.089
	Surveillance	-.005	.102	-.066	.073	.149	.144	-.006	-.086	.143	.071	1.000	-.049
	Competition	-.045	-.011	-.035	.047	.098	-.129	-.079	.115	-.023	.089	-.049	1.000

Table - 4

Analysis 1 Box's Test of Equality of Covariance Matrices

Log Determinants

Ward Method	Rank	Log Determinant
Innovator	12	-13.801
Diehard Pessimist	12	-7.201
Planner	12	-10.506
Active experimenter	12	-16.671
Blind Optimist	11	.[a]
Pooled within-groups	12	-8.606

The ranks and natural logarithms of determinants printed are those of the group covariance matrices.

a. Singular

Table 5 Test Results[a]

Box's M	848.362
Approx.	3.278
df1	234
df2	69277.536
Sig.	.000

Tests null hypothesis of equal population covariance matrices.

a. Some covariance matrices are singular and the usual procedure will not work. The non-singular groups will be tested against their own pooled within-groups covariance matrix. The log of its determinant is -7.933.

Summary of Canonical Discriminant Functions

Table 6 Eigenvalues

Function	Eigenvalue	% of Variance	Cumulative %	Canonical Correlation
1	3.299[a]	56.0	56.0	.876
2	1.557[a]	26.4	82.5	.780
3	.760[a]	12.9	95.4	.657
4	.272[a]	4.6	100.0	.462

a. First 4 canonical discriminant functions were used in the analysis.

Table 7 Wilks' Lambda

Test of Function(s)	Wilks' Lambda	Chi-square	df	Sig.
1 through 4	.041	930.401	48	.000
2 through 4	.175	506.738	33	.000
3 through 4	.447	234.040	20	.000
4	.786	69.874	9	.000

Table 8 Standardized Canonical Discriminant Function Coefficients

	Function			
	1	2	3	4
Management Skills & Capability	.351	-.019	-.211	-.069
Customer Centric	.430	-.149	.088	.140
Productive Workforce	-.188	.277	.635	.517
Constraints for Growth	-.463	-.138	-.417	.334
Accessible Location	.378	-.090	-.015	-.172
Store Specific Planogram	.217	.137	-.084	-.355
Adaptiveness	.335	.004	-.195	.135
Employees Skill	.023	-.117	.413	-.390
Supportiveness of Local Authorities	.238	-.182	-.024	.658
Convenience	-.017	-.181	-.090	.029
Surveillance	-.010	.099	.139	-.366
Competition	.151	.977	-.229	.090

Table 9 Structure Matrix

	Function			
	1	2	3	4
Customer Centric	.585*	-.144	.007	.234
Store Specific Planogram	.452*	-.023	-.080	-.115
Management Skills & Capability	.406*	-.061	.071	-.024
Accessible Location	.343*	-.011	-.058	-.172
Adaptiveness	.331*	-.049	-.255	.139

	Function			
	1	2	3	4
Convenience	.158*	-.083	-.043	.031
Competition	.094	.906*	-.204	.068
Productive Workforce	.029	.181	.729*	.411
Employees Skill	-.002	.086	.547*	-.230
Constraints for Growth	-.366	-.086	-.515*	.244
Supportiveness of Local Authorities	.392	-.157	.058	.563*
Surveillance	.120	-.015	.029	-.314*

Pooled within-groups correlations between discriminating variables and standardized canonical discriminant functions
Variables ordered by absolute size of correlation within function.

*. Largest absolute correlation between each variable and any discriminant function

Table 10 Canonical Discriminant Function Coefficients

	Function			
	1	2	3	4
	.535	-.029	-.322	-.106
	.648	-.225	.132	.210
	-.289	.426	.976	.794
	-.598	-.178	-.539	.431
	.537	-.128	-.022	-.245
	.302	.191	-.117	-.495
	.482	.006	-.279	.194
	.021	-.108	.379	-.358
	.322	-.246	-.032	.888
	-.030	-.325	-.161	.051
	-.013	.131	.184	-.486
	.186	1.205	-.283	.111
(Constant)	-9.155	-2.445	.226	-3.125

Unstandardized coefficients

Table 11 Functions at Group Centroids

Ward Method	Function			
	1	2	3	4
Innovator	-.088	1.695	.045	-.845
Diehard Pessimist	-2.807	-.137	-.747	.278
Planner	.088	-.726	1.073	.081
Active experimenter	2.468	1.706	-.367	.886
Blind Optimist	2.366	-1.904	-1.286	-.395

Unstandardized canonical discriminant functions evaluated at group means

Table 12 Classification Statistics Prior Probabilities for Groups

Ward Method	Prior	Cases Used in Analysis	
		Unweighted	Weighted
Innovator	.180	54	54.000
Diehard Pessimist	.223	67	67.000
Planner	.343	103	103.000
Active experimenter	.127	38	38.000
Blind Optimist	.127	38	38.000
Total	1.000	300	300.000

Table 13 Territorial Map

(Assuming all functions but the first two are zero)

Canonical Discriminant

Function 2

```
    -8.0    -6.0    -4.0    -2.0     .0    2.0    4.0    6.0    8.0
      +---------+---------+---------+---------+---------+---------+---------+---------+
  8.0 +       21                          14                        +
      I      211                         14                       I
      I       221                        14                       I
      I        21                        14                      I
      I         21                       14                      I
      I          21                      14                      I
  6.0 +          +   21   +       +        +      14 +       +       +        +
      I             21                   14                      I
```

```
     I        21                  14                  I
     I         211                 14                  I
     I          221                14                  I
     I           21               14                  I
 4.0 +      +       + 21     +       +     14 +      +      +      +
     I             21             14                  I
     I              21            14                  I
     I               21           14                  I
     I                21          14                  I
     I                 211         14                  I
 2.0 +      +       +      221       +    14 +      +      +      +
     I                   21     *    14  *                 I
     I                    21        14                  I
     I                     21       14                  I
     I                     2111111111111114                     I
     I                      23333333333333344                     I
  .0 +      +       +    *  +  23    +       3344      +      +      +
     I                      23          33444444444444444444444444444I
     I                      23    *      3555555555555555555555555555I
     I                      23          35                  I
     I                     23          35                   I
     I                     23          35                   I
-2.0 +      +       +       +23       +      35+ *     +      +      +
     I                    23          35                    I
     I                    23         35                     I
     I                   23         35                      I
     I                   23        35                       I
     I                   23       35                        I
-4.0 +      +       +     23        + 35     +      +      +      +
     I                  23       35                         I
     I                  23      35                          I
     I                 23      35                           I
     I                 23     35                            I
     I                 23    35                             I
-6.0 +      +       +    23 +     35 +      +      +      +      +
```

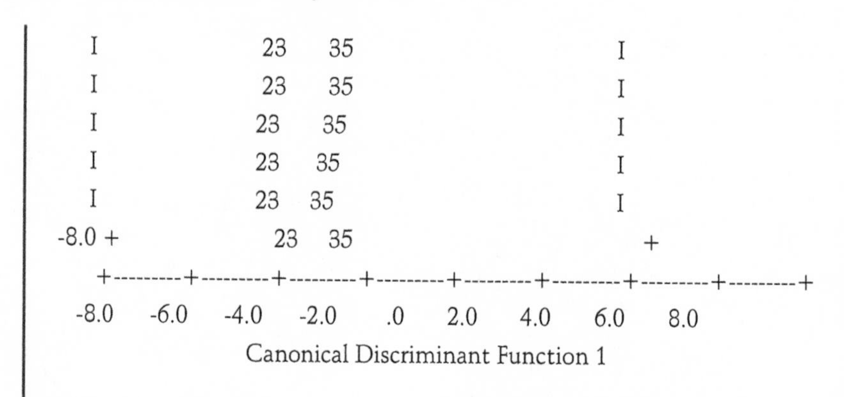

```
       I                23     35                    I
       I                23     35                    I
       I               23     35                     I
       I               23    35                      I
       I               23    35                      I
  -8.0 +                23    35                        +
       +---------+---------+---------+---------+---------+---------+---------+---------+
     -8.0      -6.0      -4.0      -2.0       .0       2.0       4.0       6.0       8.0
                         Canonical Discriminant Function 1
```

Symbols used in territorial map

```
Symbol  Group  Label
------  -----  --------------------

  1       1  Innovator
  2       2  Diehard Pessimist
  3       3  Planner
  4       4  Active experimenter
  5       5  Blind Optimist
  *          Indicates a group centroid
```

Separate-Groups Graphs

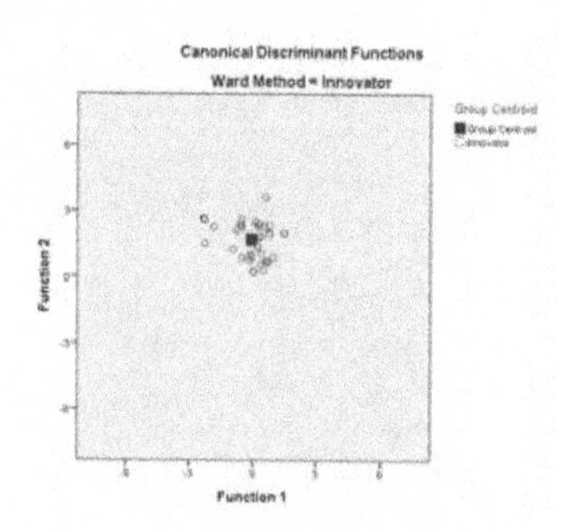

Figure - 11

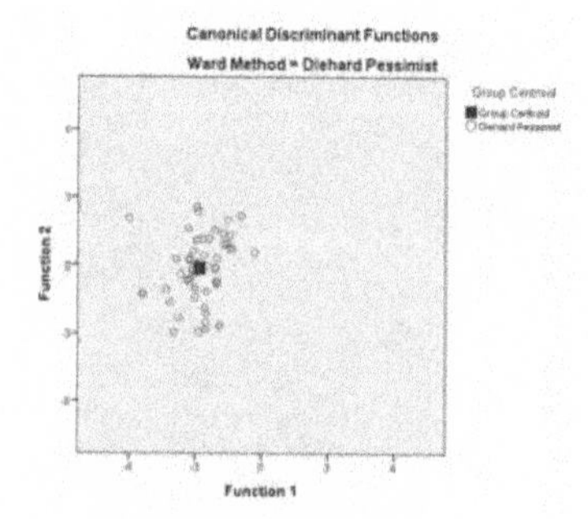

Figure - 12

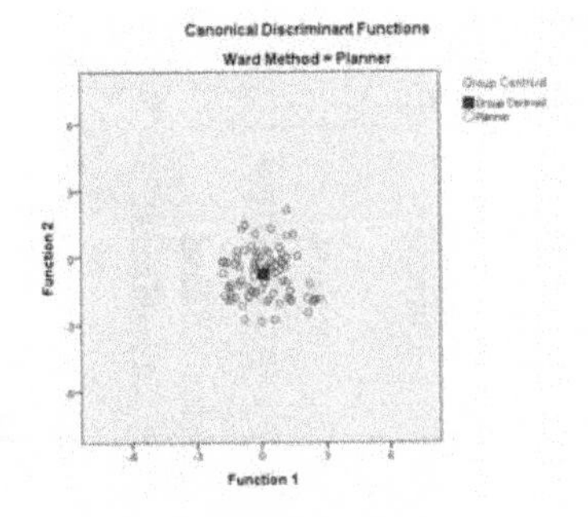

Figure - 13

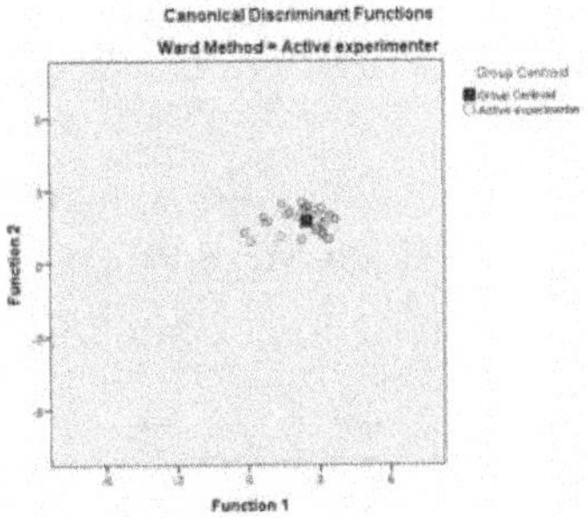

Canonical Discriminant Functions
Ward Method = Active experimenter
Group Centroid
Function 2
Function 1

Figure - 14

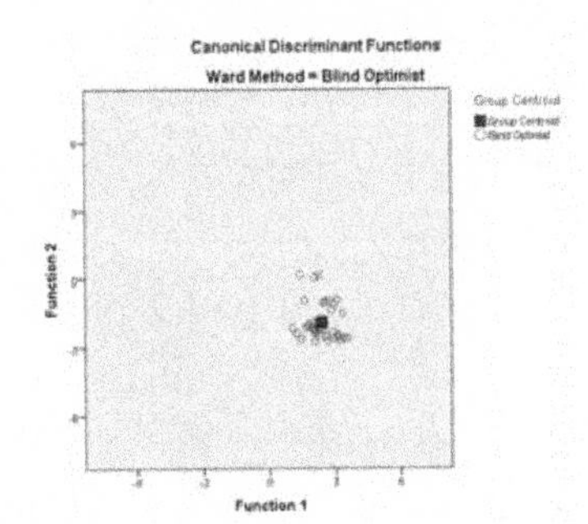

Figure - 15

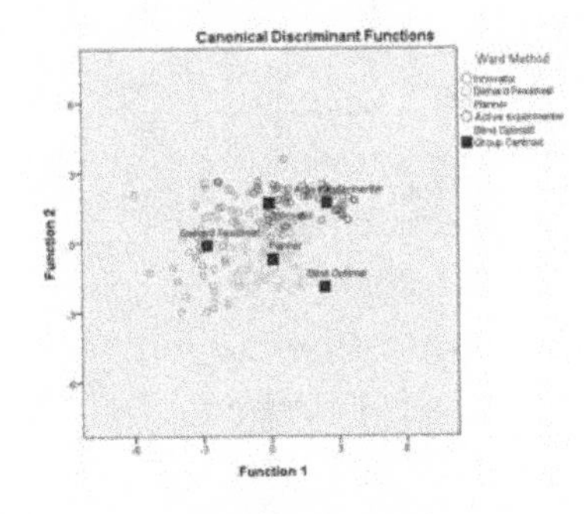

Figure - 16

Table 14 Classification Results[a,c]

		Ward Method	Predicted Group Membership: Innovator	Diehard Pessimist	Planner	Active experimenter	Blind Optimist	Total
Original	Count	Innovator	48	2	3	1	0	54
		Diehard Pessimist	4	61	2	0	0	67
		Planner	4	1	95	1	2	103
		Active experimenter	0	0	1	37	0	38
		Blind Optimist	0	0	0	1	37	38
	%	Innovator	88.9	3.7	5.6	1.9	.0	100.0
		Diehard Pessimist	6.0	91.0	3.0	.0	.0	100.0
		Planner	3.9	1.0	92.2	1.0	1.9	100.0
		Active experimenter	.0	.0	2.6	97.4	.0	100.0
		Blind Optimist	.0	.0	.0	2.6	97.4	100.0
Cross-validated[b]	Count	Innovator	48	2	3	1	0	54
		Diehard Pessimist	4	59	4	0	0	67
		Planner	4	2	93	2	2	103
		Active experimenter	0	1	3	34	0	38
		Blind Optimist	0	0	0	1	37	38
	%	Innovator	88.9	3.7	5.6	1.9	.0	100.0
		Diehard Pessimist	6.0	88.1	6.0	.0	.0	100.0
		Planner	3.9	1.9	90.3	1.9	1.9	100.0
		Active experimenter	.0	2.6	7.9	89.5	.0	100.0
		Blind Optimist	.0	.0	.0	2.6	97.4	100.0

a. 92.7% of original grouped cases correctly classified.

b. Cross validation is done only for those cases in the analysis. In cross validation, each case is classified by the functions derived from all cases other than that case.

c. 90.3% of cross-validated grouped cases correctly classified.

Results & Discussion

Group Statistics

Table 1 represents the group statistics of the variables, which provides the means and standard deviation for the groups. From the above table -1, final observations about the groups can be done and it clearly represents the groups are different from one another based on the features. Based on the loadings of features, the clusters are named as Innovator, Diehard Pessimist, Planner, Active experimenter and Blind Optimist.

Test of Equality of Group Means

Table -2 represents test of equlity of group means. F statistics determines the variable that should be included in the model and describes that when predictors are considered individually. The last column of the table is the p value corresponding to the f value, In this example, the independent variables are statistically significant from the groups.

Pooled With Group Matrices

Table 3 represents pooled within-group matrices and indicates the degree of correlation between the predictors. In this example, it can be clearly observed that there is weak correlation between the predictor variables. Thus, multi-collinearity will not be a problem.

Box's Test of Equality of Covariance Matrices

Box's test is used to determine whether two or more covariance matrices are equal. Box's M tests the null hypothesis that the covariance matrices do not differ between groups formed by the dependent. In this example, for one group "blind optimist" the rank I differs. And the log determinants ranges between -13.801 to -8.606 (table- 4). As per the rule, the log determinants should be equal. But in example the values are not significantly different from one another.

Test Results of Box's M

In this example, the log determinants is not similar and Box's M is 848.362 with F value - 3.278 which is significant at $p < .000$ (table - 5). However, with large samples, a significant result is not regarded as too important. Where three or more groups exist, and M is significant, groups with very small log determinants should be deleted from the analysis.

Summary of Canonical Discriminant Function

In the summary of canonical discriminant function, the Eigen value table provides information on each of the discriminate functions (equations) formed. The maximum number of discriminant functions formed is based on number of groups minus 1. This example contains 5 groups namely Innovator, Diehard Pessimist, Planner, Active experimenter and Blind Optimist so four function are displayed. Eigen value above 1 is considered to be good. The canonical correlation is the multiple correlations between the predictors and the discriminant function. It measures the degree of association between the discriminant scores and the groups (level of dependent variable). A high value of the canonical correlation associated with the function is .876. Square of this value is given as 76.73%, likewise for the following function the canonical correlation is .780,.657,.462 respectively i.e. 62.4%,43%, 21%. This indicates that 76.73% of the variance in the dependent variable can be attributed to this model. The Canonical Discriminant Function Coefficients below explain the factors that can be used in the different functions as predictors. 4 discriminant functions have been formed to predict the factors influencing the groups.

Wilks' Lambda

Wilks' lambda helps to test the significance of the discriminant function; chi square transformation of wilks lambda is used. A high value of chi square indicates that the function significantly differ from each other. This table -7 indicates a highly significant function (p <000) and provides the proportion of total variability not explained, i.e. it is the converse of the squared canonical correlation. It indicates that the discriminant function is statistically significant.

Standardized Canonical Discriminant Function Coefficients

Table -8 represents standardized canonical discriminant function coefficients in which the coefficients are like discriminant coefficient. This provides an index of the importance of each predictor like the standardized regression coefficient in multiple regression. The sign describes the direction of relationship with the predictor variables.

Structure Matrix

Table – 9 represents structure matrix. The loading of the structure matrix is like factor loadings, 0.30 is seen as the cut-off between important and less important variables. The pooled vales are the average of the group correlations. These structured correlation are referred as canonical loadings or discriminant loadings. The importance of predictor variables can be judged by the magnitude of correlations. Higher the value of correlations, higher is the importance of the correspondence predictor. In this example the largest absolute correlation between each variable and any discriminant function are

1. Function 1: Management Skills & Capability, Accessible Location, Store Specific Planogram, Adaptiveness,

2. Function 2: Competition.
3. Function 3: Productive Workforce, Employees Skill,
4. Function 4: Productive Workforce, Employees Skill, and Supportiveness of Local Authorities predicted as a significant factor.

Canonical Discriminant Function Coefficients

Table -10 represents canonical discriminant function coefficient table which gives an unstandarised coefficient and a constant value for the discriminant equation. After substituting the unstandarised coefficient values with the corresponding predictor and constant values, the discriminant equation can be formed.

Functions at Group Centroids

Table -11 represents function at group Centroids. A further way of interpreting discriminant analysis results is to describe each group in terms of its profile, using the group means of the predictor variables. These group means are called Centroids. In this example the Centroids ranges between -2.807 to 2.468 for function 1. Similarly the ranges can be determined for the rest of the functions.

Classification Statistics

Table -12 represents classification statistics, which is similar to percentage analysis.

Territorial Map

Table 13 represents territorial map, which shows the relative location of the boundaries of the different groups.

Separate Group Graphs

Separate group graphs are given from figure-11 to figure -15 using wards method for individual groups. Figure-15 represents collective groups of the retailers.

Classification Results

Table 14 represents the classification results of discriminant analysis. In the predictive group membership 92.3% of the original groups classified correctly. Also, through this analysis, the discriminant scores for each case explaining function classification (from 1 to 5) for each respondent are given. They are classified based on the coefficients of their predictor variables. These scores can be used for further analysis as well. With the confidence of the classification statistics derived, it can be said that the cluster segments interpreted in the study would be sufficient evidence for the potential retailers in the four cities general. The results of Cross validation shows that 90.3% of cross-validated grouped cases correctly classified.

Model - 2 Stepwise Method

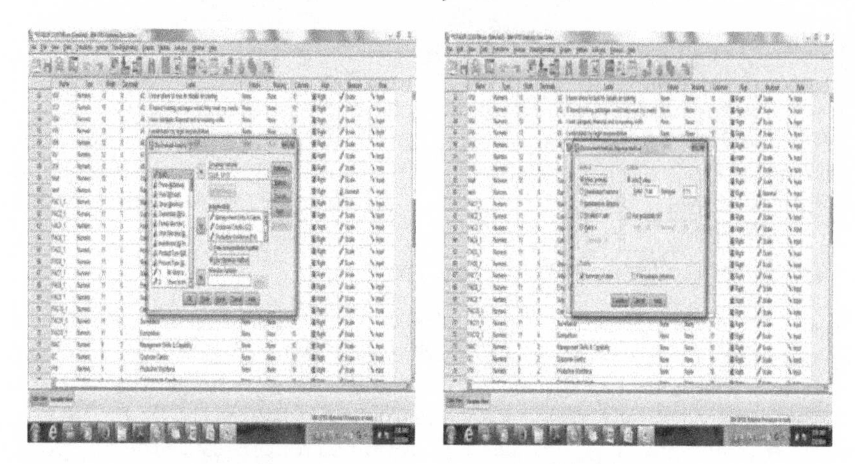

Figure - 1 **Figure - 2**

Step - 1: In discriminant analysis dialogue box, after transferring grouping variable and independent variable click on enter step wise method as given in figure -1. Then click method tab, discriminant analysis: method dialogue box will appear as given in figure -2. In methods click, wiki lambda and in display summary of steps checkbox and click continue to return to discriminant analysis dialogue box.

Step - 2: Click ok to generate output.

SPSS Output of the Discriminant Analysis

Stepwise Statistics

Table 1 Variables Entered/Removed[a,b,c,d]

Step	Entered	Wilks' Lambda											
		Statistic	df1	df2	df3	Exact F				Approximate F			
						Statistic	df1	df2	Sig.	Statistic	df1	df2	Sig.
1	Competition	.427	1	4	295.000	98.900	4	295.000	.000				
2	Customer Centric	.196	2	4	295.000	92.376	8	588.000	.000				
3	Productive Workforce	.131	3	4	295.000					74.714	12	775.497	.000
4	Constraints for Growth	.095	4	4	295.000					64.592	16	892.712	.000
5	Adaptiveness	.078	5	4	295.000					56.122	20	966.088	.000
6	Supportiveness of Local Authorities	.066	6	4	295.000					49.771	24	1012.899	.000
7	Accessible Location	.057	7	4	295.000					45.080	28	1043.427	.000
8	Management Skills & Capability	.051	8	4	295.000					41.347	32	1063.687	.000
9	Employees Skill	.047	9	4	295.000					37.918	36	1077.259	.000
10	Store Specific Planogram	.043	10	4	295.000					35.037	40	1086.334	.000

At each step, the variable that minimizes the overall Wilks' Lambda is entered.

a. Maximum number of steps is 24.
b. Minimum partial F to enter is 3.84.
c. Maximum partial F to remove is 2.71.
d. F level, tolerance, or VIN insufficient for further computation.

Table 2 Variables in the Analysis

Step		Tolerance	F to Remove	Wilks' Lambda
1	Competition	1.000	98.900	
2	Competition	1.000	98.556	.460
	Customer Centric	1.000	86.410	.427
3	Competition	.999	97.648	.306
	Customer Centric	.997	86.088	.285
	Productive Workforce	.996	36.558	.196
4	Competition	.997	97.279	.223
	Customer Centric	.997	65.475	.181
	Productive Workforce	.988	30.349	.135
	Constraints for Growth	.990	27.195	.131
5	Competition	.990	98.114	.182
	Customer Centric	.996	46.457	.127
	Productive Workforce	.984	30.260	.110
	Constraints for Growth	.977	28.406	.108
	Adaptiveness	.978	16.782	.095
6	Competition	.989	96.803	.154
	Customer Centric	.958	26.024	.090
	Productive Workforce	.978	30.163	.094
	Constraints for Growth	.977	27.308	.091
	Adaptiveness	.974	16.578	.081
	Supportiveness of Local Authorities	.949	12.631	.078
7	Competition	.979	95.566	.133
	Customer Centric	.958	21.657	.075
	Productive Workforce	.973	30.500	.082
	Constraints for Growth	.962	28.552	.080
	Adaptiveness	.965	12.689	.067
	Supportiveness of Local Authorities	.949	12.238	.067
	Accessible Location	.960	10.950	.066

Step		Tolerance	F to Remove	Wilks' Lambda
8	Competition	.978	95.334	.118
	Customer Centric	.935	14.057	.061
	Productive Workforce	.938	32.633	.074
	Constraints for Growth	.961	27.248	.070
	Adaptiveness	.965	11.552	.059
	Supportiveness of Local Authorities	.944	10.924	.058
	Accessible Location	.946	12.119	.059
	Management Skills & Capability	.915	9.330	.057
9	Competition	.959	97.867	.110
	Customer Centric	.880	13.186	.055
	Productive Workforce	.869	24.101	.062
	Constraints for Growth	.959	25.386	.063
	Adaptiveness	.945	9.603	.053
	Supportiveness of Local Authorities	.944	10.642	.053
	Accessible Location	.944	11.814	.054
	Management Skills & Capability	.801	8.822	.052
	Employees Skill	.743	6.510	.051
10	Competition	.948	99.414	.103
	Customer Centric	.867	11.770	.050
	Productive Workforce	.863	24.626	.058
	Constraints for Growth	.959	24.133	.058
	Adaptiveness	.931	8.126	.048
	Supportiveness of Local Authorities	.897	10.732	.050
	Accessible Location	.927	9.433	.049
	Management Skills & Capability	.792	7.150	.048
	Employees Skill	.725	6.488	.047
	Store Specific Planogram	.831	5.534	.047

Table 3 Variables Not in the Analysis

Step		Tolerance	Min. Tolerance	F to Enter	Wilks' Lambda
0	Management Skills & Capability	1.000	1.000	40.747	.644
	Customer Centric	1.000	1.000	86.712	.460
	Productive Workforce	1.000	1.000	37.181	.665
	Constraints for Growth	1.000	1.000	49.466	.599
	Accessible Location	1.000	1.000	29.436	.715
	Store Specific Planogram	1.000	1.000	50.363	.594
	Adaptiveness	1.000	1.000	30.947	.704
	Employees Skill	1.000	1.000	18.693	.798
	Supportiveness of Local Authorities	1.000	1.000	46.821	.612
	Convenience	1.000	1.000	6.990	.913
	Surveillance	1.000	1.000	5.568	.930
	Competition	1.000	1.000	98.900	.427
1	Management Skills & Capability	.998	.998	40.830	.275
	Customer Centric	1.000	1.000	86.410	.196
	Productive Workforce	.999	.999	36.703	.285
	Constraints for Growth	.998	.998	49.224	.256
	Accessible Location	.990	.990	29.137	.306
	Store Specific Planogram	.983	.983	52.275	.250
	Adaptiveness	.994	.994	31.505	.299
	Employees Skill	.987	.987	19.181	.339
	Supportiveness of Local Authorities	.999	.999	46.262	.262
	Convenience	.992	.992	7.587	.387
	Surveillance	.998	.998	5.681	.397

Step		Tolerance	Min. Tolerance	F to Enter	Wilks' Lambda
	Management Skills & Capability	.967	.967	11.486	.170
	Productive Workforce	.996	.996	36.558	.131
	Constraints for Growth	.998	.998	33.206	.135
	Accessible Location	.990	.990	13.741	.165
	Store Specific Planogram	.938	.938	16.027	.161
2	Adaptiveness	.993	.993	15.874	.161
	Employees Skill	.963	.963	20.160	.154
	Supportiveness of Local Authorities	.960	.960	14.108	.165
	Convenience	.982	.982	2.207	.191
	Surveillance	.987	.987	3.061	.188
	Management Skills & Capability	.934	.934	12.225	.112
	Constraints for Growth	.990	.988	27.195	.095
	Accessible Location	.987	.987	13.994	.110
	Store Specific Planogram	.931	.931	16.599	.107
3	Adaptiveness	.991	.991	15.682	.108
	Employees Skill	.866	.866	10.131	.115
	Supportiveness of Local Authorities	.954	.954	13.667	.110
	Convenience	.977	.977	2.535	.127
	Surveillance	.982	.982	3.063	.126
	Management Skills & Capability	.934	.934	10.835	.083
	Accessible Location	.969	.969	15.348	.079
	Store Specific Planogram	.930	.930	15.297	.079
	Adaptiveness	.978	.977	16.782	.078
4	Employees Skill	.865	.865	7.688	.086
	Supportiveness of Local Authorities	.953	.953	12.815	.081
	Convenience	.970	.970	2.444	.092
	Surveillance	.977	.977	3.390	.091

	Step	Tolerance	Min. Tolerance	F to Enter	Wilks' Lambda
5	Management Skills & Capability	.933	.933	9.464	.069
	Accessible Location	.960	.960	11.333	.067
	Store Specific Planogram	.912	.912	9.938	.068
	Employees Skill	.849	.849	7.485	.070
	Supportiveness of Local Authorities	.949	.949	12.631	.066
	Convenience	.966	.966	1.998	.075
	Surveillance	.977	.972	3.398	.074
6	Management Skills & Capability	.929	.929	8.191	.059
	Accessible Location	.960	.949	10.950	.057
	Store Specific Planogram	.866	.866	8.828	.059
	Employees Skill	.849	.849	7.459	.060
	Convenience	.966	.949	1.998	.064
	Surveillance	.959	.932	3.955	.063
7	Management Skills & Capability	.915	.915	9.330	.051
	Store Specific Planogram	.854	.854	6.687	.052
	Employees Skill	.849	.849	6.996	.052
	Convenience	.903	.898	1.797	.056
	Surveillance	.936	.931	3.210	.055
8	Store Specific Planogram	.851	.851	5.552	.047
	Employees Skill	.743	.743	6.510	.047
	Convenience	.900	.888	1.719	.050
	Surveillance	.936	.915	3.189	.049
9	Store Specific Planogram	.831	.725	5.534	.043
	Convenience	.893	.737	1.641	.045
	Surveillance	.933	.741	3.343	.044
10	Convenience	.890	.719	1.451	.042
	Surveillance	.927	.724	2.868	.042

Table 4 Wilks' Lambda

Step	Number of Variables	Lambda	df1	df2	df3	Exact F				Approximate F			
						Statistic	df1	df2	Sig.	Statistic	df1	df2	Sig.
1	1	.427	1	4	295	98.900	4	295.000	.000				
2	2	.196	2	4	295	92.376	8	588.000	.000				
3	3	.131	3	4	295					74.714	12	775.497	.000
4	4	.095	4	4	295					64.592	16	892.712	.000
5	5	.078	5	4	295					56.122	20	966.088	.000
6	6	.066	6	4	295					49.771	24	1012.899	.000
7	7	.057	7	4	295					45.080	28	1043.427	.000
8	8	.051	8	4	295					41.347	32	1063.687	.000
9	9	.047	9	4	295					37.918	36	1077.259	.000
10	10	.043	10	4	295					35.037	40	1086.334	.000

Results & Discussion

Many of the tables in step wise discriminant analysis are similar to independents together method. From the above results table -1 and table 4 are considered for interpretation.

Stepwise Statistics Tables

The Stepwise Statistics Table -1 represents that ten steps were taken, with each one including another variable and therefore these ten were included in the Variables in the Analysis and Wilk's Lambda tables because each was adding some predictive power to the function. In some stepwise analyses only the first one or two steps might be taken, even though there are more variables, because subsequent variables are not counted to the predictive power of the discriminant function.

Wilk's Lambda Table

Wilk's lambda table (table-4) disclosed that all the predictors include some predictive power to the discriminant function as all are significant with $p < .000$. The remaining tables providing the Discriminant Function Coefficients, Structure Matrix, Group Centroids and the classification are the same as above.

Chapter - 9

Correspondence Analysis

Learning Objectives

This chapter helps to understand the following

- Meaning of Correspondence analysis
- Development of correspondence analysis
- Importance of Correspondence analysis
- Key Terms widely used in correspondence analysis
- Types of correspondence analysis
- Procedure for Running Correspondence Analysis using SPSS
- Results and Discussion for Correspondence analysis Output

introduction

Correspondence analysis has gained an international reputation as a powerful statistical tool for the graphical analysis of contingency tables. This popularity stems from its development and application in many European countries, especially France, and its use has spread to English speaking nations such as the United States and the United Kingdom. Its growing popularity amongst statistical practitioners, and more recently those disciplines where the role of statistics is less dominant, demonstrates the importance of the continuing research and development of the methodology. The aim of this chapter is to highlight the theoretical, practical and computational issues of simple correspondence analysis and discuss its relationship with recent advances that can be used to graphically display the association in two-way categorical data.

Development of Correspondence analysis

The development of correspondence analysis techniques is to handle problems involving contingency tables are due most importantly to Karl Pearson, G. Udny Yule and R.A. Fisher (see Goodman, 1996). One of the most influential techniques developed to measure the association between two categorical variables is the Pearson chi-squared statistic. Pearson (1900) developed the ground work for the chi-squared statistic which is used to compare the observed counts with what is expected under the hypothesis of independence between the two variables.

The theoretical issues associated with correspondence analysis date back to the early 20th century and its foundation is algebraic rather than geometric. The foundation of the technique was nearly laid with the 1904 and 1906 papers of Karl Pearson, as argued by de Leeuw (1983), when he developed the correlation coefficient of a two-way contingency table using linear regression. As Pearson (1906) states:

The conception of linear regression line as giving this arrangement with the maximum Degree of correlation appears to be of considerable philosophical interest. It amounts primarily to much the same thing as saying that if we have a fine classification, we shall get the maximum correlation by arranging the arrays so that the means of the arrays fall as closely as possible on a line. De Leeuw (1983) then notes: This is exactly what correspondence analysis does. Pearson just was not familiar with singular value decomposition, although this had been discovered much earlier by Beltrami, Sylvester and Jordan.

However, the original algebraic derivation of correspondence analysis is often accredited to Hirschfeld (1935) who developed a formulation of the correlation between the rows and columns of a two-way contingency table. Others to contribute to such developments include Richardson &Kuder (1933) and Horst (1935). In fact, Horst, who discussed his findings in early 1934 before the Psychology Section of the Ohio Academy of Science, was the first to coin the term “method of reciprocal averaging”, an alternative derivation of correspondence analysis. The simplest derivation of correspondence analysis was made by the biometrician R.A. Fisher in1940 when he considered data relating to hair and eye colour in a sample of children from Caithness, Scotland. While the original development of the problem aimed at dealing with two-way contingency tables, a more complex approach dealing with multi-way contingency tables was not discussed until1941 when psychometrician Louis Guttman discussed his method, called dual (or optimal) scaling, which is now referred to as the foundation of multiple correspondence analysis. Later applications of multiple correspondence analysis were considered using the Burt matrix of Burt (1950).

In fact Guttman (1953) writes of Burt: It is gratifying to see how Professor Burt has independently arrived at much the same formulation. This convergence of thinking lends credence to the similarity of the approach. Fisher and Guttman presented essentially the same theory in the biometric and psychometric literature. Thus biometricians regard Fisher as the inventor of correspondence analysis, while psychometrician regard it as being Guttman.

In the 1940's and 1950's further advances were made to the mathematical development of correspondence, particularly in the field of psychometrics, by Guttman and his researchers. In Japan, a group of data analysts led by Chikio Hayashi also further developed Guttman's ideas, which they referred to as the quantification of qualitative data.

The 1960's saw the biggest leap in the development of correspondence analysis when it was given a geometric form by linguist Jean-Paul Benzecri and his team of researchers at the Mathematical Statistics Laboratory, Faculty of Science in Paris, France. This work culminated in two volumes on data analysis; Benźecri (1973b, 1973a). As a result the method of analyse correspondences as coined by Benzecri, is very popular in France not just among statisticians, but among researchers from most disciplines in the country. The popularity of correspondence analysis in France resulted in a journal dedicated to the development and application of the technique as well as methods of classification, Cahiers de l'Analyse des Donnees, founded by Benzecri.

In 1974, this new method was widely exposed to English speaking researchers with the popular paper by M.O. Hill (Hill, 1974). He was the first to coin the method's name correspondence analysis which is the English translation of Benzecri'sl'analyse des correspondences. Since Hill's (1974) contribution, the theory of correspondence analysis, especially its application to multivariate data, has been reinvented many times and given different names, such as homogeneity analysis (Gifi, 1990) and dual scaling (Nishisato, 1980, 1994).

What do you mean by correspondence analysis?

Correspondence analysis (CA) is a multivariate statistical technique proposed by Hirschfeld and later developed by Jean-Paul Benzécri. It is conceptually similar to principal component analysis, but applies to categorical rather than continuous data. In a similar manner to principal component analysis, it provides a means of displaying or summarising a set of data in two-dimensional graphical form.

All data should be non-negative and on the same scale for CA to be applicable, and the method treats rows and columns equivalently. It is traditionally applied to contingency tables — CA decomposes the chi-squared statistic associated with this table into orthogonal factors. Because CA is a descriptive technique, it can be applied to tables whether or not the statistic is appropriate.

Importance of Correspondence analysis

The analysis of the contingency table is a very important component of multivariate statistics with many different types of analysis dedicated solely to this type of data set. Fienberg (1982) points out that the term contingency seems to have originated with Karl Pearson (1904) who used it to describe the measure of the deviation from complete independence between the rows and columns of such a data structure. More recently, the term has come to refer to the counts and the marginal frequencies in the contingency table. As a result, a contingency table contains information which is of a discrete or categorical nature.

Some of the widely used terms while discussing correspondence analysis are as follow:

Categorical data

In statistics, categorical data is a statistical data type consisting of categorical variables, used for observed data whose value is one of a fixed number of nominal categories, or for data that has been converted into that form, for example as grouped data. More specifically, categorical data may derive from either or both of observations made of qualitative data, where the observations are summarised as counts or cross tabulations, or of quantitative data, where observations might be directly observed counts of events happening or they might counts of values that occur within given intervals. Often, purely categorical data are summarised in the form of a contingency table. However, particularly when considering data analysis, it is common to use the term "categorical data" to apply to data sets that, while containing some categorical variables, may also contain non-categorical variables.

Continuous Data:

A set of data is said to be continuous if the values / observations belonging to it may take on any value within a finite or infinite interval. You can count, order and measure continuous data. For example height, weight, temperature, the amount of sugar in an orange, the time required to run a mile.

Orthogonal factor:

"Orthogonal" has meaning for a vector space with a quadratic form Q: two vectors v and w are orthogonal if and only if Q(v,w)=0. "Orthonormal" means in addition that Q(v,v)=1=Q(w,w). Thus "orthogonal" and "orthonormal" are not synonymous, nor are they restricted to finite matrices. (E.g., v and w may be elements of a Hilbert space, such as the space of L2 complex-valued functions on R3 used in classical quantum mechanics.)

Keyword interpretations

Mass: the marginal proportions of the row variable, used to weight the point profiles when computing point distance. This weighting has the effect of compensating for unequal numbers of cases.

Scores in dimension: scores used as coordinates for points when plotting the correspondence map. Each point has a score on each dimension.

Inertia: Variance

Contribution of points to dimensions: as factor loadings are used in conventional factor analysis to ascribe meaning to dimensions, so "contribution of points to dimensions" is used to intuit the meaning of correspondence dimensions.

Contribution of dimensions to points: these are multiple correlations, which reflect how well the principal components model is explaining any given point (category).

Interpretation of Correspondence analysis using reciprocal averaging

Goals

Community ecologists frequently use CA to identify and distinguish communities on the basis of their species composition. In these studies, each species can be considered a variable, and communities our subjects, which we attempt to describe efficiently and comprehensively with abundance data for the individual species. Early attempts at defining plant communities focused on using species with known ecological characteristics (i.e., response to moisture or elevation) as indicators of certain environmental conditions and therefore of habitat types and communities. A community was therefore designated on the basis of the relative abundance of these indicator species. However, many researchers became interested in whether groups of organisms could be described as communities without reference to pre-existing knowledge of their ecological requirements.

CA accomplishes this by taking into account the relative distribution of as many species as possible, using observed associations between species, or the lack thereof, to compare or distinguish groups of organisms in different locations. Correspondence analysis constructs a score for each location sampled by weighting the abundance of species in that location. The trick is devising weights that adequately represent both the overall distribution of a species, and its co-occurrence with and abundance relative to others. Remember, the goal is to reduce the data collected on many species to a single index that accounts for as much variation in species as possible.

Many algorithms have been used and described as "correspondence analysis", but I will focus on two of the broadest, most commonly referenced ones here: "Reciprocal Averaging", and then the modern "Correspondence Analysis" described by Legendre and Legendre (1998).

Reciprocal Averaging.

Reciprocal averaging (RA) was put forth by M.O. Hill in 1973 as a way of addressing particular weaknesses of PCA (principal components analysis) in describing multivariate community data. At the time, Hill did not realize his solution was in fact another method of correspondence analysis, being developed concomitantly by other researchers in France. He later published a treatment that resolved the differences and proved mathematical relationship between the two.

A simple example will help illustrate the problems and goals of correspondence analysis and of the reason for deriving weighted averages. Consider the following artificial data taken from Pielou (1984). Species are arranged in rows, and abundance data on each is collected in five locations (quadrats):

Reciprocal averaging approaches this problem on a trial and error basis by assigning arbitrary weights to species and re-calculating them until they converge on a single number. Thus, the data themselves derive the weights, rather than being imposed by the researcher. The re-calculation is guided by a back-and forth averaging between rows (species averages) and columns (quadrats): quadrat scores become weighted averages of species, and species become weighted averages of quadrats. The back and forth averaging between row and column scores gives the name "reciprocal" to the technique. In general the procedure is as follows (Pielou gives a worked example using the above data):

1. Assign arbitrary trial values for species weights (ranging 0 to 100)
2. Calculate average abundance for each quadrat using the above weights for each species; these are trial quadrat scores.
3. Return to species data, and re-calculate species scores. Using the scores for each quadrat to weight abundance in that quadrat, calculate a weighted average for each species across the quadrats. This is your revised set of species scores.
4. Return to quadrat data, and using the new species scores as weights, calculate new quadrat scores (essentially repeat step 2).

5. Moving back and forth between species and quadrat scores, re-calculate new scores until they converge (stop changing).

Applying this procedure to our sample data yields the following one-dimensional ordination of quadrats (scores are re-scaled to fit between 0 and 100):

Quad V	Quad IV	Quad II	Quad III	Quad I
(0)	(18.6)	(52.1)	(73.0)	(100)

Due to the fact that species scores are weighted according to quadrat scores, and vice versa, the scoring system derived is a better account of similarities than our first attempt. Quadrats IV and V are accurately represented close together, each being more similar to the other than other quadrats. In addition, Quadrats I and III no longer have the same score, but are more adequately distinguished by the species weights derived by our second attempt. Thus, by accounting for species distributions within and across all quadrats, reciprocal averaging has more adequately described the variation throughout the sampled locations – a successful ordination.

Note, however, that while we have separated and ordered the quadrats, the specific ecological or environmental factor underlying this order is not apparent in our analysis; in fact, it has to be inferred from what we know. Hence, these ordination techniques are considered indirect methods of detecting underlying environmental gradients. Also, note that we derived only one axis or scoring system, and another is traditionally developed. That second axis is constructed with the constraint that it accounts for variation independent of the first.

Correspondence Analysis using Linear Algebra

Detecting and depicting structure in species distributions can also be accomplished by using a system of linear equations to solve for the "best" weights, or those that will account for the most variation. Conceptually, this is the same approach of PCA (principal components analysis), but the difference lies in how the variation in species is quantified before weights are assigned. Rather than outlining the mathematical procedure involved, I will briefly describe the conceptual goals of the technique, which are sometimes the hardest to figure out.

In both PCA and CA the weights are derived by eigenanalysis, a technique in matrix algebra that can detect systematic variation among measurements and devise linear equations to represent that systematic variation. In PCA, the matrix of species abundances is transformed into a matrix of covariances or correlations, each abundance value being replaced by a measure of its correlation (or covariance) with other abundances in other quadrats. This correlation matrix is then subjected to eigenanalysis to determine weighted combinations of species that can combine and account for variation across all the samples. In ordination terminology, the specific source of these weights, which essentially become coefficients in a linear equation, is a set of eigenvectors.

In correspondence analysis, the abundance data is transformed to a semi-chi-square statistic (I say semi, because it's never evaluated according to the chi-square distribution to test for its significance). You will recall that the goal of CA in community ecology is to use species associations to depict distance or similarity in sites, and the chi-square statistic is used to depict the degree those associations depart from independence. Each abundance value is replaced by its chi-square statistic, or in some cases a factored form of this statistic, and the variance in these chi-square "distances" is evaluated by eigenanalysis. Thus, the system of weights used to score sites or quadrats is derived from a metric of species associations, and the more these associations depart from independence, the further separated final scores will be.

CA often produces more reliable ordinations than PCA in community ecology, presumably because it better models the non-linear responses of species to environmental and ecological gradients. The linear combinations calculated in PCA can either manufacture differences where they do not exist, or likewise obscure subtle differences in species associations that separate communities. Presumably, the chi-square metric in CA preserves ecological distance by modeling differences in associations rather than abundances of single species.

But CA has a major fault that mitigates against its use, commonly referred to as the "arch" or "horseshoe" effect. When constructing a second, yet independent, axis for plotting scores, CA often introduces a spurious "arch" in the order of scores along the first axis that may not correspond to underlying ecological forces. Detrended correspondence analysis was devised in an effort to eliminate the arch, but the procedure is somewhat arbitrary and in some cases may eliminate actual underlying structure to the data. Thus some researchers suggest avoiding DCA, while others suggest CA is not valid without it. As a response, more researchers are turning to canonical correspondence analysis, which constrains the ordination to known environmental measurements and thereby eliminates some of the guesswork in inferring underlying causality.

TYPES OF CORRESPONDENCE ANALYSIS

1. Multiple correspondence analysis:

Multiple correspondence analysis (MCA) is an extension of correspondence analysis (CA) which allows one to analyse the pattern of relationships of several categorical dependent variables. As such, it can also be seen as a generalization of principal component analysis when the variables to be analyzed are categorical instead of quantitative. Because MCA has been (re)discovered many times, equivalent methods are known under several different names such as optimal scaling, optimal or appropriate scoring, dual scaling, homogeneity analysis, scalogram analysis, and quantification method. Technically MCA is obtained by using a standard correspondence analysis on an indicator matrix (i.e., a matrix whose entries are 0 or 1). The percentages of explained variance need to be corrected, and the correspondence analysis interpretation of inter-point distances needs to be adapted.

How multiple correspondence analysis is performed?

MCA is performed by applying the CA algorithm to either an indicator matrix or a Burt table formed from these variables. An indicator matrix is an individuals × variables matrix, where the rows represent individuals and the columns are dummy variables representing categories of the variables. Analyzing the indicator matrix allows the direct representation of individuals as points in geometric space. The Burt table is the symmetric matrix of all two-way crosstabulations between the categorical variables, and has an analogy to the covariance matrix of continuous variables. Analyzing the Burt table is a more natural generalization of simple correspondence analysis, and individuals or the means of groups of individuals can be added as supplementary points to the graphical display.

In the indicator matrix approach, associations between variables are uncovered by calculating the chi-square distance between different categories of the variables and between the individuals (or respondents). These associations are then represented graphically as "maps", which eases the interpretation of the structures in the data. Oppositions between rows and columns are then maximized, in order to uncover the underlying dimensions best able to describe the central oppositions in the data. As in factor analysis or principal component analysis, the first axis is the most important dimension, the second axis the second most important, and so on, in terms of the amount of variance accounted for. The number of axes to be retained for analysis is determined by calculating modified eigenvalues.

In recent years, several students of Jean-Paul Benzécri have refined MCA and incorporated it into a more general framework of data analysis known as Geometric data analysis. This involves the development of direct connections between simple correspondence analysis, principal component analysis and MCA with a form of cluster analysis known as Euclidean classification.

In the social sciences, MCA is arguably best known for its application by Pierre Bourdieu, notably in his books La Distinction, Homo Academicus and The State Nobility. Bourdieu argued that there was an internal link between his vision of the social as spatial and relational —captured by the notion of field, and the geometric properties of MCA. Sociologists following Bourdieu's work most often opt for the analysis of the indicator matrix, rather than the Burt table, largely because of the central importance accorded to the analysis of the 'cloud of individuals

When to use it

MCA is used to analyze a set of observations described by a set of nominal variables. Each nominal variable comprises several levels, and each of these levels is coded as a binary variable. For ex- ample gender, (F vs. M) is one nominal variable with two levels. The pattern for a male respondent will be 0 1 and 1 0 for a female. The complete data table is composed of binary columns with one and only one column taking the value "1" per nominal variable. MCA can also accommodate quantitative variables by recoding them as "bins." For example, a score with a range of −5 to +5 could be recoded as a nominal variable with three levels: less than 0, equal to 0, or more than 0. With this schema, a value of 3 will be expressed by the pattern 0 0 1. The coding schema of MCA implies that each row has the same total, which for CA implies that each row has the same mass.

2. Principle component analysis:

Correspondence analysis (CA) and principal component analysis (PCA) are often used to describe multivariate data. In certain applications they have been used for estimation in latent variable models. The theoretical basis for such inference is assessed in generalized linear models where the linear predictor equals is treated as a latent fixed effect. The PCA and CA eigenvectors/column scores are evaluated as estimators. With m fixed, consistent estimators cannot be obtained due to the incidental parameters problem unless sufficient "moment" conditions are imposed on. PCA is equivalent to maximum likelihood estimation for the linear Gaussian model and gives a consistent estimator of (up to a scale change) when the second sample moment of is positive and finite in the limit. It is inconsistent for Poisson and Bernoulli distributions, but when is constant, its first and/or second eigenvectors can consistently estimate (up to a location and scale change) for the quadratic Gaussian model. In contrast, the CA estimator is always inconsistent. For finite samples, however, the CA column scores often have high correlations with the , especially when the response curves are spread out relative to one another. The correlations obtained from PCA are usually weaker, although the second PCA eigenvector can sometimes do much better than the first eigenvector, and for incidence data with tightly clustered response curves its performance is comparable to that of CA. For small sample sizes, PCA and particularly CA are competitive alternatives to maximum likelihood and may be preferred because of their computational ease.

3. Correspondence mapping & cluster analysis

Correspondence is a mapping package which graphically represents relationships between variables in a given market. In the case of TGI such variables will usually be brands and/or attitudes. Correspondence analysis produces a multi-dimensional representation of a cross-tab, highlighting relationships of significance.

Cluster analysis is a statistical tool which divides TGI respondents into mutually exclusive groups – usually based on attitude – for the purpose of efficient targeting.

Together, Correspondence and Cluster allow markets to be segmented in terms of consumer attitude and motivation, thus providing a depth of insight unachievable through analysis of demographic correlations alone.

Correspondence Analysis

Correspondence analysis is to depict the relationships between two nominal variables in a correspondence table in a low-dimensional space, whereas simultaneously relating the relationships between the categories for each variable. For each variable, the distances between category points in a plot replicate the relationships between the categories with similar categories plotted close to each other. Projecting points for one variable on the vector from the origin to a category point for the other variable describe the relationship between the variables.

An analysis of contingency tables frequently includes examining row and column profiles and testing for independence via the chi-square statistic. However, the number of profiles can be rather large, and the chi-square test does not expose the dependence structure. The Crosstabs procedure offers numerous measures of association and tests of association but cannot graphically represent any relationships between the variables.

Factor analysis is a standard technique for describing relationships between variables in a low-dimensional space. However, factor analysis requires interval data, and the number of observations should be five times the number of variables. Correspondence analysis, on the other hand, presumes nominal variables and can describe the relationships between categories of each variable, as well as the relationship between the variables. In addition, correspondence analysis can be used to analyze any table of positive correspondence measures.

REQUIREMENT

In Correspondence analysis, Categorical variables to be examined are scaled nominally. For aggregated data or for a correspondence measure other than frequencies, use a weighting variable with positive similarity values.

ASSUMPTIONS

The maximum number of dimensions used in the procedure depends on the number of active rows and column categories and the number of equality constraints. If no equality constraints are used and all categories are active, the maximum dimensionality is one fewer than the number of categories for the variable with the fewest categories.

Points to remember

Correspondence analysis is appropriate for exploratory research and is not suitable for hypothesis testing. It is a form of compositional technique that requires specification of both objects and attributes to be put side by side. As well, it is considered as sensitive to outliers, which should be exterminates prior to using the technique. The number of dimensions to be retained in the solution is based on Dimensions with inertia (Eigen values) greater than .2 and/or it should be sufficient to research objectives (usually two or three dimensions). Like loadings of factor analysis, the values shows the extent of association for each category individually with each dimensions, these dimensions can be named based on disintegration of inertia measures across a dimension. These values shows the extent of association for each category individually with each dimensions

Example

A study was conducted in four metro cities of Tamilnadu to evaluate the characteristics of small retailers associated with different location in adapting to present market scenario in retailing.

Test Procedure in SPSS

Following are the steps to analyse data using Correspondence Analysis in SPSS.

Step - 1: Click Analyze > Dimension Reduction > Correspondence Analysis as shown in the figure -1, the correspondence analysis dialogue box will appear (as given in the figure - 2).

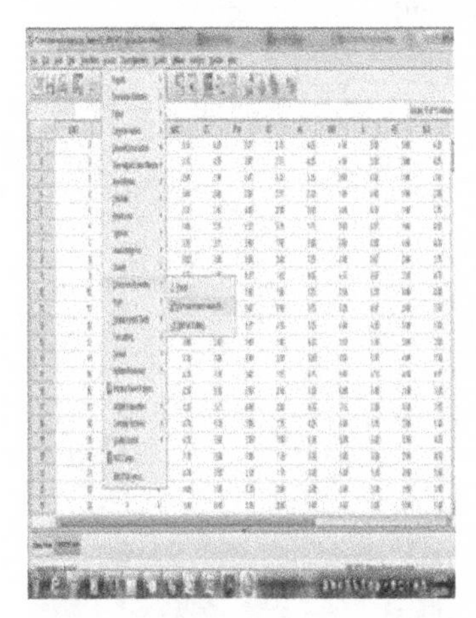

Figure - 1

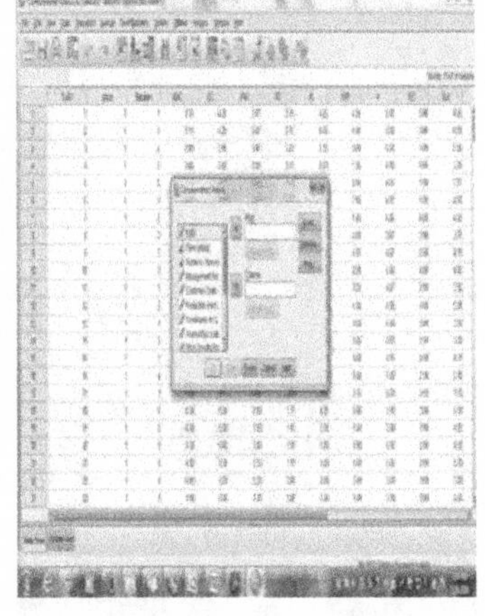

Figure - 2

Step - 2: Select nominal variable in row and column box to proceed with the analysis. In this example, place and retailer characteristics are the variables. Transfer place variable to row box and retailers characteristics to column box as given in figure -3.

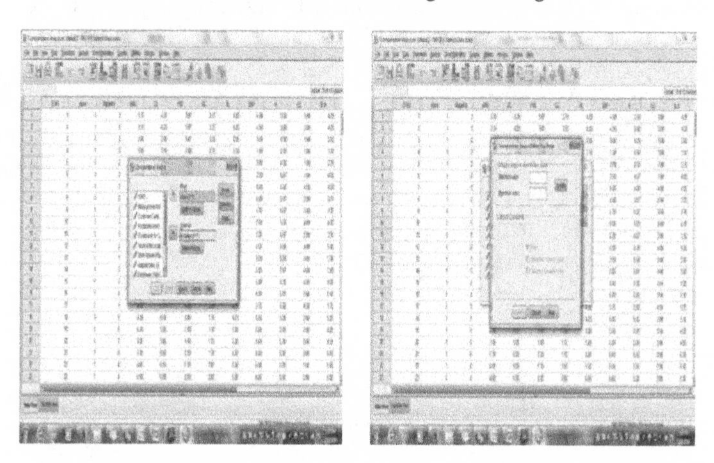

Figure - 3 **Figure - 4**

Step - 3: Then, click the define range tab below to the row box, correspondence analysis: define row range dialogue box will appear as given in figure -4 and in category range for row variable: place, Give Minimum value 1 and Maximum value 4 (depends on data), then click update tab as given in figure -5. The category constraints tab will appear, select none. Then click continue tab to return to correspondence analysis dialogue box figure -6.

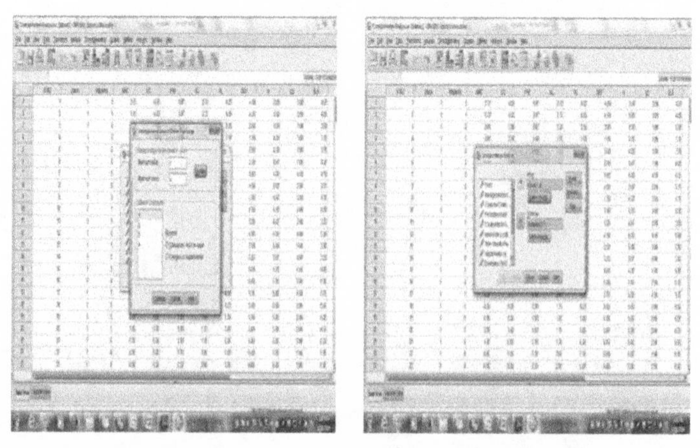

Figure - 5 **Figure - 6**

Define Row Range in Correspondence Analysis

All categories are initially unconstrained and active i.e. the researcher can constrain row categories to equal other row categories, or define a row category as supplementary.

- **Categories must be equal**. Categories must have equal scores. Use equality constraints if the obtained order for the categories is undesirable or counterintuitive. The maximum number of row categories that can be constrained to be equal is the total number of active row categories minus 1.
- **Category is supplemental**. Supplementary categories do not influence the analysis but are represented in the space defined by the active categories. Supplementary categories play no role in defining the dimensions. The maximum number of supplementary row categories is the total number of row categories minus 2.

Step - 4: Then, click the define range tab below to the column box, correspondence analysis: define column range dialogue box will appear as given in figure -7 and in category range for column variable: retailers, Give Minimum value 1 and Maximum value 5 (depends on data), then click update tab as given in figure -8. The category constraints tab will appear, select none. Then click continue tab to return to correspondence analysis dialogue box figure -9.

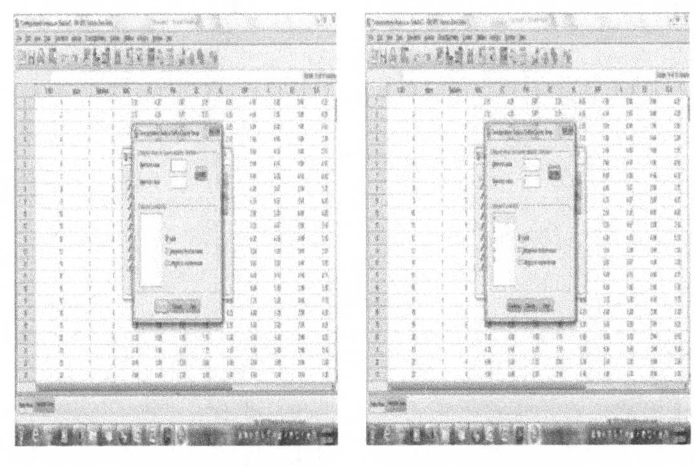

Figure - 7 **Figure - 8**

Define Column Range in Correspondence Analysis

All categories are initially unconstrained and active. The user can constrain column categories to equal other column categories, or define a column category as supplementary.

- ***Categories must be equal:*** Categories must have equal scores. Use equality constraints if the obtained order for the categories is undesirable or counterintuitive. The maximum number of column categories that can be constrained to be equal is the total number of active column categories minus 1.
- ***Category is supplemental***: Supplementary categories do not influence the analysis but are represented in the space defined by the active categories. Supplementary categories play no role in defining the dimensions. The maximum number of supplementary column categories is the total number of column categories minus 2.

Step - 5: Click model tab in correspondence analysis dialogue box, a correspondence analysis: model dialogue box will appear as given in figure -10. Give 2 in dimensional solution area. Select chi square in distance measure. By default in standardized model, row and column means are removed. In normalization method, select symmetrical then click continue (figure -10) to return to correspondence analysis dialogue box.

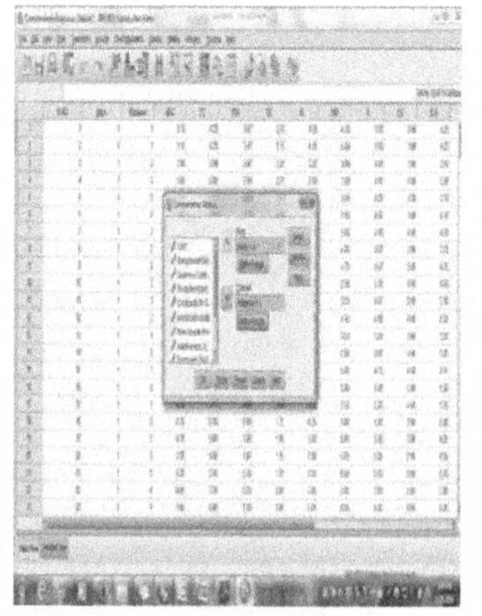

Figure - 9

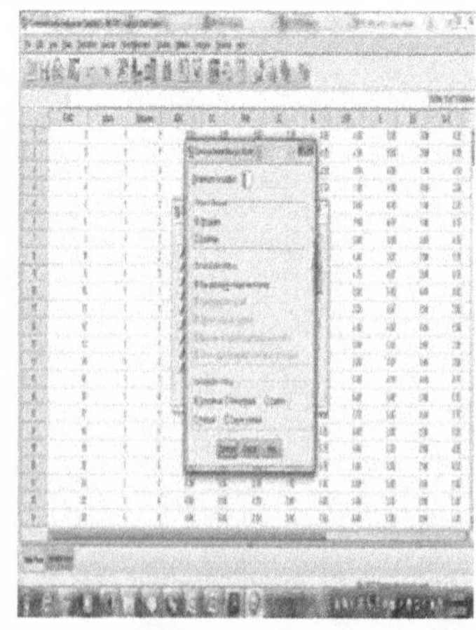

Figure - 10

Correspondence Analysis Model

A dimension in solution specifies the number of dimensions. In general, choose as few dimensions as needed to explain most of the variation. The maximum number of dimensions depends on the number of active categories used in the analysis and on the equality constraints. The maximum number of dimensions is the smaller of

- The number of active row categories minus the number of row categories constrained to be equal, plus the number of constrained row category sets.
- The number of active column categories minus the number of column categories constrained to be equal, plus the number of constrained column category sets.

Distance Measures enables the rows and columns of the correspondence table. It contains following alternatives

- ***Chi-square distance measure*** uses for a weighted profile distance, where the weight is the mass of the rows or columns. This measure is required for standard correspondence analysis.
- ***Euclidean distance measure*** employs the square root of the sum of squared differences between pairs of rows and pairs of columns.

In Standardization Method measures, any one of the following options to be selected

- ***Row and column means are removed***, in which both the rows and columns are centered. This method is required for standard correspondence analysis.
- ***Row means are removed***, in which only the rows are centered.
- ***Column means are removed***, in which only the columns are centered.
- ***Row totals are equalized and means are removed***, in which before centering the rows, the row margins are equalized.
- ***Column totals are equalized and means are removed*** in which before centering the columns, the column margins are equalized.

In **Normalization Method** any one of the following options to be selected

- ***Symmetrical normalization method*** is used to examine the differences or similarities between the categories of the two variables. For each dimension, the row scores are the weighted average of the column scores divided by the matching singular value, as well as the column scores are the weighted average of row scores divided by the matching singular value.

- ***Principal normalization method*** is used to examine differences between categories of either or both variables instead of differences between the two variables. The distances between row points and column points are approximations of the distances in the correspondence table according to the selected distance measure.
- ***Row principal normalization method*** is used to examine differences or similarities between categories of the row variable. The distances between row points are approximations of the distances in the correspondence table according to the selected distance measure. The row scores are the weighted average of the column scores.
- ***Column principal normalization method*** is used to examine differences or similarities between categories of the column variable. The distances between column points are approximations of the distances in the correspondence table according to the selected distance measure. The column scores are the weighted average of the row scores.
- ***Custom normalization method*** is useful for making tailor-made biplots. It is important to specify a value between –1 and 1. A value of –1 corresponds to column principal. A value of 1 corresponds to row principal. A value of 0 corresponds to symmetrical. All other values spread the inertia over both the row and column scores to varying degrees.

Step - 6: In correspondence analysis dialogue box, click statistics tab. Correspondence analysis: statistics dialogue box will appear as given in figure -11. Select correspondence table, overview of row points, overview of column points, row profiles, and column profiles check box. As well in confidence statistics, select row points and column point check box as given in figure -12. Then click continue to return to correspondence analysis dialogue box.

Correspondence Analysis Statistics

- ***Correspondence table*** is a cross tabulation of the input variables with row and column marginal totals.
- ***Overview of row points*** generates is for each row category, the scores, mass, inertia, contribution to the inertia of the dimension, and the contribution of the dimension to the inertia of the point.

- ***Overview of column points*** generates is for each column category, the scores, mass, inertia, contribution to the inertia of the dimension, and the contribution of the dimension to the inertia of the point.
- ***Row profiles*** enables distribution across the categories of the column variable.
- ***Column profiles*** enable distribution across the categories of the row variable.
- ***Permutations of the correspondence table*** reorganized such that the rows and columns are in increasing order according to the scores on the first dimension. Optionally, the researcher can specify the maximum dimension number for which permuted tables will be produced. A permuted table for each dimension from 1 to the number specified is produced.
- ***Confidence Statistics for Row points*** includes standard deviation and correlations for all non supplementary row points.
- ***Confidence Statistics for Column points*** includes standard deviation and correlations for all non supplementary column points.

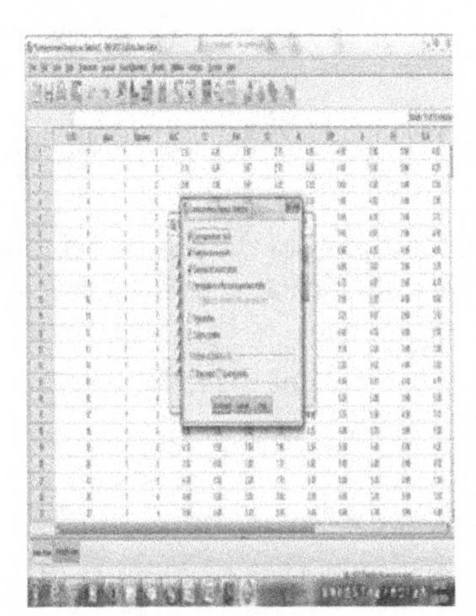

Figure - 11

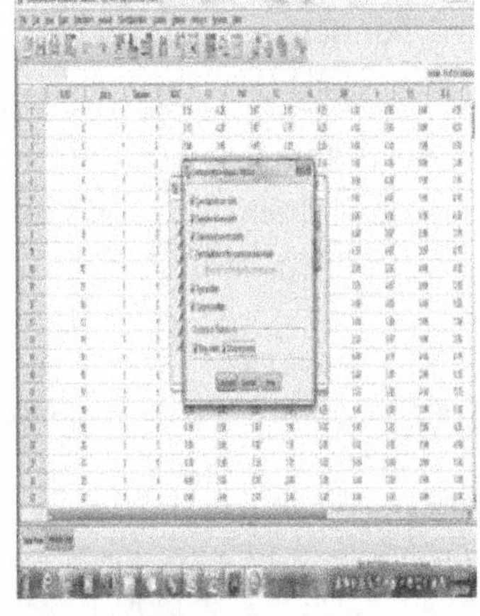

Figure - 12

Step - 7: In correspondence analysis dialogue box, click plots tab. Correspondence analysis: plots tab will appear as given in figure-13. In scatterplots, click biplot, row points, column points check box.

In line plots, click transformed row categories, transformed column categories check box. Select display all dimensions in the solution in plot dimension and click continue to return to correspondence analysis dialogue box (given in figure -14).

Step - 8: Then click ok tab in correspondence analysis dialogue box to generate output.

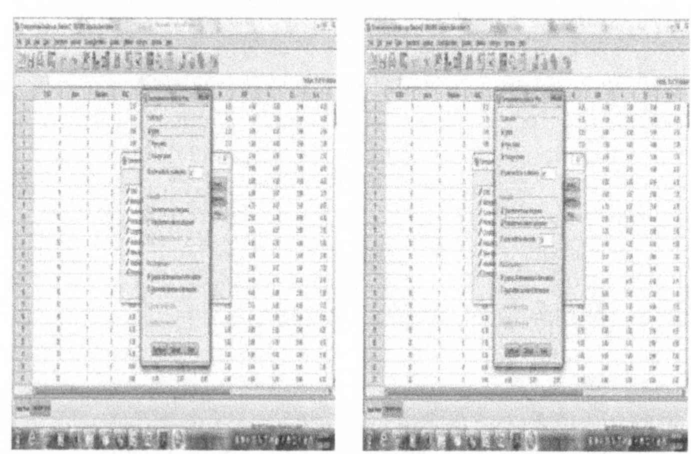

Figure - 13 **Figure - 14**

Correspondence Analysis Plots

Scatter plots generate a matrix of all pair wise plots of the dimensions. In SPSS, scatter plots include:

- Biplot generates a matrix of joint plots of the row and column points; if principal normalization is chosen, the biplot is not available.
- A Row point creates a matrix of plots of the row points.
- A column point generates a matrix of plots of the column points.

Optionally, SPSS helps to value the label width of the scatter plots. This value must be a non-negative integer less than or equal to 20. For every dimensions Line plots will be produced.

SPSS Output of the Correspondence Analysis

Table 1 Correspondence Table

Place	Retailers Characteristics					
	Innovator	**Diehard Pessimist**	**Planner**	**Active experimenter**	**Blind Optimist**	**Active Margin**
Chennai	13	17	20	13	16	79
Coimbatore	17	21	26	6	4	74
Madurai	15	19	24	8	6	72
Trichy	9	10	33	11	12	75
Active Margin	54	67	103	38	38	300

Table 2 Row Profiles

Place	Retailers Characteristics					
	Innovator	**Diehard Pessimist**	**Planner**	**Active experimenter**	**Blind Optimist**	**Active Margin**
Chennai	.165	.215	.253	.165	.203	1.000
Coimbatore	.230	.284	.351	.081	.054	1.000
Madurai	.208	.264	.333	.111	.083	1.000
Trichy	.120	.133	.440	.147	.160	1.000
Mass	.180	.223	.343	.127	.127	

Table 3 Column Profiles

Place	Retailers Characteristics					
	Innovator	Diehard Pessimist	Planner	Active experimenter	Blind Optimist	Mass
Chennai	.241	.254	.194	.342	.421	.263
Coimbatore	.315	.313	.252	.158	.105	.247
Madurai	.278	.284	.233	.211	.158	.240
Trichy	.167	.149	.320	.289	.316	.250
Active Margin	1.000	1.000	1.000	1.000	1.000	

Table 4 Summary

Dimension	Singular Value	Inertia	Chi Square	Sig.	Proportion of Inertia		Confidence Singular Value	
					Accounted for	Cumulative	Standard Deviation	Correlation
								2
1	.231	.053			.721	.721	.053	-.038
2	.143	.021			.277	.998	.057	
3	.013	.000			.002	1.000		
Total		.074	22.253	.035[a]	1.000	1.000		

a. 12 degrees of freedom

Table 5 Overview Row Points[a]

Place	Mass	Score in Dimension		Inertia	Contribution				
					Of Point to Inertia of Dimension		Of Dimension to Inertia of Point		
		1	2		1	2	1	2	Total
Chennai	.263	.432	-.522	.022	.212	.500	.525	.475	1.000
Coimbatore	.247	-.616	.058	.022	.404	.006	.992	.006	.997
Madurai	.240	-.349	-.040	.007	.126	.003	.976	.008	.984
Trichy	.250	.488	.531	.024	.257	.491	.577	.423	1.000
Active Total	1.000			.074	1.000	1.000			

a. Symmetrical normalization

Table 6 Overview Column Points[a]

Retailers Characteristics	Mass	Score in Dimension		Inertia	Contribution				
					Of Point to Inertia of Dimension		Of Dimension to Inertia of Point		
		1	2		1	2	1	2	Total
Innovator	.180	-.456	-.209	.010	.162	.055	.884	.115	.999
Diehard Pessimist	.223	-.474	-.323	.015	.217	.163	.776	.224	1.000
Planner	.343	.015	.517	.013	.000	.640	.001	.999	1.000
Active experimenter	.127	.511	-.168	.008	.143	.025	.924	.062	.986
Blind Optimist	.127	.934	-.365	.028	.477	.118	.912	.086	.998
Active Total	1.000			.074	1.000	1.000			

a. Symmetrical normalization

Table 7 Confidence Row Points

Place	Standard Deviation in Dimension		Correlation
	1	2	1-2
Chennai	.335	.190	.790
Coimbatore	.145	.296	.269
Madurai	.194	.280	.106
Trichy	.330	.225	-.868

Table 8 Confidence Column Points

Retailers Characteristics	Standard Deviation in Dimension		Correlation
	1	2	1-2
Innovator	.147	.199	-.732
Diehard Pessimist	.200	.188	-.889
Planner	.315	.108	-.053
Active experimenter	.301	.451	.121
Blind Optimist	.285	.420	.553

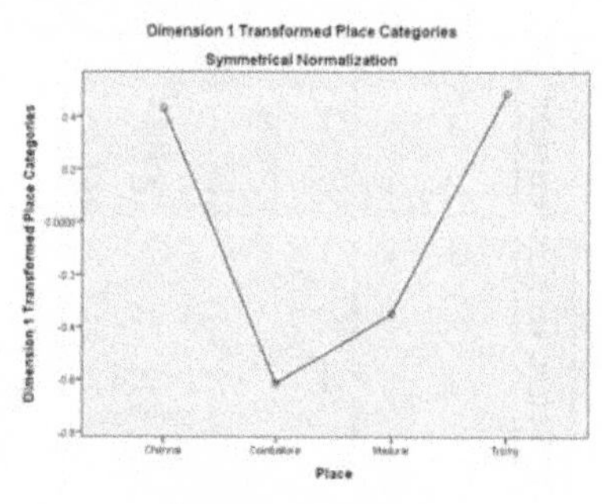

Figure - 15

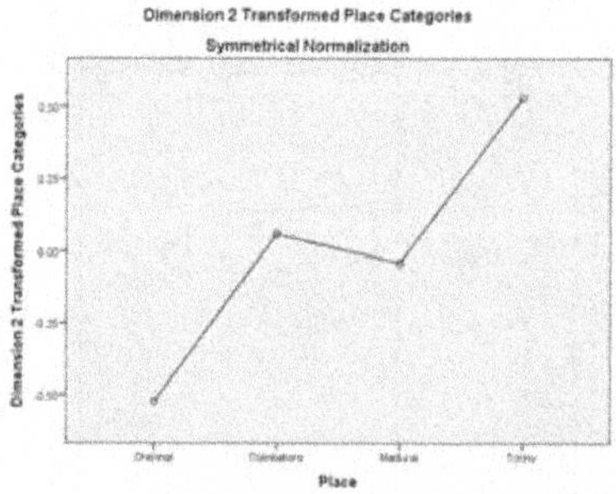

Figure - 16

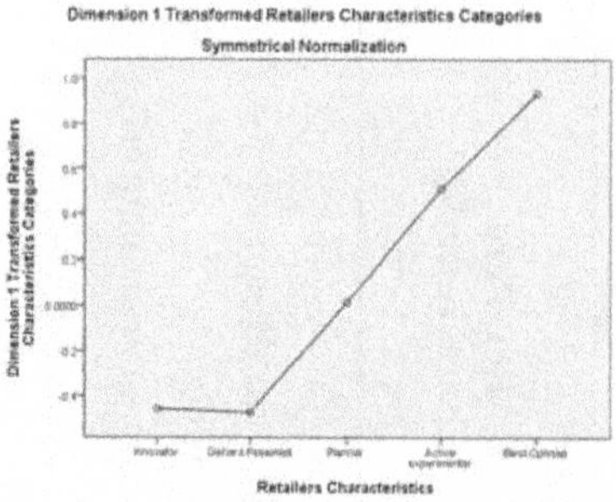

Figure - 17

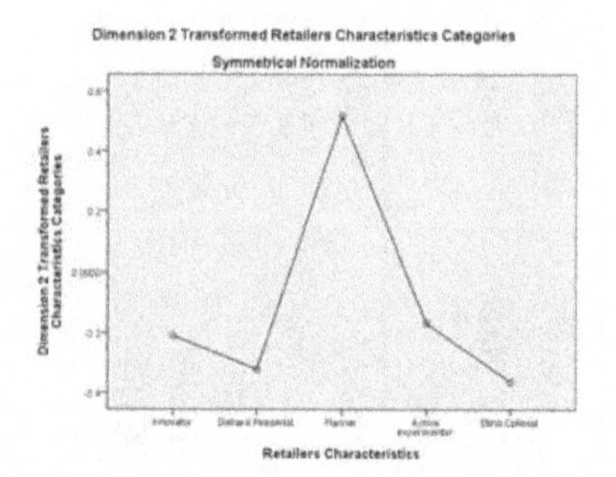

Figure - 18

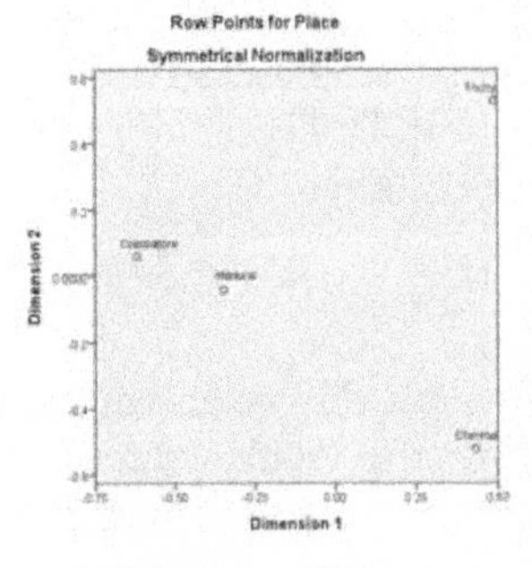

Figure - 19

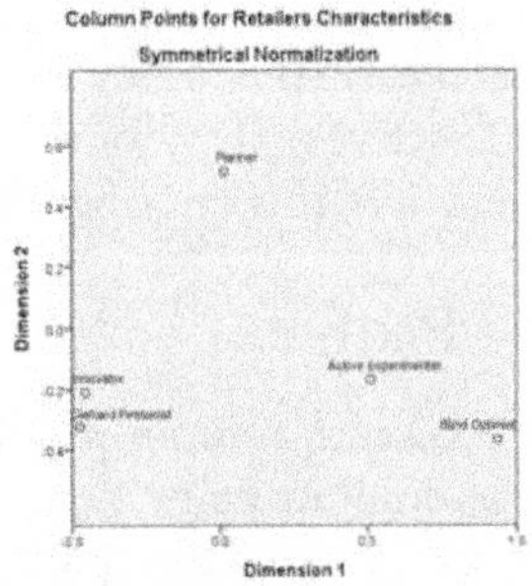

Figure - 20

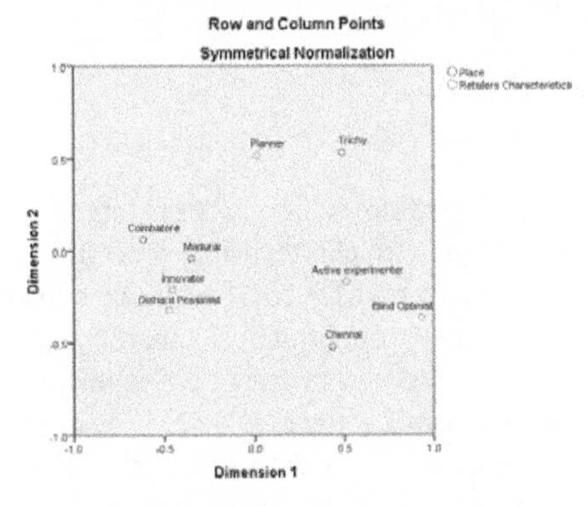

Figure - 21

Results & Discussion

Correspondence Table

SPSS will generate a tabulation table called correspondence table as given in table -1. The data assimilated based on the data entered for place and retailer characteristics. The above table includes the frequency of row category "place" and column category "retailer characteristics based on summation of each row and column categories called as Active margin. In this example, in place across all retailer characteristics, Chennai retailers are sum to 79. Similarly, summation of planner across each category of the place is 103. The correspondence table is similar to cross tabulation.

Row Profiles

Table -2 that represents row profile provides the weighted frequency of each of the row point such as the total for the whole row sum to 1. The row profiles are calculated by taking each row point and dividing it by the respective active margin for the respective row. In this example, for the place Chennai and the retailer characteristics innovator, the frequency (as given in correspondence table) is 13. Active margin for that respective row is 79. Therefore 13/79 =.16445 i.e. .165. Likewise it is done for all categories in the table.

Column Profiles

Table -3 represents column profile, which provides the weighted frequency of each of the column point such as the total for the whole column sum to 1. Similar to row profile, the column profile table values are calculated. From the correspondence table, the place – trichy with the retailer characteristics active experimenter, the frequency given in correspondence table is 11. Active margin for the respective column is 38. Therefore 11/38=.289. Likewise it is done for all categories in the table.

Summary

Table – 4 represents summary table, which is considered to be a most important table in correspondence analysis. In this example correspondence analysis is done based on chi-square statistics to text for total variance explained, along with the associated probability. It is important to check the model is significant or not in the summary table; our model is significant at 5% level and the chi square value is 22.253 with degree of freedom 12, p value is .035. It is important to note that SPSS has generated three dimensions to explain the model. Correspondence analysis only produces dimensions that can be interpreted rather than including all dimensions that explain something about the model.

For this reason, inertia does not always add upto 100%. The inertia column gives the total variance explains by each dimension in the model. In our model, the total inertia is 7.4%. This represents the model explains are 7% place and retailer characteristics or vice versa. The association is weak; still, it is significant as indicated in the chi-square statistics.

Each dimension is listed according to the amount of variance explained in the model. Dimension 1 explains the most variance in the model 5.3% of the 7.4% of the variance accounted. Furthermore, dimension 2 explains 2.1% of the total 7.4% of variance accounted for. Dimension 3 is 0% of the total variance accounted and dropped from further analysis. The singular value column gives the square roots of the Eigen values, which describes "the maximum canonical correlation between the categories of the variable analysis for any dimensions. In correspondence analysis, Eigen value and inertia are similar.

The values in the Proportion of Inertia column give the percent of variance that each dimension explains of the total variance explained by the model. In this example, Dimension 1 explains approximately 72% of the total 7.4% of variance explained in the model. Furthermore, Dimension 2 explains approximately 27.7% of the 7.4% of variance explained in the model. Dimension 3 explains too little of the total variance explained to be kept for further analysis. In essence, this example dictates that there are two dimensions that can explain the most variance between place and retailer characteristics. Some research questions may reveal that three dimensions are necessary to explain most of the variance.

Overview Row Points

The Overview Row Points (Table 5) gives information on how each of the row points is plotted in the final biplot. The 'Mass' column in this table indicates the proportion of each age group with respect to all age groups in the analysis. The column 'Score in Dimension' indicates the coordinates in each dimension (1 and 2) where each row category will be situated on the biplot. Inertia again reflects variance. The 'Contribution' column reflects how well each of the points load onto each of the dimensions, as well as how well the extraction of dimensions explains each of the points. In this example, for Chennai respondents 21% for dimension 1 and 50% for dimension 2 which is nominal in both the cases. It can also be seen that the extraction of Dimension 1 explains 52.5% of the variance for Chennai retailers, whereas the extraction of Dimension 2 explains around 47.5% of the variance in the Chennai retailers group.

Overview Column Point

As discussed in the overview row points, the overview column points (table -6) give the same information for the plotting of column points on the biplot. In this example, the retailer characteristics of diehard pessimist, dimension -1 21.7% and for dimension -2 it is 16%. Furthermore the dimension1 explains 77.6% of the variance and dimension- 2, 22.4% across the retailer groups.

Confidence Row Points & Confidence Column Points

Tables 7 and 8, Confidence Row Points and Confidence Column Points provide the standard deviations of row and column scores in each dimension, which is used to evaluate the accuracy of the estimates of points on their axes, much like confidence intervals are used in other statistical analyses.

As well the data's are graphically represented in transformed row categories and transformed column categories biplot chart. The chi square statistics reveals the strength of trends within data, which is based on the point distance of the categories. Distance between row and column points are interpreted differently. In this example, symmetrical normalization method is used, that helps for a general comparison between standardized row and column data.

Correspondence analysis can also be done in Euclidean distance measures. The example is follows as

Model -2 – Euclidean – Correspondence Analysis

Step - 5: Click model tab in correspondence analysis dialogue box, a correspondence analysis: model dialogue box will appear as given in figure -10. Give 2 in dimensional solution area. Select Euclidean in distance measure. By default in standardized model, row and column means are removed. In normalization method, select symmetrical then click continue (figure -1) to return to correspondence analysis dialogue box.

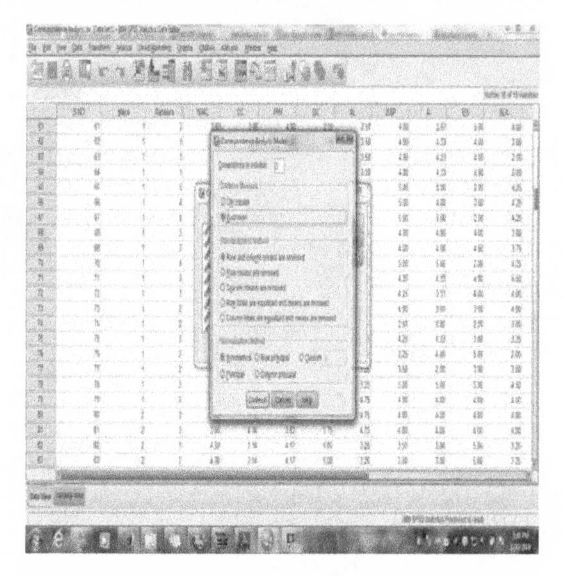

Figure - 1

Table 1 Summary

Dimension	Singular Value	Inertia	Proportion of Inertia		Confidence Singular Value	
			Accounted for	Cumulative	Standard Deviation	Correlation
						2
1	.207	.043	.598	.598	.015	.107
2	.170	.029	.401	.998	.020	
3	.010	.000	.002	1.000		
Total		.072	1.000	1.000		

Table 2 Overview Row Points[a]

Place	Mass	Score in Dimension		Inertia	Contribution				
					Of Point to Inertia of Dimension		Of Dimension to Inertia of Point		
		1	2		1	2	1	2	Total
Chennai	.250	.226	-.670	.022	.061	.662	.122	.878	1.000
Coimbatore	.250	-.533	.203	.016	.343	.061	.892	.106	.998
Madurai	.250	-.319	.035	.005	.123	.002	.978	.010	.987
Trichy	.250	.626	.432	.028	.473	.275	.719	.281	1.000
Active Total	1.000			.072	1.000	1.000			

a. Symmetrical normalization

Table 3 Overview Column Points[a]

Ward Method	Mass	Score in Dimension		Inertia	Contribution				
					Of Point to Inertia of Dimension		Of Dimension to Inertia of Point		
		1	2		1	2	1	2	Total
Innovator	.200	-.459	-.030	.009	.204	.001	.996	.004	.999
Diehard Pessimist	.200	-.619	-.145	.017	.370	.025	.957	.043	1.000
Planner	.200	.229	.787	.023	.051	.730	.094	.906	1.000
Active experimenter	.200	.303	-.205	.005	.089	.049	.719	.268	.987
Blind Optimist	.200	.545	-.406	.018	.287	.195	.686	.312	.998
Active Total	1.000			.072	1.000	1.000			

a. Symmetrical normalization

Table 4 Confidence Row Points

Place	Standard Deviation in Dimension		Correlation
	1	2	1-2
Chennai	.309	.100	.943
Coimbatore	.106	.195	.880
Madurai	.074	.103	.313
Trichy	.190	.225	-.987

Table 5 Confidence Column Points

Ward Method	Standard Deviation in Dimension		Correlation
	1	**2**	**1-2**
Innovator	.028	.177	-.553
Diehard Pessimist	.075	.223	-.949
Planner	.376	.092	-.925
Active experimenter	.116	.182	.642
Blind Optimist	.195	.216	.890

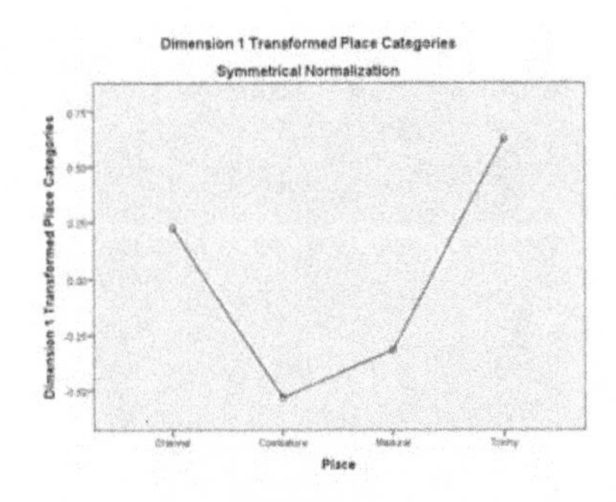

Figure -2

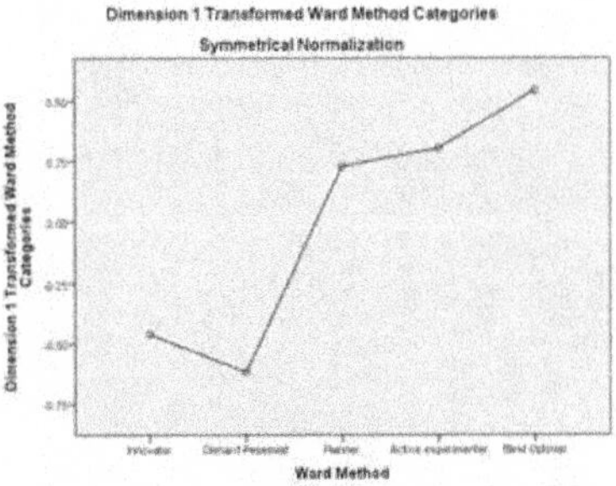

Figure -4

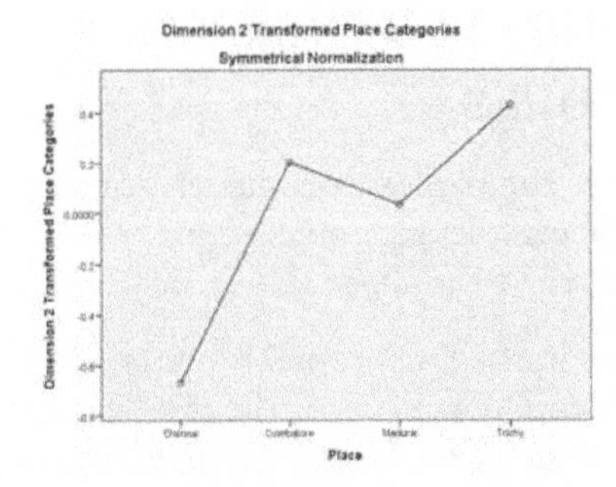

Figure -3

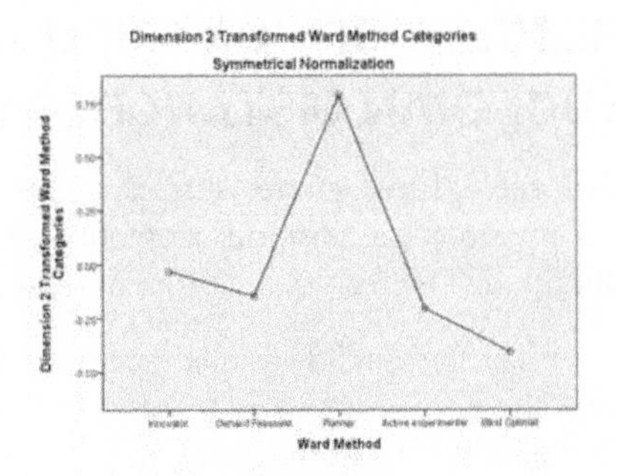

Figure -5

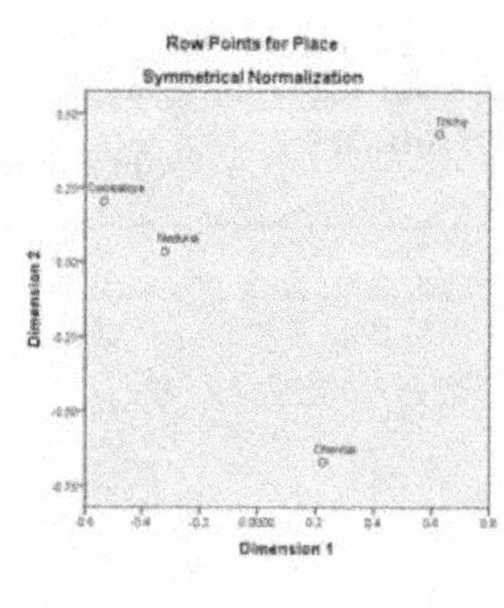

Figure -6

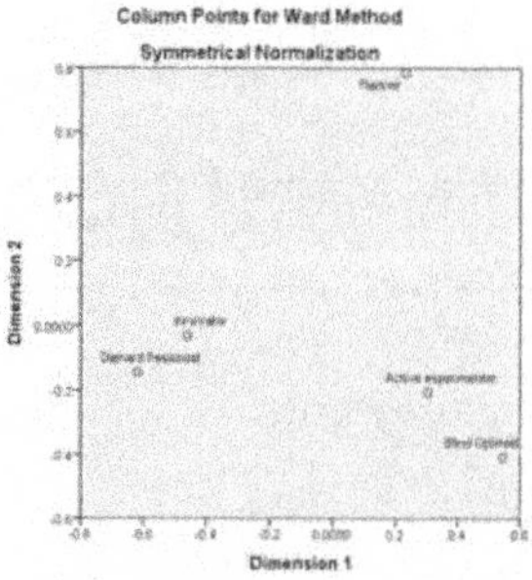

Figure -7

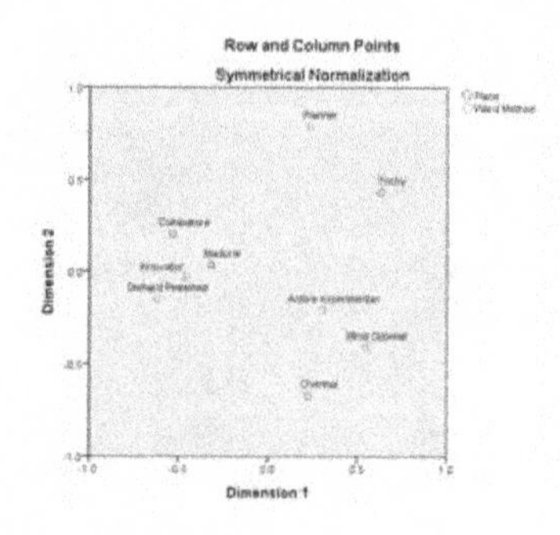

Figure -8

RESULTS OF CORRESPONDENCE ANALYSIS UNDER EUCLIDIAN DISTANCE MEASURES

The table 1 represents summary table, in the earlier example, the observed chi-square value is significant but in Euclidean distance no chi square value is obtained. The results are more or less similar like the chi-square model.

For the Euclidean distance, the values in the Proportion of Inertia column give the percent of variance that each dimension explains of the total variance explained by the model. In this example, Dimension 1 explains approximately 59.8% of the total 7.4% of variance explained in the model. Furthermore, Dimension 2 explains approximately 40.1% of the 7.4% of variance explained in the model. Dimension 3 explains too little of the total variance explained to be kept for further analysis. In essence, this example dictates that there are two dimensions that can explain the most variance between place and retailer characteristics. Some research questions may reveal that three dimensions are necessary to explain most of the variance. The correlation between the two values is .107.

Table -2 & Table -3 represents overview of row point and overview of column point; even though the Euclidean distance and Chi-square measure are similarly computed, they tend to represent the row and column scores different when under normalization.

Table- 4 & Table- 5 represents confidence row point and confidence column point, which provide the standard deviations of row and column scores in each dimension It is used to evaluate the accuracy of the estimates of points on their axes, much like confidence intervals are used in other statistical analyses.

The biplot for correspondence analysis of Chi-square distance and Euclidean distance tend give different representations of the row and column scores.

Chapter - 10

Multidimensional Scaling

Learning Objectives

This chapter helps to understand the following

- Meaning of Multidimensional Scaling
- Concepts of multidimensional scaling
- Uses of Multidimensional scaling
- Advantage and Disadvantage of Multidimensional scaling
- Difference between multidimensional scaling with factor analysis and cluster analysis
- Requirements for multidimensional scaling
- Steps in Conducting Multidimensional Scaling
- Types of multidimensional scaling
- Algorithm of Multidimensional Scaling
- Running Multidimensional Scaling Using SPSS (Alscal)
- Results and Discussion of Multidimensional Scaling Output (Alscal)

Meaning of Multidimensional Scaling

Multidimensional scaling (**MDS**) is a procedure that helps in reducing the data that can be used to analyze the similarity or dissimilarity among them. Here objects are represented as two dimensional spaces. The points are closely observed to identify the dissimilarities. The concept of grouping of dissimilarities in a data is found in both MDS and cluster analysis so they have a common perspective. This is a more subjective concept.

Co-ordinates of consumers are disclosed on the basis of their preference or perception through an ideal point model. It helps in simplification of data. A proper insight is provided before carrying on psycho-attitudinal surveys.

Let's take the example may demonstrate the logic of an MDS analysis; take a matrix of distances between major Indian cities from a map. On analysing the matrix it would be helpful to specify the distance to be reproduced on two dimensions. MDS analysis helps to infer the output in a two dimension map.

Objects are arranged in a particular number of dimensions to reproduce the observed distance for which the above step has to be followed. This will help to explain the underlying dimensions. The distance of north and south, east and west can be explained in terms of geographical dimensions.

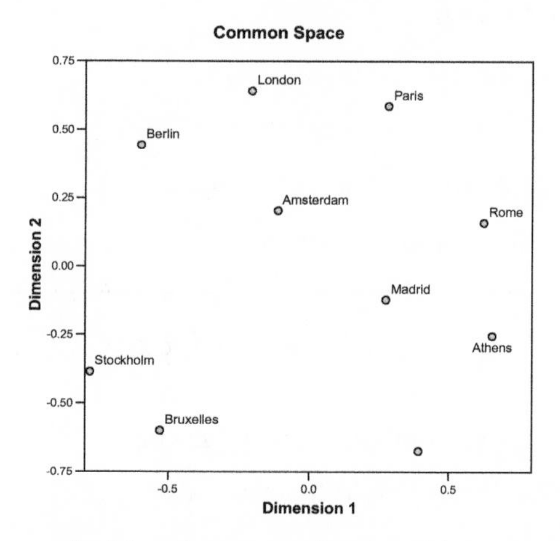

- Cities are ranked by the respondents, without any specifications to priority.
- Adequate number of responses will reflect perceived similarities between cities.
- The distances represented graphically reflect dissimilarities
- The main perceived difference can be understood if the 2 dimensions can be labelled according to some criterion.

Concepts of multidimensional scaling

- Proximities are used to identify how similar or how different two objects could be.
- Geometric configuration of points is produced with the help of proximities in a two-dimensional space as output.

- The distance between derived distance and 2 proximity in each dimension is evaluated through a measure called stress.
- Plotting stress values will help to identify the appropriate number of dimensions to locate objects.

Uses of Multidimensional scaling

1. ***Social sciences***- Group of objects with similar functions like tastes, colours, sounds, people, nations, facial expressions of emotions etc are found out using Multidimensional scaling .
2. ***Psychology***
    - Perception – the dimensions would be based on confusion errors with the perceptual similarity
    - Cognition – MDS is used based on item recall, sorting task, etc.
3. ***Political science***–in political science the multidimensional scaling is used to analyse the dimensions of voting behaviour of people, legislative bills, and court decisions.
4. ***Sociology*** – the basis of decision would be on relationship, social network analysis such similarity or dissimilarity data which is often called "proximity data"
5. ***Archaeology*** – Most of the similarities of artifacts are found in archaeological sites like the similarity of two digging sites.
6. ***Chemistry*** – In chemistry multidimensional scaling can be used for spatial structure of molecules.
7. ***Graph layout*** – To find positions of nodes in abstract graphs, an active area at the intersection of discrete mathematics and network visualization connect with edges. From graphs one obtains dissimilarities by computing shortest-path lengths for all pairs of vertices
8. ***Dimension reduction*** – To calculate distances in high-Dspace, in dimension reduction, the distance scaling is a nonlinear competitor of principal components map to low-D space.

Advantages of multidimensional scaling

1. The relationship between the objects are denoted by multi dimension scaling.
    - For example- A multidimensional scale of city distances is formed for the places in India. Chennai would be close to Mumbai than to Delhi.

2. Multidimensional Scaling would help to analyse how closely different values are related.
3. When there is large amount of data presented in a tabular form, Multidimensional Scaling helps in simplification of the tables.
4. Multidimensional Scaling makes the assessing of data more feasible by immediately assessing relationships even when presented with table of 10000 fig.
5. Multidimensional Scaling is used in various subjects like psychology, graphing subjects etc.
6. The relationship importance can be figured by the researchers & hence this finds more weightage.
7. This method is extremely useful in psychological subject where the quantity of data tends to be at high volume and have many different aspects.
8. Identify dimensions by which objects are perceived or evaluated
9. MDS helps to position the objects with respect to dimensions

Disadvantages of multidimensional scaling

1. The method does not deal in real numbers. Multidimensional Scaling between Chennai, Mumbai, and Kolkata is roughly similar to Delhi, Ahmadabad, Bangalore even though the actual figures are profoundly different.
2. Multidimensional Scaling uses complex formulas to convert raw figures to a Multidimensional Scaling scale.
3. It is easy to analyse the relationships but difficult to compute the table.
4. The process is very time consuming.
5. Additional layer of subjectivity is created by this method.
6. Two data may be formed one being modelling tabled data and the other original data, so the decision as to which data will be used for computation will be a question and this may sometime affect the accuracy.
7. Perceptual mapping has not been shown to be reliable across different methods
8. The interpretation of dimensions is difficult in MDS and when more than two or three dimensions are needed, usefulness is reduced

Difference between multidimensional scaling and factor analysis

Multi Dimensions Scaling	Factor Analysis
This is used by the researcher to identify the similarities or dissimilarities between investigated objects.	This is used to describe variability among observed and correlated variables.
Any kind of (dis)similarity matrix in adds to correlation matrix.	Where the similarities between objects are expressed in correlation matrix.
It uses a function minimization algorithm.	This is related to principal component analysis.
It arranges objects in a space with particular dimension.	Its main purpose is grouping of variables.
It assumes the concept of minimalist any pt must be at least an close to itself as it is to any other point.	This is a form of simplification.

Difference between multidimensional scaling and cluster analysis

Multi Dimensions Scaling	Cluster Analysis
This is a technique to identify the underlying dimensions that the respondents use to evaluate the objectives.	This is the task of grouping a set of objectives, so that the objectives are all similar to each other when compared to those in other groups.
It is often used in marketing researches.	It is used in several fields like machine learning, image analysis, and information retrieval.
Multi Dimensions Scaling helping to determine how many dimensions they might use.	This helps to derive segments.
It is based on comparison of object.	Cluster analysis is an interactive process of knowledge discovery.
It helps to analyse the similarity or dissimilarity matrix.	It helps to analyse objects that are similar to one another.
There are all two types of dimensions objective as perceived.	There cannot be one specific algorithm. It requires various algorithms.

Requirements for multidimensional scaling

1. Related objects

The most important requirement of Multidimensional scaling is that the objects must be related, so as to find the similarity or dissimilarity between them. The major function of multidimensional scaling is to find the similarity or dissimilarity. Eg, red and pink are more similar than red and green. In the above example the points as red and pink are located closer in space than red and green.

2. Criteria

It represents inter-object dissimilarities by inter-point distances. The distances can be calculated using the Euclidean model. The promising feature about Euclidean model is that there is invariance over the choice of origin and co-ordinate axes. The origin is generally placed in centroid of the object configuration. The configuration is rotated in such a way that the axes denote the meaningful attributes. However weighted Euclidean model does not allow rotation of axes without changing the inter-point distances. When the data is observed there can probably be certain errors, so exact representation is not possible, hence best approximation of the (dis)similarity of data is observed. The notion to be more validated has to be introduced to an index that measures the goodness of agreement between the observed and the distance model. This index acts as a criterion to be optimized in multi dimensional scaling. There are two broad class of goodness criteria

- Least squares
- Maximum likelihood

The latter provides with statistical inference capabilities. The former is used more predominantly for simplicity and flexibility methods that are interactive approximations to final solution, which is obtained by gradually improving the goodness of fit.

3. Scales in transformations

Scale levels represent approximate relationship that may hold between observed dissimilarity data and distances. The scale levels are very important because multi dimensional scaling procedure apply to only data measured on certain scale levels. There are 4 scale levels: ratio, interval, nominal and ordinal.

Researcher may not know the priori exact scale liners, practically the researcher may start with weaker assumptions, once the result is generated. It might show that stronger measurement assumption would be the correct assumption, this way we can get more reliable results avoiding unaffordable assumptions.

4. Selection of Dimensions

The most important decision in multi dimensional scaling is the selection of dimension. It refers to co-ordinates required to fix a point in the spatial representation of objects. The considerations to be taken into amount are that the configuration derived for the object must fix to the data reasonably but not fit well, when the dimensionality of the solution space increases automatically. The practical strategy used to determine the adequate number of dimension is the data has to be analysed under varied dimensionalities. The interpretability of derived dimensions is also an important consideration. Uninterruptable dimensions are useless hence retaining is of no use.

5. Model selection

Model selection will basically depend on the similarity of data, whether the data is one set or group of subjects when no systematic individual difference exists. In such cases, Euclidean distance model can be fitted simultaneously or a single Euclidean model is fit to average (dis)similarity data. The assumption of no systematic individual differences is not true. In such a situation the matrix may be analysed separately which will yield many configuration. The commonality and individual differences are captured by the multidimensional scaling. It represents a common object that applies to all individuals. In few cases the subject background information such as gender, age, level of education may show different patterns.

6. Unfolding analysis

Preference judgements have more individual differences. This is commonly analyzed using unfolding analysis. This analysis assumes to have an ideal point. The distances of ideal point and objects points are inversely proportional to subject preferences. The ideal and object points are designed in a way that preference value of objects show a decreasing function. When the object point is closer to the ideal point the object would be more preferred by the subject. This is a useful analysis specially for marketing research. This would be helpful in understanding patterns of individual differences and relationships to product features.

7. Software

There are few software's that can be used for the computation of multi dimensional scaling. They are as follows

PREFSCAL

Unfolding analysis is generally difficult to generate. With the advent of PREFSCAL, this incorporated a penalty term which overcomes the difficulty. The scale helps to unfold the proximity data, which helps to find the least squares representation of the rows and columns of objects in a low dimensional scale. The algorithm generally minimizes stress & convergence in both neutral & non-metric data under a variety of constraints. The virtual relations between 2 sets can be found using a common quantitative scale.

The PREFSCALE produces a two-dimensional metric input, which consists of one or more rectangular matrices with proximities. The classical scaling start is considered to the initial configuration. The fit and stress values are a part of output.

- This requires minimum of at least 2 objects
- Similarly requires 2 variables
- The data must contain at least 2 cases
- This will not work with SPLIT FILE

INDSCAL

INDSCAL finds the relationship between subjects, and the matrixes in 2nd and 3rd ways of matrix are identical. This scale is used to find the relationship between subjects. This is a special form of CANDECOM, i.e. the second in third ways of matrix are identical. This is a dimensional model; the optimality of the solution will be destroyed by any rotation of the axes of group space. The distance of private spares as compared to group spaces are simple Euclidean, in group it is weighted Euclidean. Interpretation can be readily yielded of the dimension are resolved. The input matrix has to be considered to scales products in the beginning. The data has to be normalized by scaling each subject influence.

ALSCAL

This uses an alternating least squares algorithm. The ALSCAL include 4 models.

- Data Input

  This will help to read two or three way data; in other words, it can read inline data matrices when the matrix is in the form of square or rectangle. It can read with the help of SHAPE subcommand.

- Methodological Assumptions

  Data can be commanded based on the condition, i.e., it can be row conditional, matrix conditional or unconditional.

- Model Selection

  There are various models to ALSCAL wherein the perfect combination of ALSCAL has to be selected.

- Output

  Output can be produced either with raw or scaled input data, missing value patterns, normalized data, with means. Data can be transformed using PRINT or PLOT commands.

Limitations

1. ALSCAL allows the usage of only 100 values.
2. A maximum of 6 dimensions can be scaled.
3. It cannot recognize data weights created by weight commands.
4. ALSCAL analyses cannot include more than 32767 values in each of input matrix.

Steps in Conducting Multidimensional Scaling

- Problem identification
- Data collection
- Procedure for MDS
- Dimension decision
- Labelling & interpretation
- Assessment

Problem identification

- The purpose /problem has to be identified
- Analysis has to include or other stimuli so selection has to be made.
- The number of brands to be included should be based on the statement of marketing research problem.

Data collection

- *Direct Approaches* - Data is gathered on the basis of how the respondents judge similarities/ dissimilarities of various brands, using their own criteria also known as similarity judgements.
- *Derived Approaches* - The data is collected based on the ratings they give to the brand/stimuli on the attributes using Likert scale.

A similarity measure is derived for each pair of brands/stimuli if attribute ratings are obtained.

Procedure for MDS

- The nature of input data is a determining factor analyses will require both kinds of data (perception/preferences)
- Input data are ordinal is the assumption of non-metric multidimensional scaling but result in metric output
- Input data are metric in non-metric multidimensional scaling; a strong relationship between input and output data is maintained.
- The most important factor is whether the information will be collected at individual level or at an aggregate level.

Dimension decision

- Particular number of dimensions can be sought from previous or past research
- Interpretation of configuration or maps is difficult in case of having more than 3 dimensions
- Two-dimension maps are convenient than involving more dimensions.
- Dimensionality can be determined through statistical approach
- Labelling and interpretation
- Researcher-supplied attributes may be collected using statistical methods such as regression.
- The criteria used for decision making has to be identified
- The spatial maps of the respondents may be exposed and could be labelled
- The subjective dimensions can be interpreted with the help of objective characteristics of the brand.

Assessment

- The proportion of variance of the optimally scaled data can be accounted by MDS procedure
- Stress values indicate the quality of solution when the values are less than 10% it is considered acceptable

- In an aggregate analysis, original data is split into 2 or more parts stimuli has to be selectively eliminated in the input data.
- Error term has to be added to the input data. The resulting data are given for analysis or comparison.

Types of multidimensional scaling

1. Classical Multidimensional Scaling
2. Metric Multidimensional Scaling
3. Non-Metric Multidimensional Scaling
4. Replicated Multidimensional scaling
5. Weighted Multidimensional scaling

The types of Multidimensional Scaling is as follows

1. Classical Multidimensional Scaling

It displays the distance in a data as a geometrical picture. It is based on the self evident principle. The Classic multidimensional scaling organises objectives that give dissimilarities, and co-ordinates them together to prevent a loss function called strain. It is also known as principal coordinate's analysis.

Example: In a map showing a number of cities one might be interested in knowing the distance between them. This can be found easily by measuring through a rules, but the classical multidimensional scaling addresses to the inverse of this situation, i.e.- is it possible to obtain the map having only the distances. Classical Multi Dimension Scaling uses the Euclidean distances. In applications of multidimension scaling the data are not distances but rather proximity data. The proximity data when applied performs like real measured distances. Classical multidimension scaling avoids extractive procedure and provides only analytical solution.

2. Metric Multi Dimension Scaling

Metric Multi Dimension Scaling is superior scaling method as compared to classic multi dimension scaling. This generalises the loss function by assigning the input of known distance with weights. The multi dimension scaling transforms distance matrix into cross product by factorising the matrix into a mathematical object.

Example- In a study 6 subjects were scanned while they were watching pictures of 8 categories (faces, houses, cats chairs, shoes, scissors, bottle & mixed images). Distance matrix was calculated for each subject on the basis of level of prediction. The distances were computed in 2 form. The first distance correlated was the average matrix; on the other hand, direct matrix was also calculated. The two matrix were used to transform a cross product matrix, which also assisted in the projecting a set of supplementary observations. The variables reflect the perception of the respondent in relation to distance. In this the researcher may apply Factor Analysis or Principal Component Analysis to reduce dimension coordinates. Principal Component Analysis or Factor Analysis are best representation of data matrix.

3. Non-Metric Multidimensional Scaling

Multi Dimension Scaling finds the relationship between the dissimilarities in the item matrix. It takes for consideration statistics that includes data which has no characteristics & ordered sets. The non-metric scaling typically uses isotonic regression which includes finding weighted least squares. Non-metric multidimensional scaling is more adequate in the case of weaker measurements mainly in dissimilarity. It derives Euclidean distance approximation using only ordinal information from the data.

Non-metric Multidimensional Scaling has an interactive algorithms. The positions of samples at the initial stages are adjusted to minimise stress until any other interactions achieve a sufficient improvement. In Non metric Multi Dimension Scaling, the initial position is sensitive and sometimes settles into an optimum level which is not a best solution. This problem can be minimised with the help of multiple random starts. There can never be a guarantee that the best solution is found but when the majority of the results achieve low levels of stress, it would then be certain that a good solution has been found.

The other approach could be using of principle component analysis in the initial position, as the principle component analysis is the optimal geometric projection. The variables in non-metric reflects ranking; this will not basically assess the distance between the first and second object.

4. Replicated Multidimensional scaling

This kind of multidimensional scaling helps in analysing several matrices of similar data simultaneously rather than describing one.

5. Weighted multidimensional scaling

The distance model was generalised such that several similarity matrix could be assumed. Individual difference in bias only gets recorded in replicated multi dimensional scaling. In weighted multi dimensional scaling the individual difference gets accounted by considering cognitive process. This is otherwise known as individual differences scaling. There are specific parameters in process regarding the individual variations. Larger the weight the dimension would be very important and vice versa.

Algorithm of Multidimensional Scaling

KYST is the most commonly used algorithm for multidimensional computations of the process. It is very time consuming without computer software. The steps are:

1. The data has to be transformed or standardised using a data matrix.
2. Matrix of dissimilarities has to be made between objects. On the other hand matrix of similarities can also be used; it will have no effect in further steps.
3. Number of dimensions for scaling, which will act as a compromise between dissimilarities and distance among the objects must be calculated. The dimension must have way for simple interpretation. The suspected numbers for the underlying eco-gradients has to be identified. The dissimilarities between objects and distance between objects must be noted closely as possible. When there are more dimensions a clear matrix can be forced. It is better to usually have 2-4 dimensions.
4. Objects must be arranged either at random or more commonly by using coordinate from principal component analysis. The object must be arranged in the K-dimensional space in starting configuration. The objects arranged must be arranged based on the chosen number of dimensions. As mentioned earlier this will depend on the starting configuration & must be generated randomly.
5. Once the objects are plotted, comparison has to be done with Bray-Curtin dissimilarities. The Krukats stress value can be used to generate the strength of relationship. When the values are lower the matrix indicated better match this can measure "badness of fit".
6. The location of objects must be moved interactively so that the match among the inter-object distance in the configuration and actual dissimilarities improves the K-dimensional space.

The final configuration or position plot is achieved when the interactive moving of objects has no longer improved the match between the distance among the inter objects in the configuration and dissimilarities.

Multidimensional Scaling

Multidimensional scaling attempts to find the structure in a set of distance measures between objects or cases. This task is accomplished by assigning observations to specific locations in a conceptual space (usually two- or three-dimensional) such that the distances between points in the space match the given dissimilarities as closely as possible. In many cases, the dimensions of this conceptual space can be interpreted and used to further understand the data.

If there is objectively measured variables, use multidimensional scaling as a data reduction technique (the Multidimensional Scaling procedure will compute distances from multivariate data, if necessary). Multidimensional scaling can also be applied to subjective ratings of dissimilarity between objects or concepts. Additionally, the Multidimensional Scaling procedure can handle dissimilarity data from multiple sources, with multiple raters or questionnaire respondents.

In this example, the consumers are rated the store attributes according to the level of agreement, to check which one placing highest position in the mind.

Statistics: For each model: data matrix, optimally scaled data matrix, S-stress (Young's), stress (Kruskal's), RSQ, stimulus coordinates, average stress and RSQ for each stimulus (RMDS models). For individual difference (INDSCAL) models: subject weights and weirdness index for each subject. For each matrix in replicated multidimensional scaling models: stress and RSQ for each stimulus. Plots: stimulus coordinates (two- or three-dimensional), scatterplot of disparities versus distances.

Data: If the data are dissimilarity data, all dissimilarities should be quantitative and should be measured in the same metric. If the data are multivariate data, variables can be quantitative, binary, or count data. Scaling of variables is an important issue—differences in scaling may affect the solution. If the variables have large differences in scaling (for example, one variable is measured in dollars and the other variable is measured in years), consider standardizing them (this process can be done automatically by the Multidimensional Scaling procedure).

Assumptions: The Multidimensional Scaling procedure is relatively free of distributional assumptions. Be sure to select the appropriate measurement level (ordinal, interval, or ratio) in the Multidimensional Scaling Options dialog box so that the results are computed correctly.

Related procedures: If the goal is data reduction, an alternative method to consider is factor analysis, particularly if the variables are quantitative. If there is a need to identify groups of similar cases, consider supplementing the multidimensional scaling analysis with a hierarchical or *k*-means cluster analysis.

Example - Multidimensional Scaling (ALSCAL)

Test Procedure in SPSS

Following are the steps to analyse data using Discriminant Analysis in SPSS.

Step - 1: Click Analyze > Scale > Multidimensional Scaling (ALSCAL) as shown in the figure -1, the Multidimensional Scaling dialogue box will appear (as given in the figure - 2).

- Select at least four numeric variables for analysis.
- In the Distances group, select either Data are distances or Create distances from data.
- If needed Select Create distances from data, also select a grouping variable for individual matrices. The grouping variable can be numeric or string.

Optionally

- Specify the shape of the distance matrix when data are distances.
- Specify the distance measure to use when creating distances from data.

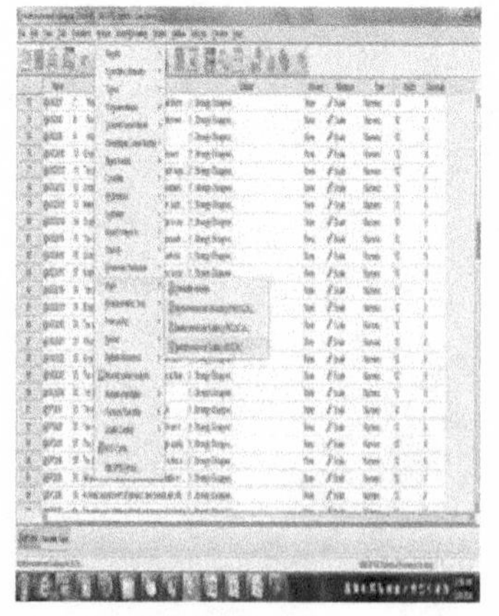

Figure - 1

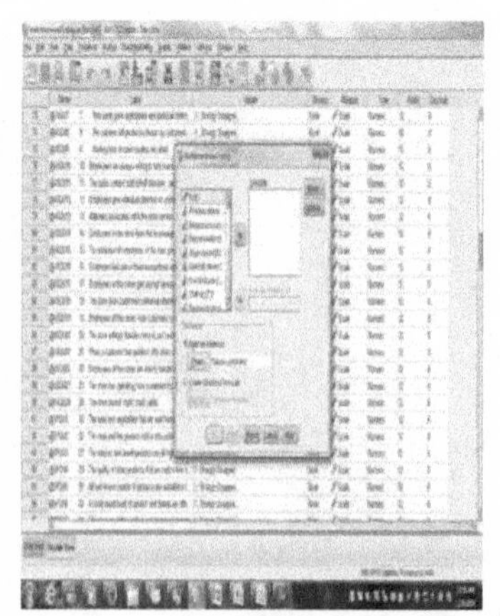

Figure - 2

Step - 2: Transfer the variables to variables box by dragging or by using right arrow button in multidimensional scaling dialogue box as given in figure -3.

Step - 3: Click model tab in multidimensional scaling, a dialogue box will appear as given in figure -4. Select level of measurement as interval followed by dimensions minimum 2 and maximum 2 and scaling model Euclidean distance as given in figure -5 and click continue to return to multidimensional dimension box.

Multidimensional Scaling Model

Correct estimation of a multidimensional scaling model depends on aspects of the data and the model itself.

- **Level of Measurement:** Allow to specify the level of data. Alternatives are Ordinal, Interval, or Ratio. If the variables are ordinal, selecting Untie tied observations requests that the variables be treated as continuous variables, so that ties (equal values for different cases) are resolved optimally.
- **Conditionality:** Allow to specify which comparisons are meaningful. Alternatives are Matrix, Row, or Unconditional.
- **Dimensions:** Allow to specify the dimensionality of the scaling solution(s). One solution is calculated for each number in the range. Specify integers between 1 and 6; a minimum of 1 is allowed only if Euclidean distance model is selected as the scaling. For a single solution, specify the same number for minimum and maximum.
- **Scaling Model**: Allow to specify the assumptions by which the scaling is performed. Available alternatives are Euclidean distance or Individual differences Euclidean distance (also known as INDSCAL). For the Individual differences Euclidean distance model, select Allow negative subject weights, if appropriate for the data.

Multidimensional Scaling Create Measure

Multidimensional scaling uses dissimilarity data to create a scaling solution. If the data are multivariate data (values of measured variables), it is required to create dissimilarity data in order to compute a multidimensional scaling solution. Specify the details of creating dissimilarity measures from the data.

Measure: Allow to specify the dissimilarity measure for analysis. Select one alternative from the Measure group corresponding to the type of data, and then choose one of the measures from the drop-down list corresponding to that type of measure. Available alternatives are:

- **Interval**: Euclidean distance, Squared Euclidean distance, Chebychev, Block, Minkowski, or Customized.
- **Counts:** Chi-square measure or Phi-square measure.
- **Binary:** Euclidean distance, Squared Euclidean distance, Size difference, Pattern difference, Variance, or Lance and Williams.

Create Distance Matrix: Allows to choose the unit of analysis. Alternatives are Between variables or Between cases.

Transform Values: In certain cases, such as when variables are measured on very different scales to standardize values before computing proximities (not applicable to binary data).

Choose a standardization method from the Standardize drop-down list. If no standardization is required, choose None.

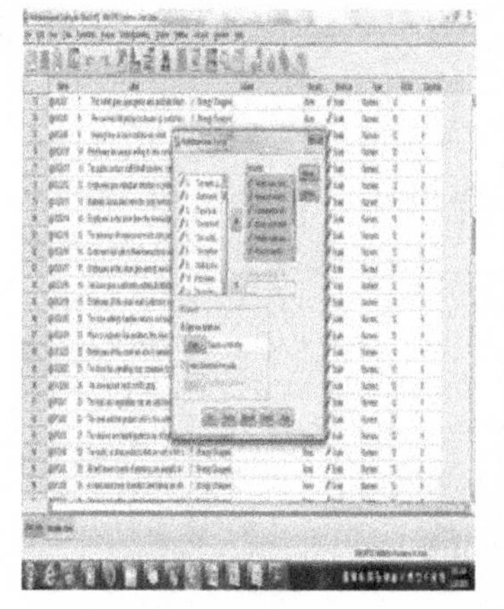

Figure - 3

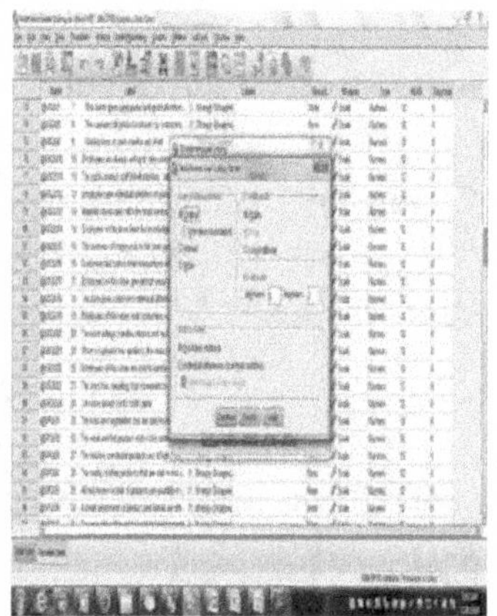

Figure - 4

Step - 4: In multidimensional scaling dialogue box, click option tab, a dialogue box will appear as given in figure -6. Select group plots, individual subject plots, data matrix, model and options summary checkbox as given in figure -7, click continue to return to multidimensional scaling dialogue box.

Multidimensional Scaling Options

Specify options for multidimensional scaling analysis.

Display: Allows to select various types of output. Available options are Group plots, Individual subject plots, Data matrix, and Model and options summary.

Criteria: Allow to determine when iteration should stop. To change the defaults, enter values for S-stress convergence, Minimum s-stress value, and Maximum iterations.

Treat distances less than n as missing. Distances that are less than this value are excluded from the analysis.

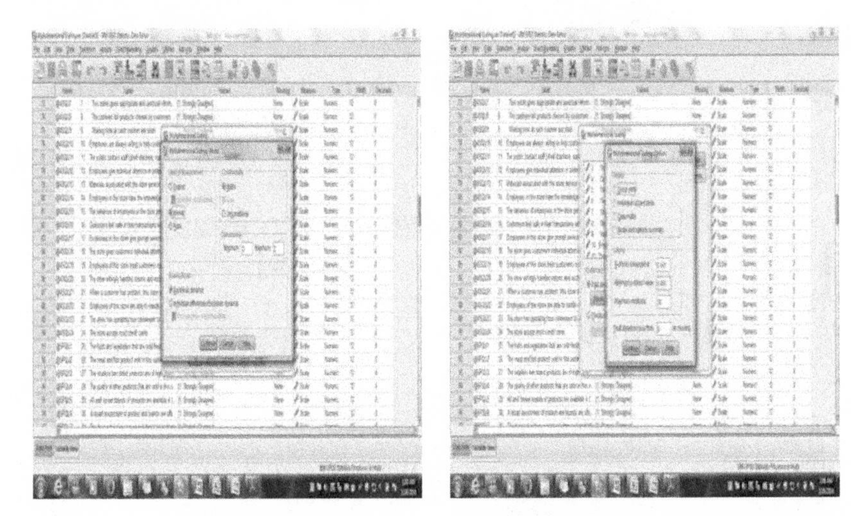

Figure - 5 **Figure - 6**

Step - 5: in multidimensional scaling dialogue box, select - create distance from data, measures as Euclidian distances a dialogue box will appear as given in figure -9. In that select measures as interval, followed by create distance matrix by variables and click continue.

Multidimensional Scaling Shape of Data

If the active dataset represents distances among a set of objects or represents distances between two sets of objects, specify the shape of data matrix in order to get the correct results.

Note: it is not required to select Square symmetric if the Model dialog box specifies row conditionality

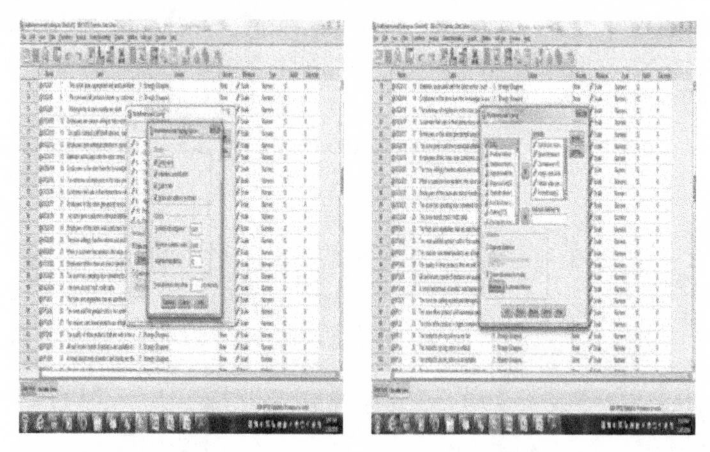

Figure - 7 Figure - 8

Step - 6: click ok in multidimensional scaling dialogue box to generate output.

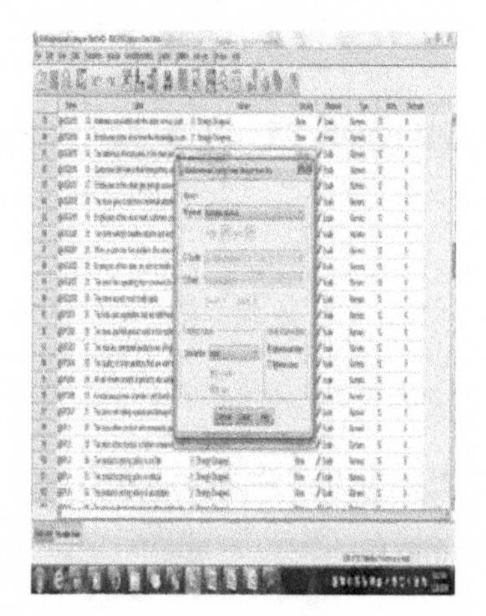

Figure - 9

SPSS Output of the Multidimensional Scaling Analysis

Table 1 Alscal Procedure Options

Data Options-

Number of Rows (Observations/Matrix). 10
Number of Columns (Variables) . . . 10
Number of Matrices 1
Measurement Level . . . . Interval
Data Matrix Shape . . . . Symmetric
Type . . . . Dissimilarity
Approach to Ties . . . . Leave Tied
Conditionality . . . . Matrix
Data Cutoff at 000000

Model Options-

Model . . . . Euclid
Maximum Dimensionality 2
Minimum Dimensionality 2
Negative Weights . . . . Not Permitted

Output Options-

Job Option Header . . . . Printed
Data Matrices . . . . Printed
Configurations and Transformations . Plotted
Output Dataset . . . . Not Created
Initial Stimulus Coordinates . . . Computed

Algorithmic Options-

Maximum Iterations 30
Convergence Criterion 00100
Minimum S-stress 00500
Missing Data Estimated by . . . . Ulbounds

Raw (unscaled) Data for Subject 1

	1	2	3	4	5	6	7	8	9	10
1	.000									
2	19.287	.000								
3	21.494	20.248	.000							
4	22.891	20.688	21.260	.000						
5	25.000	24.597	23.979	20.273	.000					
6	26.608	23.065	24.207	21.119	22.338	.000				
7	25.318	23.896	22.694	24.187	24.698	22.068	.000			
8	26.495	24.454	24.454	25.807	25.318	24.042	22.956	.000		
9	25.534	23.664	24.083	21.166	24.515	22.847	23.259	21.448	.000	
10	28.583	26.173	25.515	26.058	26.458	25.942	23.833	23.601	21.000	.000

Iteration history for the 2 dimensional solution (in squared distances)

Young's S-stress formula 1 is used.

Iteration	**S-stress**	**Improvement**
1	.24137	
2	.21641	.02497
3	.21456	.00185
4	.21453	.00003

Iterations stopped because
S-stress improvement is less than .001000

Stress and squared correlation (RSQ) in distances

RSQ values are the proportion of variance of the scaled data (disparities)
in the partition (row, matrix, or entire data) which
is accounted for by their corresponding distances.
Stress values are Kruskal's stress formula 1.

For matrix
Stress = .20393 RSQ = .76529

Configuration derived in 2 dimensions
Stimulus Coordinates

Dimension

Stimulus Number	Stimulus Name	1	2
1	Product_	1.9220	-.7687
2	Service_	1.0083	-.4291
3	Location	.8492	-.5876
4	Atmosphe	.8008	.6403
5	Salesper	.5205	1.4863
6	Quality_	-.1840	1.2695
7	Promotio	-.6431	-.5082
8	Parking	-1.3041	-.7190
9	Credit_C	-.8569	.0437
10	Competit	-2.1126	-.4272

Optimally scaled data (disparities) for subject 1

	1	2	3	4	5	6	7	8	9	10
1	.000									
2	.000	.000								
3	.950	.414	.000							

4	1.552	.603	.850	.000						
5	2.460	2.286	2.020	.424	.000					
6	3.152	1.627	2.119	.789	1.314	.000				
7	2.597	1.984	1.467	2.110	2.330	1.197	.000			
8	3.104	2.225	2.225	2.807	2.597	2.047	1.580	.000		
9	2.690	1.885	2.065	.809	2.251	1.533	1.710	.930	.000	
10	4.003	2.965	2.682	2.915	3.087	2.866	1.957	1.857	.737	.000

Abbreviated Name	**Extended** Name
Atmosphe	Atmosphere
Competit	Competitive_Price
Credit_C	Credit_Card_Facility
Product_	Product_Variety
Promotio	Promotion
Quality_	Quality_Product
Salesper	Salesperson
Service_	Service_Speed

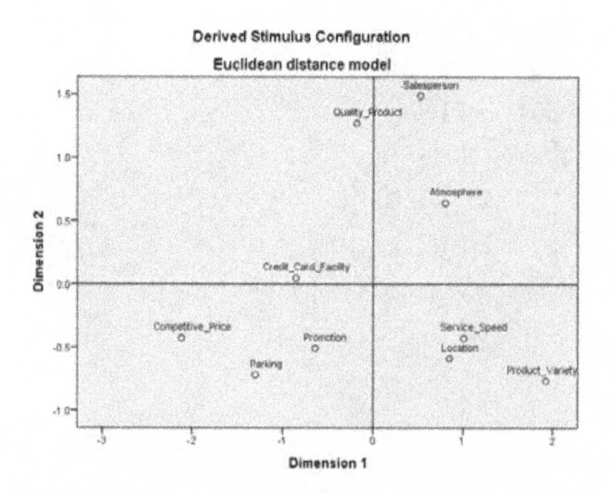

Figure - 1

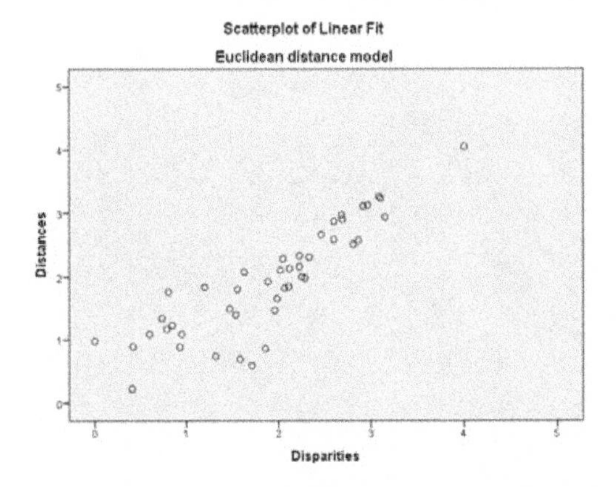

Figure - 2

Results and Discussion

The results of the ALSCAL analysis revealed the consumer perception about organised retail store attributes with two dimension solution on squared matrix. This two-dimensional MDS solution was attained with a Kruskal's stress value of .20393 and RSQ value is .76529. Stress is a lack-of-fit index, values of the stress index higher than .10 may indicate a poor solution and a lack of fit of the MDS model to the data. RSQ is an index of fit that indicates how well the MDS model corresponds to the input data (Hair, Anderson, & Tatham, 1987). The Kruskal's stress value and RSQ index indicate an adequate fit and high level of correspondence of the MDS model to the data. A three-dimensional solution, even though it had slightly lower stress and higher fit, was rejected as being not readily interpretable. The two-dimensional solution thus presents an accurate and reliable visual depiction of the data obtained from consumer perception towards store attributes.

The figure -1 depicts the dimension - 1(horizontally) containing 6 dimensions such as competitive pricing, parking, promotion, speedy service, location and product value named as "Convenience". Dimension -2 vertically named as "Adaptability".

The stress and other goodness of fit measures are used to provide an indication of how well the final configuration fits the original proximities of data. The Euclidean distance model provides a good interpretation of the distance of the attributes after the rotation of the distance. This roughly represents the mind mapping of consumer's perception towards store attributes.

Chapter - 11

Neural Network

Learning Objectives

This chapter helps to understand the following

- Overview of Neural Network
- Meaning of Neural Network
- Basic Concept of Neural Networks
- Elements of Neural Networks
- Process Development of Neural Network Analysis
- Purpose of Neural Networks
- The Neural Networks Model
- Application of Neural Networks
- Advantages and Limitation of Neural Networks
- Neural Network – Multilayer Perceptron - Key Terms
- Procedure for running Neural Network – Multilayer Perceptron
- Results and Discussion of Neural Network – Multilayer Perceptron Output
- Neural Network – Radical Basis Function - Key Terms
- Procedure for running Neural Network – Radical Basis Function
- Results and Discussion of Neural Network – Radical Basis Function Output

Overview of Neural Networks

Following are the various historical backgrounds of neural networks:-

1. **First Attempts:** There were some initial simulations using formal logic. McCulloch and Pitts (1943) developed models of neural networks based on their understanding of neurology. Two groups (Farley and Clark, 1954; Rochester, Holland, Haibit and Duda, 1956). The first group (reserchers) maintained closed contact with neuroscientists at McGill University. So whenever their models did not work, they consulted the neuroscientists. This interaction established a multidisciplinary trend which continues to the present day.

2. **Promising & Emerging Technology:** Not only was neuroscience influential in the development of neural networks, but psychologists and engineers also contributed to the progress of neural network simulations. Rosenblatt (1958) stirred considerable interest and activity in the field when he designed and developed the Perceptron. The Perceptron had three layers with the middle layer known as the association layer. This system could learn to connect or associate a given input to a random output unit.Another system was the ADALINE (Adaptive Linear Element) which was developed in 1960 by Windrow and Hoff (of Stanford University). The ADALINE was an analogue electronic device made from simple components. The method used for learning was different to that of the Perceptron; it employed the Least-Mean-Squares (LMS) learning rule.

3. **Period of Frustration & Disrepute:** In 1969 Minsky and Papert wrote a book in which they generalised the limitations of single layer Perceptrons to multilayered systems. In the book they said: "... our intuitive judgment that the extension (to multilayer systems) is sterile". The significant result of their book was to eliminate funding for research with neural network simulations. The conclusions supported the disenchantment of researchers in the field. As a result, considerable prejudice against this field was activated.

4. **Innovation:** Although public interest and available funding were minimal, several researchers continued working to develop neuromorphically based computaional methods for problems such as pattern recognition. During this period several paradigms were generated which modern work continues to enhance. Grossberg's (Steve Grossberg and Gail Carpenter in 1988) influence founded a school of thought which explores resonating algorithms. They developed the ART (Adaptive Resonance Theory) networks based on biologically

plausible models. Anderson and Kohonen developed associative techniques independent of each other. Klopf (A. Henry Klopf) in 1972, developed a basis for learning in artificial neurons based on a biological principle for neuronal learning called heterostasis. Werbos (Paul Werbos 1974) developed and used the back-propagation learning method, however several years passed before this approach was popularized. Back-propagation nets are probably the most well known and widely applied of the neural networks today. In essence, the back-propagation net. is a Perceptron with multiple layers, a different threshold function in the artificial neuron, and a more robust and capable learning rule. Amari (A. Shun-Ichi 1967) was involved with theoretical developments: he published a paper which established a mathematical theory for a learning basis (error-correction method) dealing with adaptive pattern classification. While Fukushima (F. Kunihiko) developed a step wise trained multilayered neural network for interpretation of handwritten characters. The original network was published in 1975 and was called the Cognitron.

5. **Re-Emergence:** Progress during the late 1970s and early 1980s was important to the re-emergence on interest in the neural network field. Several factors influenced this movement. For example, comprehensive books and conferences provided a forum for people in diverse fields with specialized technical languages, and the response to conferences and publications was quite positive. The news media picked up on the increased activity and tutorials helped disseminate the technology. Academic programs appeared and courses were introduced at most major Universities (in US and Europe). Attention is now focused on funding levels throughout Europe, Japan and the US and as this funding becomes available, several new commercial with applications in industry and financial institutions are emerging.

6. **Today:** Significant progress has been made in the field of neural networks-enough to attract a great deal of attention and fund further research. Advancement beyond current commercial applications appears to be possible, and research is advancing the field on many fronts. Neurally based chips are emerging and applications to complex problems are developing. Clearly, today is a period of transition for neural network technology.

Meaning of Neural Networks

Neural Networks

A neural network can approximate a wide range of predictive models with minimal demands on model structure and assumption. The form of the relationships is determined during the learning process. If a linear relationship between the target and predictors is appropriate, the results of the neural network should closely approximate those of a traditional linear model. If a nonlinear relationship is more appropriate, the neural network will automatically approximate the "correct" model structure.

The trade-off for this flexibility is that the neural network is not easily interpretable. If you are trying to explain an underlying process that produces the relationships between the target and predictors, it would be better to use a more traditional statistical model. However, if model interpretability is not important, you can obtain good predictions using a neural network.

Neural networks can be considered as nonlinear function approximating tools (i.e., linear combinations of nonlinear basis functions), where the parameters of the networks should be found by applying optimization methods.

Neural networks can learn to recognize patterns:-

- By repeated exposure to many different examples.
- Patterns or salient characteristics whether they are hand written characters, profitable loans or good trading decisions.
- Data that is inexact and incomplete.

Neural networks find this relationship through a learning cycle where many hundreds of samples are presented repeatedly to the network. Neural network cannot guarantee an optimal solution to a problem. However, properly configured and trained neural networks can often make consistently good classifications, generalizations or decisions in a statistical sense.

Basic Concept of Neural Networks

Back Propagation: - The best known learning algorithm in neural computing. Learning is done by comparing computed outputs to desired outputs of historical cases.

Network Structure:- It consists of three different layers namely,

- Input Layer
- Intermediate(Hidden Layer)

- Output Layer

Transformation Function: - This is otherwise known as "activation function". Maps the summation (combination) function onto a narrower range (0 to 1 or -1 to 1) to determine whether or not an output is produced (neuron fires). The transformation occurs before the output reaches the next level in the network.

Sigmoid (logical activation) Function: - an S-shaped transfer function in the range of zero to one –exp(x)/ (1-exp(x))

Threshold Value: - it is sometimes used instead of a transformation function. A hurdle value for the output of a neuron to trigger the next level of neurons. If an output value is smaller than the threshold value, it will not be passed to the next level of neurons.

Elements of Neural Networks

The key element of this network analysis is the novel structure of the **Network Information Processing System** that includes the following processing elements:-

- Inputs
- Outputs
- Connection weights
- Summation function

It is composed of a large number of highly interconnected processing elements (neurons) working in units to solve specific problems. Neural networks, like people, learn by example. The other element of NN is **Network Architecture** includes:-

- **Hidden layers** and
- **Parallel processing**

Process Development of Neural Network Analysis

The following are the various stages of development process of neural networks:-

1. **Data Collection and Preparation:** The data used for training and testing must include all the attributes that are useful for solving the problem. Recall the bankruptcy prediction problem we modeled using logistic regression -- The same data can be used to train a neural network:

- working capital/total assets (WC/TA)
- retained earnings/total assets (RE/TA)
- earnings before interest and taxes/total assets (EBIT/TA)
- market value of equity/total debt (MVE/TD)
- sales/total assets (S/TA)

2. **Selection of network structure:** Determination of:
   - Input nodes
   - Output nodes
   - Number of hidden layers
   - Number of hidden nodes.
     i For the bankruptcy problem, we have one hidden layer. The Bankruptcy problem has ten nodes in the hidden layer – sometimes one might experiment with the number of nodes.
3. **Learning algorithm selection:** Identify a set of connection weights that best cover the training data and have the best predictive accuracy.
4. **Network training:** An iterative process that starts from a random set of weights and gradually enhances the fitness of the network model and the known data set
   - The iteration continues until the error sum is converged to below a preset acceptable level.
5. **Testing:** Black-box testing- Comparing test results to actual results. The test plan should include routine cases as well as potentially problematic situations. If the testing reveals large deviations, the training set must be reexamined, and the training process may have to be repeated. Might compare NN results with other methods such as logistic regression.
6. **Implementation of an NN:** Implementation often requires interfaces with other computer-based information systems and user training. Ongoing monitoring and feedback to the developers are recommended for system improvements and long-term success.It is important to gain the confidence of users and management early in the deployment to ensure that the system is accepted and used properly.

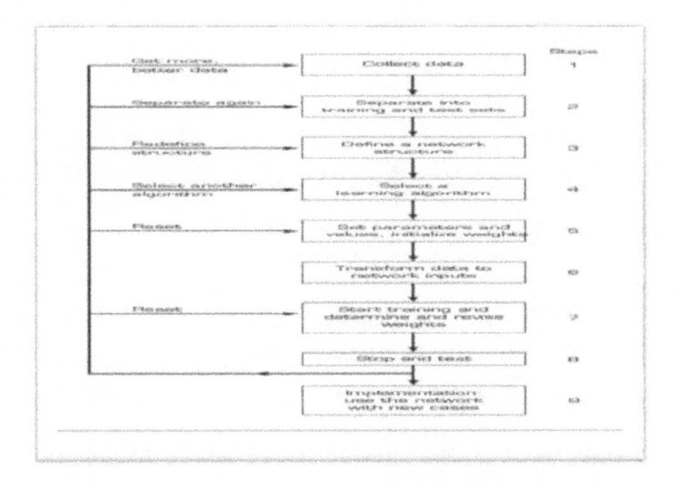

Figure 1 DEVELOPMENT PROCESS OF NN

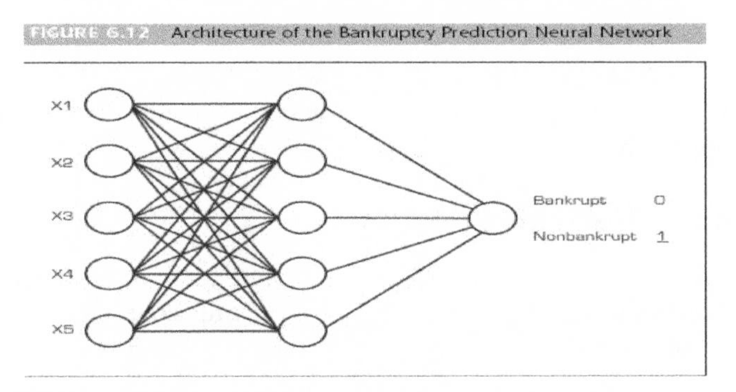

Figure 2 Architecture of Bankruptcy Prediction NN

Purpose of Neural Networks

Neural networks, with their remarkable ability to derive meaning from complicated or imprecise data, can be used to extract patterns and detect trends that are too complex to be noticed by either humans or other computer techniques. A trained neural network can be thought of as an "expert" in the category of information it has been given to analyse. This expert can then be used to provide projections given new situations of interest and answer "what if" questions.Other advantages include:

1. Adaptive learning: An ability to learn how to do tasks based on the data given for training or initial experience.
2. Self-Organisation: NN can create its own organisation or representation of the information it receives during learning time.

3. Real Time Operation: NN computations may be carried out in parallel, and special hardware devices are being designed and manufactured which take advantage of this capability.
4. Fault Tolerance via Redundant Information Coding: Partial destruction of a network leads to the corresponding degradation of performance. However, some network capabilities may be retained even with major network damage.

Neural networks are universal approximates, and they work best if the system you are using them to model has a high tolerance to error. One would therefore not be advised to use a neural network to balance one's cheque book! However they work very well for:

- capturing associations or discovering regularities within a set of patterns;
- where the volume, number of variables or diversity of the data is very great;
- the relationships between variables are vaguely understood; or,
- the relationships are difficult to describe adequately with conventional approaches.

The Neural Networks Model

Neural networks are simple models of the way the nervous system operates. The basic units are neurons, which are typically organized into layers, as shown in the following figure.

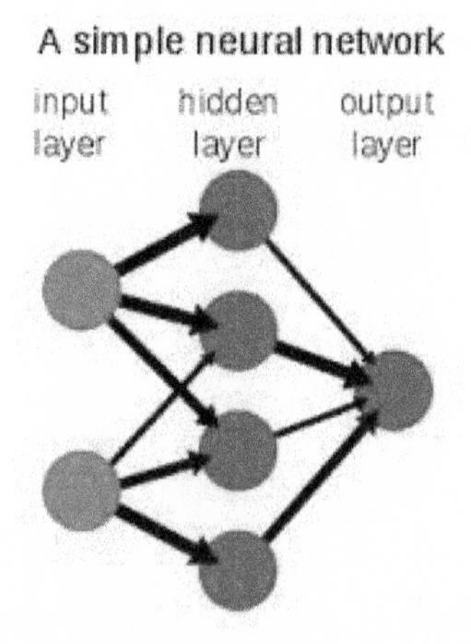

Structure of a neural network

A neural network is a simplified model of the way the human brain processes information. It works by simulating a large number of interconnected processing units that resemble abstract versions of neurons.

The processing units are arranged in layers. There are typically three parts in a neural network: an input layer, with units representing the input fields; one or more hidden layers; and an output layer, with a unit or units representing the target field(s). The units are connected with varying connection strengths (or weights). Input data are presented to the first layer, and values are propagated from each neuron to every neuron in the next layer. Eventually, a result is delivered from the output layer.

The network learns by examining individual records, generating a prediction for each record, and making adjustments to the weights whenever it makes an incorrect prediction. This process is repeated many times, and the network continues to improve its predictions until one or more of the stopping criteria have been met.

Initially, all weights are random, and the answers that come out of the net are probably nonsensical. The network learns through training. Examples for which the output is known are repeatedly presented to the network, and the answers it gives are compared to the known outcomes. Information from this comparison is passed back through the network, gradually changing the weights. As training progresses, the network becomes increasingly accurate in replicating the known outcomes. Once trained, the network can be applied to future cases where the outcome is unknown.

The type of model determines how the network connects the predictors to the targets through the hidden layer(s). The multilayer perceptron (MLP) allows for more complex relationships at the possible cost of increasing the training and scoring time. The radial basis function (RBF) may have lower training and scoring times, at the possible cost of reduced predictive power compared to the MLP.

Hidden Layers. The hidden layer(s) of a neural network contains unobservable units. The value of each hidden unit is some function of the predictors; the exact form of the function depends in part upon the network type. A multilayer perceptron can have one or two hidden layers; a radial basis function network can have one hidden layer.

- Automatically compute number of units. This option builds a network with one hidden layer and computes the "best" number of units in the hidden layer.
- Customize number of units. This option allows you to specify the number of units in each hidden layer. The first hidden layer must have at least one unit. Specifying 0 units for the second hidden layer builds a multilayer perceptron with a single hidden layer.

The following are various neural networks models:-

Feed-Forward networks

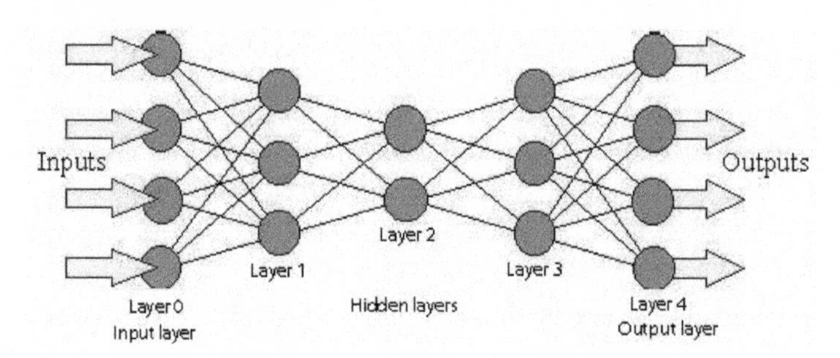

Figure 1 A feed-forward network

Feed-forward networks have the following characteristics:

- Perceptrons are arranged in layers, with the first layer taking in inputs and the last layer producing outputs. The middle layers have no connection with the external world, and hence are called hidden layers.
- Each perceptron in one layer is connected to every perceptron on the next layer. Hence information is constantly "fed forward" from one layer to the next, and this explains why these networks are called feed-forward networks.
- There is no connection among perceptrons in the same layer.

Multilayered Feed Forward Neural Networks

According to a research study, approximately 95% of reported neural network business application studies utilize multilayered feed forward neural networks (MFNNs) with the back propagation learning rule. This type of neural network is popular because of its broad applicability to many problem domains of relevance to business: principally prediction, classification, and modelling. MFNNs are appropriate for solving problems that involve learning the relationships between a set of inputs and known outputs. They are a supervised learning technique in the sense that they require a set of training data in order to learn the relationships.

Hopfield Networks

A neural network model of interest is the Hopfield network (Hopfield, 1982). John Hopfield showed in a series of papers in the 1980s how highly interconnected networks or nonlinear neurons can be extremely effective in computing. These networks provided a rapid computed solution for problems stated in terms of desired optima, often subject to constraints.

A general Hopfield network is a single large layer of neurons with total interconnectivity—that is, each neuron is connected to every other neuron. In addition, the output of each neuron may depend on its previous values. One use of Hopfield networks has been in solving constrained optimization problems, such as the classic traveling salesman problem (TSP). In this type of application, each neuron represents the desirability of a city n being visited in position mof a TSP tour. Interconnection weights are specified, representing the constraints of feasible solution to the TSP (e.g., forcing a city to appear in a tour only once). An energy function is specified, which represents the objective of the model solution process (e.g., minimize totaldistance in the TSP tour) and is used in determining when to stop the neural network evolution to a final state. The network starts with random neuron values and, using the stated interconnection weights, the neuron values are updated over time. Gradually, the neuron values stabilize, evolving into a final state (as driven by the global energyfunction) that represents a solution to the problem. At this point in the network evolution, the value of neuron (n,m) represents whether city n should be in location mof the TSP tour. While Hopfield and Tank (1985) and others claimed great success in solving the TSP, further research has shown those claims to be somewhat premature.

Nonetheless, this novel approach to a classic problem offers promise for optimization problems, especially when technology allows for taking advantage of the inherent parallelism of neural networks.

Hopfield networks are distinct from feedforward networks because the neurons are highly interconnected, weights between neurons tend to be fixed, and there is no training per se. The complexity and challenge in using a Hopfield network for optimization problems is in the correct specification of the interconnection weights and the identification of the proper global energy function to drive the network evolution process.

Self-Organising Networks

Kohonen's network, also known as a self-organizing network is another neural network model. Such networks learn in an unsupervised mode. The biological basis of these models is the conjecture that some organization takes place in the human brain when an external stimulus is provided. Kohonen's algorithm forms "feature maps," where neighborhoods of neurons are constructed. These neighborhoods are organized such that topologically close neurons are sensitive to similar inputs into the model.

Self-organizing maps, or self-organizing feature maps, can sometimes be used to develop some early insight into the data. For example, self-organizing maps could learn to identify clusters of data so that an analyst could build more refined models for each subset/cluster. In cases in which the analyst does not have a good idea of the number of classes or output or actual output class for any given pattern, the self-organizing map scan work well.

The SOM algorithm exhibits the following properties:

1. Approximation of the continuous input space by the weight vectors of the discrete lattice.
2. Topological ordering exemplified by the fact that the spatial location of a neuron in the lattice corresponds to a particular feature of the input pattern.
3. The feature map computed by the algorithm reflects variations in the statistics of the input distribution.
4. SOM may be viewed as a nonlinear form of principal components analysis.

Multilayer Perceptrons and Back-Propagation Learning

The back-propagation algorithm has emerged as the workhorse for the design of a special class of layered feedforward networks known as multilayer perceptrons (MLP). A multilayer perceptron has an input layer of source nodes and an output layer of neurons (i.e., computation nodes); these two layers connect the network to the outside world. In addition to these two layers, the multilayer perceptron usually has one or more layers of hidden neurons, which are so called because these neurons are not directly accessible. The hidden neurons extract important features contained in the input data.

Radial Basis Function Network

Another popular layered feedforward network is the radial-basis function (RBF) network which has important universal approximation properties (Park and Sandberg 1993), RBF networks use memory-based learning for their design. Specifically, learning is viewed as a curve-fitting problem in high-dimensional space (Broomhead and Lowe 1989; Poggio and Girosi 1990):

1. Learning is equivalent to finding a surface in a multidimensional space that provides a best fit to the training data.
2. Generalization (i.e., response of the network to input data not seen before) is equivalent to the use of this multidimensional surface to interpolate the test data.

RBF networks differ from multilayer perceptrons in some fundamental respects:

- RBF networks are local approximators, whereas multilayer perceptrons are global approximators.
- RBF networks have a single hidden layer, whereas multilayer perceptrons can have any number of hidden layers.
- The output layer of a RBF network is always linear, whereas in a multilayer perceptron it can be linear or nonlinear.
- The activation function of the hidden layer in an RBF network computes the Euclidean distance between the input signal vector and parameter vector of the network, whereas the activation function of a multilayer perceptron computes the inner product between the input signal vector and the pertinent synaptic weight vector.

The use of a linear output layer in an RBF network may be justified in light of Cover's theorem on the separability of patterns. According to this theorem, provided that the transformation from the input space to the feature (hidden) space is nonlinear and the dimensionality of the feature space is high compared to that of the input (data) space, then there is a high likelihood that a nonseparable pattern classification task in the input space is transformed into a linearly separable one in the feature space.

Design methods for RBF networks include the following:

1. Random selection of fixed centers (Broomhead and Lowe1998)
2. Self-organized selection of centers (Moody and Darken 1989)
3. Supervised selection of centers (Poggio and Girosi 1990)
4. Regularized interpolation exploiting the connection between an RBF network and the Watson–Nadaraya regression kernel(Yee 1998).

Other neural network models

There are many other different types of neural network models, each with their own purpose and application areas. Most of these are extensions of the three main models we have discussed here. Their potential application to problems of concern to the business world and the operations researcher is unclear, but they are referenced here for completeness. These other neural network models include adaptive resonance networks, modular networks, neocognitron, and brain-state-in-a-box, to name just a few.

Application of Neural Networks

Neural network applications abound in almost all business disciplines as well as in virtually all other functional areas. Business applications of neural networks included finance, firm failure prediction, time series forecasting, and so on. New applications of neural networks are emerging in health care, security, and so on. Neural networks are best at identifying patterns or trends in data; they are well suited for prediction or forecasting needs including:

- Sales forecasting
- Industrial process control
- Customer research
- Data validation
- Risk management
- Target marketing
- Pattern Recognition (reading zip codes)
- Signal Filtering (reduction of radio noise)
- Data Segmentation (detection of seismic onsets)
- Data Compression (TV image transmission)
- Database Mining (marketing, finance analysis)
- Adaptive Control (vehicle guidance)
- Time Series Prediction (forecast the short and long time evolution)
- Speech Generation (training for pronunciation and writing English text)
- Speech Recognition (speech convention into written text by Markon model using some symbols)
- Autonomous Vehicle Navigation (vision based and robot guidance method)
- Handwriting Recognition(hidden layer are used for reducing the free parameter and develop the writing skills)
- Robotics(Finger print system, Bio-Metrics, Robots which behave just like human)

Guidelines for using neural networks

1. Try the best existing method first
2. Get a big training set
3. Try a network without hidden units

4. Use a sensible coding for input variables
5. Consider methods of constraining network
6. Use a test set to prevent over-training
7. Determine confidence in generalization through cross-validation

Advantages and Limitation of Neural Networks

The following are the advantages of neural networks:-

- Several NN models available to choose in particular problem.
- They are very fast.
- Increase accuracy and result in cost saving.
- Represent any function and are called as "universal approximation".
- NN are able to learn and represent example by back propagation.

The following are the limitations of neural networks:-

- Low learning rate: - problem require large but complex network.
- Forgot full: - forgot old data and training new ones.
- Imprecision: - not provide precise numerical answer.
- Black box approach: - we cannot see physical part of training the data transfer.
- Limited flexibility: - implemented only one system available.
- Sample size has to be large.
- Requires lot of trial and error so training can be time consuming.

Neural Network in SPSS

_SPSS Neural Networks provides a complementary approach to the data analysis techniques available in SPSS Statistics Base and its modules. From the familiar SPSS Statistics interface, you can "mine" your data for hidden relationships, using either the Multilayer Perceptron (MLP) or Radial Basis Function (RBF) procedure.

Both of these are supervised learning techniques – that is, they map relationships implied by the data. Both use feed-forward architectures, meaning that data moves in only one direction, from the input nodes through the hidden layer or layers of nodes to the output nodes.

Your choice of procedure will be influenced by the type of data you have and the level of complexity you seek to uncover. While the MLP procedure can find more complex relationships, the RBF procedure is generally faster.

With either of these approaches, the procedure operates on a training set of data and then applies that knowledge to the entire dataset, and to any new data.

Control the process from start to finish

After selecting a procedure, you specify the dependent variables, which may be scale, categorical or a combination of the two. You adjust the procedure by choosing how to partition the dataset, what sort of architecture you want and what computation resources will be applied to the analysis.

Finally, you choose whether you want to display results in tables or graphs, save optional temporary variables to the active dataset and/or export models in XML-based file format to score future data.

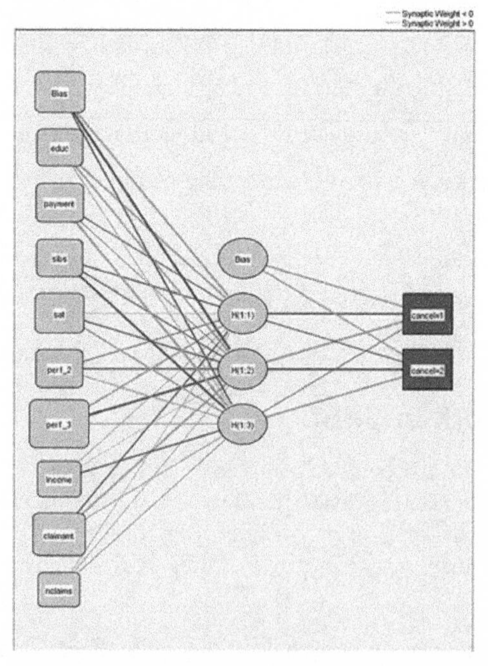

In an MLP procedure like the one shown here, nodes in the input and output layers are connected to nodes in one or more hidden layers.

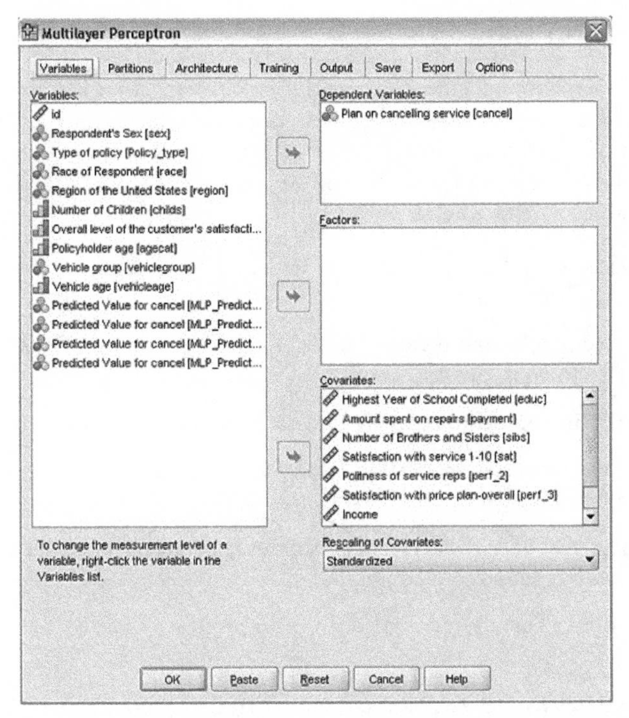

Just as you do when using SPSS Statistics Base or other modules, from the dialog boxes in SPSS Neural Networks, you select the variables that you want to include in your model.

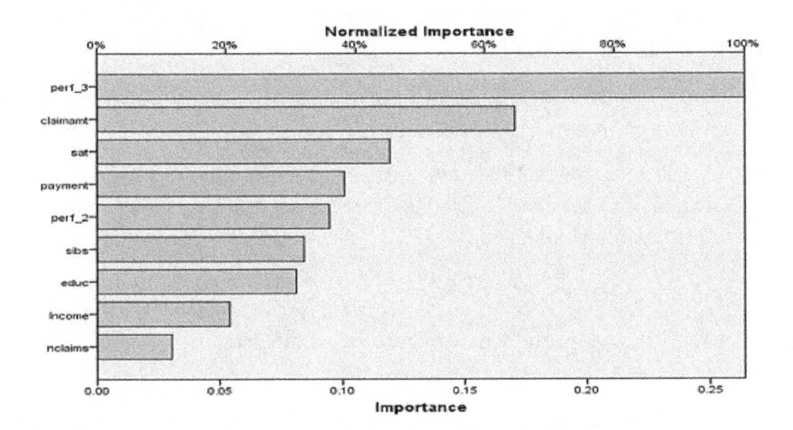

The results of exploring data with SPSS Neural Networks can be shown in a variety of graphic formats. This simple bar chart is one of many options.

Features and Benefits

SPSS Statistics Base is statistical analysis software that delivers the core capabilities you need to take the analytical process from start to finish. It is easy to use and includes a broad range of procedures and techniques to help you increase revenue, outperform competitors, conduct research and make better decisions.

SPSS Statistics Base provides essential statistical analysis tools for every step of the analytical process.

- **A comprehensive range of statistical procedures** for conducting accurate analysis.
- **Built-in techniques** to prepare data for analysis quickly and easily.
- **Sophisticated reporting functionality** for highly effective chart creation.
- **Powerful visualization capabilities** that clearly show the significance of your findings.
- **Support for all types of data** including very large data sets.

A comprehensive range of statistical procedures

- Carry out a wide range of descriptive procedures including cross tabulations, frequencies, compare means and correlation.
- Predict numerical outcomes and identify groups using factor analysis, cluster analysis, linear regression, ordinal regression, discriminant analysis and Nearest Neighbor analysis.
- Apply Monte Carlo simulation techniques to build better models and assess risk when inputs are uncertain.
- Use SPSS Statistics Base with other modules, such as SPSS Regression and SPSS Advanced Statistics, to more accurately identify and analyze complex relationships.

Built-in techniques

- Identify and eliminate duplicate cases and restructure your data files prior to analysis.
- Set up data dictionary information (for example, value labels and variable types) and use it as a template to prepare all of your data for analysis faster.

- Open multiple data sets within a single session to save time and condense steps.

Sophisticated reporting functionality

- Create commonly used charts, such as scatterplot matrices, histograms and population pyramids, more easily.
- Drag and drop variables and elements onto a chart creation canvas and preview the chart as it is being built.
- Build a chart once, and then use those specifications to create hundreds more just like it.

Powerful visualization capabilities

- Distribute and manipulate information for ad hoc decision-making using report online analytical processing (OLAP) technology.
- Create high-end charts and graphs to aid analysis and reporting and identify new insights in your data.
- Use pre-built map templates to generate a geographic or demographic analysis that can provide critical information for decision making.
- Quickly change information and statistics in graphs for new levels of understanding, and convert a table to a graph with just a few mouse clicks.

Support for all types of data

- Access, manage and analyze any kind of data set including survey data, corporate databases, data downloaded from the web and Cognos Business Intelligence data.
- Eliminate variability in data due to language-specific encoding and view, analyze and share data written in multiple languages using built-in Unicode support.

Neural Network

MULTILAYER PERCEPTRON

Key terms

Before proceeding with the analysis part, review the key terms to develop an understanding of the concepts and terminologies used and associated with the chapter.

MULTILAYER PERCEPTRON

The Multilayer Perceptron (MLP) method produces an analytical model for one or more dependent (target) variables based on the values of the predictor variables.

VARIABLE TAB OF MULTILAYER PERCEPTRON DIALOG BOX

- The dependent variables may be nominal, ordinal or scale.
- The predictor variables can be specified as factors i.e. categorical or covariates i.e. scale.
- The Variables tab optionally provides the method for rescaling covariates. The choices are:
    - **Standardized**: Subtract the mean and divide by the standard deviation, (x−mean)/s.
    - **Normalized**: Subtract the minimum and divide by the range, (x−min)/(max−min). Normalized values fall between 0 and 1.
    - **Adjusted Normalized**: Adjusted version of subtracting the minimum and dividing by the range, [2*(x−min)/ (max−min)] −1. Adjusted normalized values fall between −1 and 1.
    - **None:** No rescaling of covariates.

PARTITION TAB OF MULTILAYER PERCEPTRON DIALOG BOX

Partition Dataset: This group specifies the method of partitioning the active dataset into training, testing, and holdout samples. The training sample comprises the data records used to train the neural network; some percentage of cases in the dataset must be assigned to the training sample in order to obtain a model. The testing sample is an independent set of data records used to track errors during training in order to prevent overtraining. It is highly recommended to create a training sample, and network training will generally be most efficient if the testing sample is smaller than the training sample. The holdout sample is another independent set of data records used to assess the final neural network; the error for the holdout sample gives an "honest" estimate of the predictive ability of the model because the holdout cases were not used to build the model.

- ***Randomly assign cases based on relative number of cases***: Specify the relative number (ratio) of cases randomly assigned to each sample (training, testing, and holdout). The % column reports the percentage of cases that will be assigned to each sample based on the relative numbers which have to be specified. For example, specifying 7, 3, 0 as the relative numbers for training, testing, and holdout samples corresponds to 70%, 30%, and 0%. Specifying 2, 1, 1 as the relative numbers corresponds to 50%, 25%, and 25%; 1, 1, 1 corresponds to dividing the dataset into equal thirds among training, testing, and holdout.
- ***Use partitioning variable to assign cases:*** Specify a numeric variable that assigns each case in the active dataset to the training, testing, or holdout sample. Cases with a positive value on the variable are assigned to the training sample, cases with a value of 0, to the testing sample, and cases with a negative value, to the holdout sample. Cases with a system-missing value are excluded from the analysis. Any user-missing values for the partition variable are always treated as valid.

Note: Using a partitioning variable will not guarantee identical results in successive runs of the procedure.

ARCHITECTURE TAB OF MULTILAYER PERCEPTRON DIALOG BOX

The Architecture tab is used to specify the structure of the network. The procedure can select the "best" architecture automatically, or one can specify a custom architecture. Automatic architecture selection builds a network with one hidden layer. Specify the minimum and maximum number of units allowed in the hidden layer, and the automatic architecture selection computes the "best" number of units in the hidden layer. Automatic architecture selection uses the default activation functions for the hidden and output layers.

Custom architecture selection gives expert control over the hidden and output layers and can be most useful in advance to tweak the results of the Automatic architecture selection.

Hidden Layers contains unobservable network nodes (units). Each hidden unit is a function of the weighted sum of the inputs. The function is the activation function, and the values of the weights are determined by the estimation algorithm. If the network contains a second hidden layer, each hidden unit in the second layer is a function of the weighted sum of the units in the first hidden layer. The same activation function is used in both layers.

Number of Hidden Layers: A multilayer perceptron can have one or two hidden layers. Activation Function. The activation function "links" the weighted sums of units in a layer to the values of units in the succeeding layer.

- Hyperbolic tangent takes real-valued arguments and transforms them to the range (–1, 1). When automatic architecture selection is used, this is the activation function for all units in the hidden layers.
- Sigmoid takes real-valued arguments and transforms them to the range (0, 1).

The number of units in each hidden layer can be specified explicitly or determined automatically by the estimation algorithm.

Output Layer contains the target (dependent) variables.

Activation Function: The activation function "links" the weighted sums of units in a layer to the values of units in the succeeding layer.

- **Identity** takes real-valued arguments and returns them unchanged. When automatic architecture selection is used, this is the activation function for units in the output layer if there are any scale-dependent variables.
- **Softmax** takes a vector of real-valued arguments and transforms it to a vector whose elements fall in the range (0, 1) and sum to 1. Softmax is available only if all dependent variables are categorical. When automatic architecture selection is used, this is the activation function for units in the output layer if all dependent variables are categorical.
- **Hyperbolic tangent** takes real-valued arguments and transforms them to the range (–1, 1).

- **Sigmoid** takes real-valued arguments and transforms them to the range (0, 1).

Rescaling of Scale Dependent Variables

These controls are available only if at least one scale-dependent variable has been selected.

- **Standardized:** Subtract the mean and divide by the standard deviation, (x−mean)/s.
- **Normalized:** Subtract the minimum and divide by the range, (x−min)/(max−min). Normalized values fall between 0 and 1. This is the required rescaling method for scale-dependent variables if the output layer uses the sigmoid activation function. The correction option specifies a small number ε that is applied as a correction to the rescaling formula; this correction ensures that all rescaled dependent variable values will be within the range of the activation function. In particular, the values 0 and 1, which occur in the uncorrected formula when x takes its minimum and maximum value, define the limits of the range of the sigmoid function but are not within that range. The corrected formula is [x−(min−ε)]/[(max+ε)−(min−ε)]. Specify a number greater than or equal to 0.
- **Adjusted Normalized:** Adjusted version of subtracting the minimum and dividing by the range, [2*(x−min)/(max−min)]−1. Adjusted normalized values fall between −1 and 1. This is the required rescaling method for scale-dependent variables if the output layer uses the hyperbolic tangent activation function. The correction option specifies a small number ε that is applied as a correction to the rescaling formula; this correction ensures that all rescaled dependent variable values will be within the range of the activation function. In particular, the values −1 and 1, which occur in the uncorrected formula when x takes its minimum and maximum value, define the limits of the range of the hyperbolic tangent function but are not within that range. The corrected formula is {2*[(x−(min−ε))/((max+ε)−(min−ε))]}−1. Specify a number greater than or equal to 0.
- **None**: No rescaling of scale-dependent variables.

TRAINING TAB OF MULTILAYER PERCEPTRON DIALOG BOX

The Training tab is used to specify how the network should be trained. The type of training and the optimization algorithm determine which training options are available.

Type of Training: The training type determines how the network processes the records. Select one of the following training types:

- **Batch:** Updates the synaptic weights only after passing all training data records; that is, batch training uses information from all records in the training dataset. Batch training is often preferred because it directly minimizes the total error; however, batch training may need to update the weights many times until one of the stopping rules is met and hence may need many data passes. It is most useful for "smaller" datasets.
- **Online:** Updates the synaptic weights after every single training data record; that is, online training uses information from one record at a time. Online training continuously gets a record and updates the weights until one of the stopping rules is met. If all the records are used once and none of the stopping rules is met, then the process continues by recycling the data records. Online training is superior to batch for "larger" datasets with associated predictors; that is, if Multilayer Perceptron there are many records and many inputs, and their values are not independent of each other, then online training can more quickly obtain a reasonable answer than batch training.
- **Mini-batch:** Divides the training data records into groups of approximately equal size, then updates the synaptic weights after passing one group; that is, mini-batch training uses information from a group of records. Then the process recycles the data group if necessary. Mini-batch training offers a compromise between batch and online training, and it may be best for "medium-size" datasets. The procedure can automatically determine the number of training records per mini-batch, or one has to specify an integer greater than 1 and less than or equal to the maximum number of cases to store in memory. The user can set the maximum number of cases to store in memory on the Options tab.

Optimization Algorithm: This is the method used to estimate the synaptic weights.

- **Scaled conjugate gradient:** The assumptions that justify the use of conjugate gradient methods apply only to batch training types, so this method is not available for online or mini-batch training.
- **Gradient descent:** This method must be used with online or mini-batch training; it can also be used with batch training.

Training Options: The training options allow fine-tuning the optimization algorithm. In general, there is no need to change these settings unless the network runs into problems with estimation. Training options for the scaled conjugate gradient algorithm include:

- **Initial Lambda:** The initial value of the lambda parameter for the scaled conjugate gradient algorithm. Specify a number greater than 0 and less than 0.000001.
- **Initial Sigma**: The initial value of the sigma parameter for the scaled conjugate gradient algorithm. Specify a number greater than 0 and less than 0.0001.
- **Interval Center and Interval Offset:** The interval center (a0) and interval offset (a) define the interval [a0−a, a0+a], in which weight vectors are randomly generated when simulated annealing is used. Simulated annealing is used to break out of a local minimum, with the goal of finding the global minimum, during application of the optimization algorithm. This approach is used in weight initialization and automatic architecture selection. Specify a number for the interval center and a number greater than 0 for the interval offset.

Training options for the gradient descent algorithm include:

- **Initial Learning Rate:** The initial value of the learning rate for the gradient descent algorithm. A higher learning rate means that the network will train faster, possibly at the cost of becoming unstable. Specify a number greater than 0.
- **Lower Boundary of Learning Rate**: The lower boundary on the learning rate for the gradient descent algorithm. This setting applies only to online and mini-batch training. Specify a number greater than 0 and less than the initial learning rate.
- **Momentum:** The initial momentum parameter for the gradient descent algorithm. The momentum term helps to prevent instabilities caused by a too-high learning rate. Specify a number greater than 0.
- **Learning rate reduction, in Epochs:** The number of epochs (p), or data passes of the training sample, required to reduce the initial learning rate to the lower boundary of the learning rate when gradient

descent is used with online or mini-batch training. This gives control of the learning rate decay factor $\beta = (1/pK)*\ln(\eta0/\eta low)$, where $\eta 0$ is the initial learning rate, ηlow is the lower bound on the learning rate, and K is the total number of mini-batches (or the number of training records for online training) in the training dataset. Specify an integer greater than 0.

OUTPUT TAB OF MULTILAYER PERCEPTRON DIALOG BOX

Network Structure: Displays summary information about the neural network.

- **Description**. Displays information about the neural network, including the dependent variables, number of input and output units, number of hidden layers and units, and activation functions.
- **Diagram.** Displays the network diagram as a non-editable chart. Note that as the number of covariates and factor levels increases, the diagram becomes more difficult to interpret.
- **Synaptic weights.** Displays the coefficient estimates that show the relationship between the units in a given layer to the units in the following layer. The synaptic weights are based on the training sample even if the active dataset is partitioned into training, testing, and holdout data.

Note that the number of synaptic weights can become rather large and that these weights are generally not used for interpreting network results.

Network Performance: Displays results used to determine whether the model is "good". Note: Charts in this group are based on the combined training and testing samples or only on the training sample if there is no testing sample.

Model summary: Displays a summary of the neural network results by partition and overall, including the error, the relative error or percentage of incorrect predictions, the stopping rule used to stop training, and the training time.

The error is the sum-of-squares error when the identity, sigmoid, or hyperbolic tangent activation function is applied to the output layer. It is the cross-entropy error when the softmax activation function is applied to the output layer.

Relative errors or percentages of incorrect predictions are displayed depending on the dependent variable measurement levels. If any dependent variable has scale measurement level, then the average overall relative error (relative to the mean model) is displayed. If all dependent variables are categorical, then the average percentage of incorrect predictions is displayed. Relative errors or percentages of incorrect predictions are also displayed for individual dependent variables.

- **Classification results:** Displays a classification table for each categorical dependent variable by partition and overall. Each table gives the number of cases classified correctly and incorrectly for each dependent variable category. The percentage of the total cases that were correctly classified is also reported.
- **ROC curve:** Displays an ROC (Receiver Operating Characteristic) curve for each categorical dependent variable. It also displays a table giving the area under each curve. For a given dependent variable, the ROC chart displays one curve for each category. If the dependent variable has two categories, then each curve treats the category at issue as the positive state versus the other category. If the dependent variable has more than two categories, then each curve treats the category at issue as the positive state versus the aggregate of all other categories.
- **Cumulative gains chart:** Displays a cumulative gains chart for each categorical dependent variable. The display of one curve for each dependent variable category is the same as for ROC curves.
- **Lift chart**: Displays a lift chart for each categorical dependent variable. The display of one curve for each dependent variable category is the same as for ROC curves.
- **Predicted by observed chart**: Displays a predicted-by-observed-value chart for each dependent variable. For categorical dependent variables, clustered box plots of predicted pseudo-probabilities are displayed for each response category, with the observed response category as the cluster variable. For scale-dependent variables, a scatter plot is displayed.
- **Residual by predicted chart**: Displays a residual-by-predicted-value chart for each scale-dependent variable. There should be no visible patterns between residuals and predicted values. This chart is produced only for scale-dependent variables.

Case processing summary: Displays the case processing summary table, which summarizes the number of cases included and excluded in the analysis, in total and by training, testing, and holdout samples.

Independent variable importance analysis: Performs a sensitivity analysis, which computes the importance of each predictor in determining the neural network. The analysis is based on the combined training and testing samples or only on the training sample if there is no testing sample. This creates a table and a chart displaying importance and normalized importance for each predictor. Note that sensitivity analysis is computationally expensive and time-consuming if there are large numbers of predictors or cases.

SAVE TAB OF MULTILAYER PERCEPTRON DIALOG BOX

The Save tab is used to save predictions as variables in the dataset.

- **Save predicted value or category for each dependent variable.** This saves the predicted value for scale-dependent variables and the predicted category for categorical dependent variables.
- **Save predicted pseudo-probability or category for each dependent variable.** This saves the predicted pseudo-probabilities for categorical dependent variables. A separate variable is saved for each of the first n categories, where n is specified in the Categories to Save column.

Names of Saved Variables: Automatic name generation ensures that to keep all the work done. Custom names allow to discard/replace results from previous runs without first deleting the saved variables in the Data Editor.

Probabilities and Pseudo-Probabilities

Categorical dependent variables with softmax activation and cross-entropy error will have a predicted value for each category, where each predicted value is the probability that the case belongs to the category.

Categorical dependent variables with sum-of-squares error will have a predicted value for each category, but the predicted values cannot be interpreted as probabilities. The procedure saves these predicted pseudo-probabilities even if any are less than 0 or greater than 1, or the sum for a given dependent variable is not 1.

The ROC, cumulative gains, and lift charts (see Output on p. 14) are created based on pseudo-probabilities. In the event that any of the pseudo-probabilities are less than 0 or greater than 1, or the sum for a given variable is not 1, they are first rescaled to be between 0 and 1 and to sum to 1. Pseudo-probabilities are rescaled by dividing by their sum. For example, if a case has predicted pseudo-probabilities of 0.50, 0.60, and 0.40 for a three-category dependent variable, then each pseudo-probability is divided by the sum 1.50 to get 0.33, 0.40, and 0.27. If any of the pseudo-probabilities are negative, then the absolute value of the lowest is added to all pseudo-probabilities before the above rescaling. For example, if the pseudo-probabilities are -0.30, 0.50, and 1.30, then first add 0.30 to each value to get 0.00, 0.80, and 1.60. Next, divide each new value by the sum 2.40 to get 0.00, 0.33, and 0.67.

EXPORT TAB OF MULTILAYER PERCEPTRON DIALOG BOX

The Export tab is used to save the synaptic weight estimates for each dependent variable to an XML (PMML) file. This model file is used to apply the model information to other data files for scoring purposes. This option is not available if split files have been defined.

OPTION TAB OF MULTILAYER PERCEPTRON DIALOG BOX

User-Missing Values: Factors must have valid values for a case to be included in the analysis. These controls allow deciding whether user-missing values are treated as valid among factors and categorical dependent variables.

Stopping Rules: These are the rules that determine when to stop training the neural network. Training proceeds through at least one data pass. Training can then be stopped according to the following criteria, which are checked in the listed order. In the stopping rule definitions that follow, a step corresponds to a data pass for the online and mini-batch methods and an iteration for the batch method.

- **Maximum steps without a decrease in error:** The number of steps to allow before checking for a decrease in error. If there is no decrease in error after the specified number of steps, then training stops. Specify an integer greater than 0. It also specify which data sample is used to compute the error. Choose automatically uses the testing sample if it exists and uses the training sample otherwise. Note that batch training guarantees a decrease in the training sample error after each data pass; thus, this option applies only to batch training if a testing sample exists. Both training and test data checks the error for each of these samples; this option applies only if a testing sample exits.

Note: After each complete data pass, online and mini-batch training require an extra data pass in order to compute the training error. This extra data pass can slow training considerably, so it is generally recommended that supply a testing sample and select Choose automatically in any case.

- **Maximum training time.** Choose whether to specify a maximum number of minutes for the algorithm to run. Specify a number greater than 0.
- **Maximum Training Epochs.** The maximum number of epochs (data passes) allowed. If the maximum number of epochs is exceeded, then training stops. Specify an integer greater than 0.
- **Minimum relative change in training error.** Training stops if the relative change in the training error compared to the previous step is less than the criterion value. Specify a number greater than 0. For online and mini-batch training, this criterion is ignored if only testing data is used to compute the error.
- **Minimum relative change in training error ratio.** Training stops if the ratio of the training error to the error of the null model is less than the criterion value. The null model predicts the average value for all dependent variables. Specify a number greater than 0. For online and mini-batch training, this criterion is ignored if only testing data is used to compute the error.

Maximum cases to store in memory: This controls the following settings within the multilayer perceptron algorithms. Specify an integer greater than 1.

- In automatic architecture selection, the size of the sample used to determine the network architecture is min (1000, memsize), where memsize is the maximum number of cases to store in memory.
- In mini-batch training with automatic computation of the number of mini-batches, the number of mini-batches is min(max(M/10,2),memsize), where M is the number of cases in the training sample.

SPSS

Test Procedure in SPSS

In this example, A supermarket studied the importance of store attributes such as Personalised services, Customised services, pricing policy, product quality and store environment based on consumer preference.

Following are the steps to analyse data using neural network –Multilayer Perceptron in SPSS.

Step - 1: Click Analyze > Neural Network > Multilayer Perceptron as shown in the figure -1, the Multilayer perceptron dialogue box will appear (as given in the figure - 2).

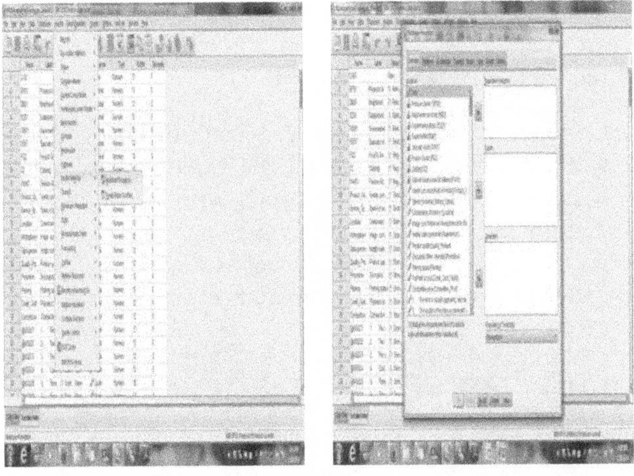

Figure - 1 **Figure - 2**

Step - 2: From the Multilayer Perceptron (MLP) dialog, select the variables to include in the model. Transfer the dependent variable i.e. consumer preference into dependent variables box and transfer the independent variables such as Store_Environment, Personalised_ Services, Customised_Services, Pricing_Policy and Product_Quality in covariates box and select standardized in Rescale of covariates box by drag-and-dropping the variables into their respective boxes or by using the SPSS Right Arrow Button. The result is shown below in figure -3.

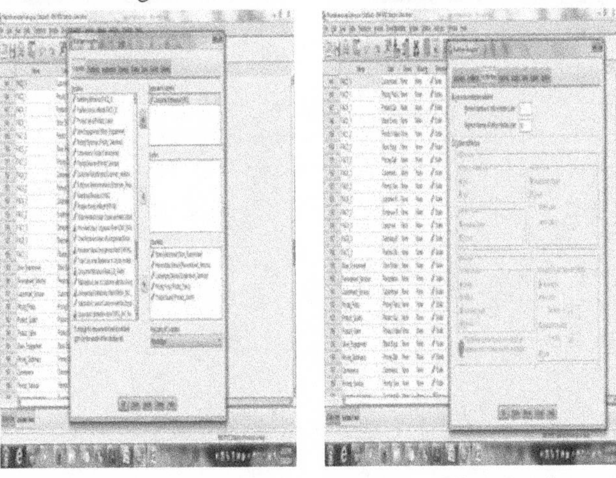

Figure -3 **Figure -4**

Step - 3: Click partition tab in multilayer perceptron dialog, in automatic architecture selection, replace 2 in maximum number of units in hidden layer as shown in figure 4.

Step - 4: Go to output tab in multilayer perceptron dialog, the default image given in figure -6. Select all the three in network structure i.e. descriptive, diagram and synaptic weight followed by in network performance select model summary, predicted observed chart, residual by predicted chart. As well as select case processing summary and independent variable importance analysis as given in figure -7.

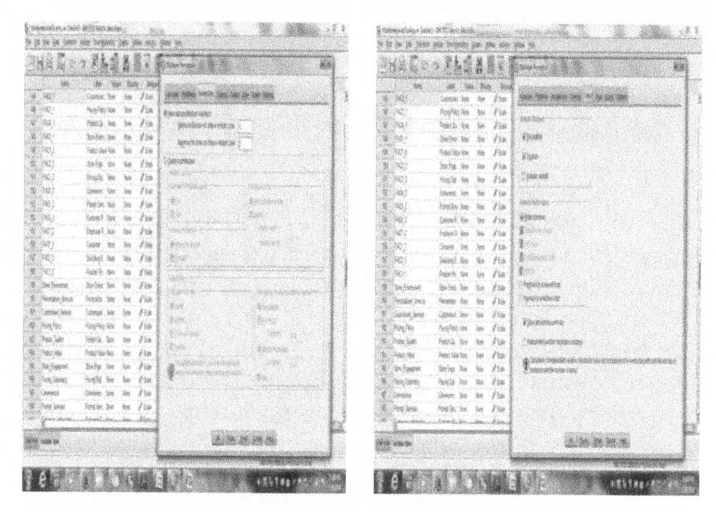

Figure - 5 **Figure - 6**

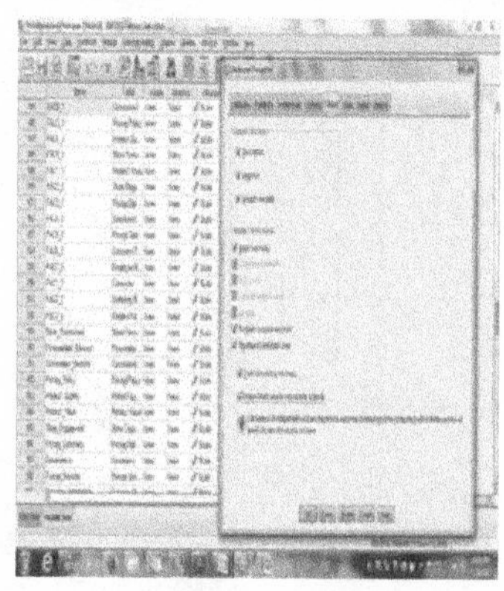

Figure - 7

Step - 5: Click the OK button to generate the output.

SPSS Output of the Neural Network – Multilayer Perceptron

Table 1 Case Processing Summary

		N	Percent
Sample	Training	381	73.8%
	Testing	135	26.2%
Valid		516	100.0%
Excluded		0	
Total		516	

Table 2 Network Information

Input Layer	Covariates	1	Store Environment
		2	Personalized Service
		3	Customized Service
		4	Pricing Policy
		5	Product Quality
	Number of Units[a]		5
	Rescaling Method for Covariates		Standardized
Hidden Layer(s)	Number of Hidden Layers		1
	Number of Units in Hidden Layer 1[a]		2
	Activation Function		Hyperbolic tangent
Output Layer	Dependent Variables	1	Consumer Preference
	Number of Units		1
	Rescaling Method for Scale Dependents		Standardized
	Activation Function		Identity
	Error Function		Sum of Squares

a. Excluding the bias unit

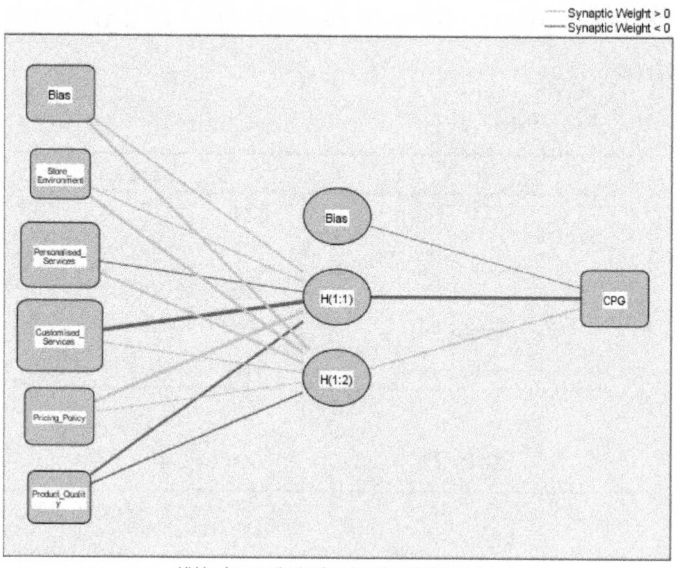

Hidden layer activation function: Hyperbolic tangent

Output layer activation function: Identity

Figure - 8

Table 3 Model Summary

Training	Sum of Squares Error	172.456
	Relative Error	.908
	Stopping Rule Used	1 consecutive step(s) with no decrease in error[a]
	Training Time	0:00:00.07
Testing	Sum of Squares Error	52.163
	Relative Error	.861

Dependent Variable: Consumer Preference

a. Error computations are based on the testing sample.

Table 4 Parameter Estimates

Predictor		Predicted		
		Hidden Layer 1		Output Layer
		H(1:1)	H(1:2)	CPG
Input Layer	(Bias)	.387	.447	
	Store_Environment	.272	.451	
	Personalised_ Services	-.115	.397	
	Customised_Services	-.586	.146	
	Pricing_Policy	.480	.074	
	Product_Quality	-.391	-.243	
Hidden Layer 1	(Bias)			-.003
	H(1:1)			-.412
	H(1:2)			.346

Table 5 Independent Variable Importance

	Importance	Normalized Importance
Store Environment	.091	27.0%
Personalized Service	.270	80.3%
Customized Service	.336	100.0%
Pricing Policy	.181	53.7%
Product Quality	.122	36.2%

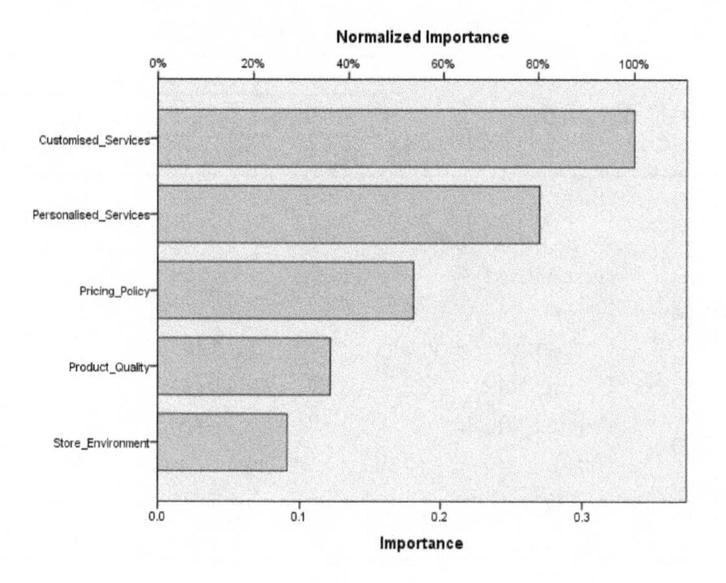

Figure - 9

Results and Discussion

Case Processing Summary

The case processing summary shows that 381 cases were assigned to the training sample, 135 to the testing sample. No cases were excluded from the analysis.

Network Information

Table -2 displays the network information. the input layer are the covariates such as store environment, personalized services, customized services, pricing policy and product quality, the number of units excluding the bias unit is 5. In rescale methods for covariates are standardized. The number of hidden layers is 1 and the number or units in the hidden layer is 2. The activation function is based on hyperbolic tangent with dependent variable - consumer preference. The output layer is with 1 unit based on standardized method of rescaling scale dependents. Activation function is identity and error function is sum of squares.

Hidden Layer Activation Function: Hyperbolic Tangent

Output Layer Activation Function: Identity

Model Summary

Table 3 displays information about the results of training and applying the MLP network to the holdout sample. Sum-of-squares error is displayed because the output layer has scale dependent variables. This is the error function that the network tries to minimize during training. One consecutive step with no decrease in error was used as stopping rule. The relative error for each scale-dependent variable is the ratio of the sum-of-squares error for the dependent variable to the sum-of-squares error for the "null" model, in which the mean value of the dependent variable is used as the predicted value for each case. There appears to be more error in the predictions of consumer preference in organised retail in the training and holdout samples.

The average overall relative errors are fairly constant across the training (0.908),and testing (0.861) samples, which give us some confidence that the model is not over trained and that the error in future cases, scored by the network will be close to the error reported in this table.

Parameter Estimates

Independent Variable Importance

Table -5 displays the independent variable importance based on covariates with the dependent variable. In this example, the consumer preference is measured based on store attributes, in which the most dominating variable is customized services (100%), followed by personalized services (80.3%), pricing policy (53.7%), product quality (36.2%) and the last influencing variable is store environment (27%).

Normalized Importance

Figure – 9 displays the normalized importance chart of the variable based on independent variable importance.

RADICAL BASIS FUNCTION

The Radial Basis Function (RBF) procedure produces a predictive model for one or more dependent (target) variables based on values of predictor variables.

VARIABLE TAB OF RADICAL BASIS FUNCTION DIALOG BOX

- The dependent variables can be nominal, ordinal or scale
- The Variables tab optionally provides to change the method for rescaling covariates. The choices are:

- **Standardized:** Subtract the mean and divide by the standard deviation, (x−mean)/s.
- **Normalized**: Subtract the minimum and divide by the range, (x−min)/(max−min). Normalized values fall between 0 and 1.
- **Adjusted Normalized**: Adjusted version of subtracting the minimum and dividing by the range, [2*(x−min)/(max−min)]−1. Adjusted normalized values fall between −1 and 1.
- **None**: No rescaling of covariates.

PARTITION TAB OF RADICAL BASIS FUNCTION DIALOG BOX

Partition Dataset

This group specifies the method of partitioning the active dataset into training, testing, and holdout samples. The training sample comprises the data records used to train the neural network; some percentage of cases in the dataset must be assigned to the training sample in order to obtain a model. The testing sample is an independent set of data records used to track errors during training in order to prevent overtraining. It is highly recommended that create a training sample, and network training will generally be most efficient if the testing sample is smaller than the training sample. The holdout sample is another independent set of data records used to assess the final neural network; the error for the holdout sample gives an "honest" estimate of the predictive ability of the model because the holdout cases were not used to build the model.

- Randomly assign cases based on relative number of cases. Specify the relative number (ratio) of cases randomly assigned to each sample (training, testing, and holdout). The % column reports the percentage of cases that will be assigned to each sample based on the relative numbers specified. For example, specifying 7, 3, 0 as the relative numbers for training, testing, and holdout samples corresponds to 70%, 30%, and 0%. Specifying 2, 1, 1 as the relative numbers corresponds to 50%, 25%, and 25%; 1, 1, 1 corresponds to dividing the dataset into equal thirds among training, testing, and holdout.
- Use partitioning variable to assign cases. Specify a numeric variable that assigns each case in the active dataset to the training, testing, or holdout sample. Cases with a positive value on the variable are assigned to the training sample, cases with a value of 0, to the testing sample, and cases with a negative value, to the holdout sample. Cases with a system-missing value are excluded from the analysis. Any user-missing values for the partition variable are always treated as valid.

ARCHITECTURE TAB OF RADICAL BASIS FUNCTION DIALOG BOX

The Architecture tab is used to specify the structure of the network. The procedure creates a neural network with one hidden "radial basis function" layer; in general, it will not be necessary to change these settings.

Number of Units in Hidden Layer: There are three ways of choosing the number of hidden units.

1. Find the best number of units within an automatically computed range. The procedure automatically computes the minimum and maximum values of the range and finds the best number of hidden units within the range. If a testing sample is defined, then the procedure uses the testing data criterion: The best number of hidden units is the one that yields the smallest error in the testing data. If a testing sample is not defined, then the procedure uses the Bayesian information criterion (BIC): The best number of hidden units is the one that yields the smallest BIC based on the training data.
2. Find the best number of units within a specified range. It helps to give own range, and the procedure will find the "best" number of hidden units within that range. As before, the best number of hidden units from the range is determined using the testing data criterion or the BIC.
3. Use a specified number of units to override the use of a range and specify a particular number of units directly.

Activation Function for Hidden Layer: The activation function for the hidden layer is the radial basis function, which "links" the units in a layer to the values of units in the succeeding layer. For the output layer, the activation function is the identity function; thus, the output units are simply weighted sums of the hidden units.

- **Normalized radial basis function**: Uses the softmax activation function so the activations of all hidden units are normalized to sum to 1.
- **Ordinary radial basis function**: Uses the exponential activation function so the activation of the hidden unit is a Gaussian "bump" as a function of the inputs.

Overlap Among Hidden Units. The overlapping factor is a multiplier applied to the width of the radial basis functions. The automatically computed value of the overlapping factor is 1+0.1d, where d is the number of input units (the sum of the number of categories across all factors and the number of covariates).

OUTPUT TAB OF RADICAL BASIS FUNCTION DIALOG BOX

Network Structure: Displays summary information about the neural network.

- **Description**. Displays information about the neural network, including the dependent variables, number of input and output units, number of hidden layers and units, and activation functions.
- **Diagram**: Displays the network diagram as a non-editable chart. Note that as the number of covariates and factor levels increases, the diagram becomes more difficult to interpret.
- **Synaptic weights**: Displays the coefficient estimates that show the relationship between the units in a given layer to the units in the following layer. The synaptic weights are based on the training sample even if the active dataset is partitioned into training, testing, and holdout data. Note that the number of synaptic weights can become rather large, and these weights are generally not used for interpreting network results.

Network Performance: Displays results used to determine whether the model is "good." Note: Charts in this group are based on the combined training and testing samples or only the training sample if there is no testing sample.

- **Model summary:** Displays a summary of the neural network results by partition and overall, including the error, the relative error or percentage of incorrect predictions, and the training time. The error is the sum-of-squares error. In addition, relative errors or percentages of incorrect predictions are displayed, depending on the dependent variable measurement levels. If any dependent variable has scale measurement level, then the average overall relative error (relative to the mean model) is displayed. If all dependent variables are categorical, then the average percentage of incorrect predictions is displayed. Relative errors or percentages of incorrect predictions are also displayed for individual dependent variables.
- **Classification results:** Displays a classification table for each categorical dependent variable. Each table gives the number of cases classified correctly and incorrectly for each dependent variable category. The percentage of the total cases that were correctly classified is also reported.
- **ROC curve:** Displays an ROC (Receiver Operating Characteristic) curve for each categorical dependent variable. It also displays a table giving the area under each curve. For a given dependent variable, the

ROC chart displays one curve for each category. If the dependent variable has two categories, then each curve treats the category at issue as the positive state versus the other category. If the dependent variable has more than two categories, then each curve treats the category at issue as the positive state versus the aggregate of all other categories.

- **Cumulative gains chart**: Displays a cumulative gains chart for each categorical dependent variable. The display of one curve for each dependent variable category is the same as for ROC curves.
- **Lift chart:** Displays a lift chart for each categorical dependent variable. The display of one curve for each dependent variable category is the same as for ROC curves.
- **Predicted by observed chart:** Displays a predicted-by-observed-value chart for each dependent variable. For categorical dependent variables, clustered box plots of predicted pseudo-probabilities are displayed for each response category, with the observed response category as the cluster variable. For scale dependent variables, a scatter plot is displayed.
- **Residual by predicted chart:** Displays a residual-by-predicted-value chart for each scale dependent variable. There should be no visible patterns between residuals and predicted values. This chart is produced only for scale dependent variables. Case processing summary. Displays the case processing summary table, which summarizes the number of cases included and excluded in the analysis, in total and by training, testing, and holdout samples.

Independent variable importance analysis: Performs a sensitivity analysis, which computes the importance of each predictor in determining the neural network. The analysis is based on the combined training and testing samples or only the training sample if there is no testing sample. This creates a table and a chart displaying importance and normalized importance for each predictor. Note that sensitivity analysis is computationally expensive and time-consuming if there are large number of predictors or cases.

SAVE TAB OF RADICAL BASIS FUNCTION DIALOG BOX

The Save tab is used to save predictions as variables in the dataset.

- **Save predicted value or category for each dependent variable:** This saves the predicted value for scale dependent variables and the predicted category for categorical dependent variables.

- **Save predicted pseudo-probability for each dependent variable:** This saves the predicted pseudo-probabilities for categorical dependent variables. A separate variable is saved for each of the first n categories, where n is specified in the Categories to Save column.

Names of Saved Variables: Automatic name generation ensures that keep all the work done. Custom names allow to discard or replace results from previous runs without first deleting the saved variables in the Data Editor.

Probabilities and Pseudo-Probabilities: Predicted pseudo-probabilities cannot be interpreted as probabilities because the Radial Basis Function procedure uses the sum-of-squares error and identity activation function for the output layer. The procedure saves these predicted pseudo-probabilities even if any are less than 0 or greater than 1 or the sum for a given dependent variable is not 1.

The ROC, cumulative gains, and lift charts (see Output on p. 29) are created based on pseudo-probabilities. In the event that any of the pseudo-probabilities are less than 0 or greater than 1 or the sum for a given variable is not 1, they are first rescaled to be between 0 and 1 and to sum to 1. Pseudo-probabilities are rescaled by dividing by their sum. For example, if a case has predicted pseudo-probabilities of 0.50, 0.60, and 0.40 for a three-category dependent variable, then each pseudo-probability is divided by the sum 1.50 to get 0.33, 0.40, and 0.27. If any of the pseudo-probabilities are negative, then the absolute value of the lowest is added to all pseudo-probabilities before the above rescaling. For example, if the pseudo-probabilities are –0.30, .50, and 1.30, then first add 0.30 to each value to get 0.00, 0.80, and 1.60. Next, divide each new value by the sum 2.40 to get 0.00, 0.33, and 0.67.

EXPORT TAB OF RADICAL BASIS FUNCTION DIALOG BOX

The Export tab is used to save the synaptic weight estimates for each dependent variable to an XML (PMML) file. This model file can be used to apply the model information to other data files for scoring purposes. This option is not available if split files have been defined.

EXPORT TAB OF RADICAL BASIS FUNCTION DIALOG BOX

User-Missing Values: Factors must have valid values for a case to be included in the analysis. These controls allow to decide whether user-missing values are treated as valid among factors and categorical dependent variables.

SPSS

Test Procedure in SPSS

Following are the steps to analyse data using neural network – Radical basis function in SPSS.

Step - 1: Click Analyze > Neural Network > Radical basis function as shown in the figure -1, the Radical basis function dialogue box will appear (as given in the figure - 2).

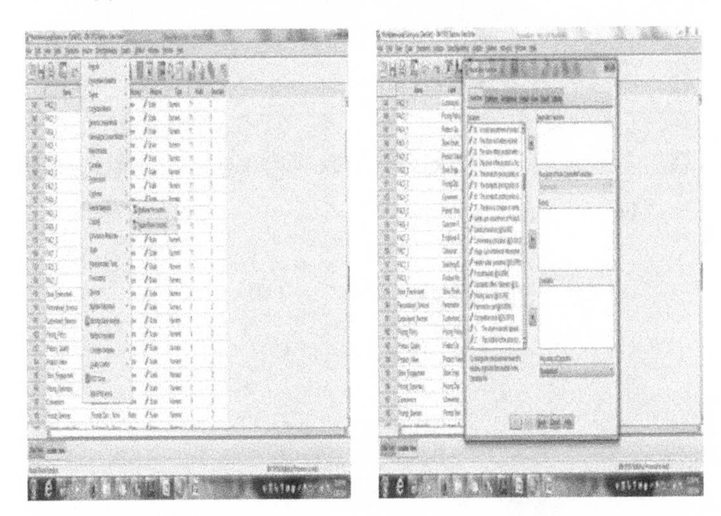

Figure - 1 **Figure - 2**

Step - 2 Go to output tab in Radical basis function (RBF), the default image given in figure -4. Select all the three in network structure i.e. descriptive, diagram and synaptic weight followed by in network performance select model summary, predicted observed chart, residual by predicted chart. As well as select case processing summary and independent variable importance analysis as given in figure -5.

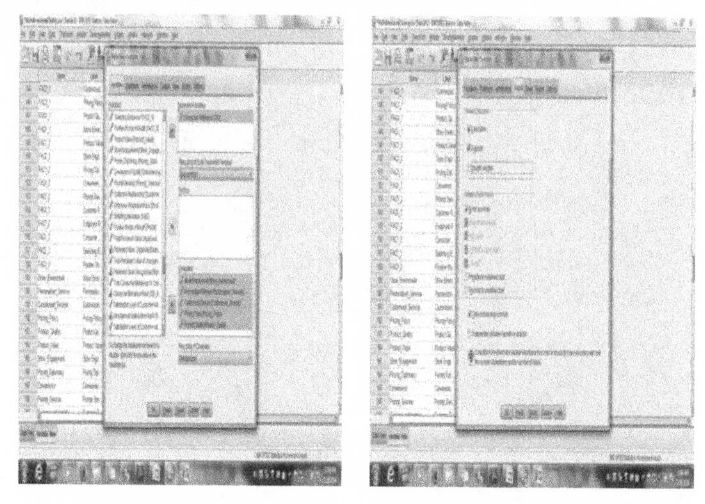

Figure - 3 **Figure - 4**

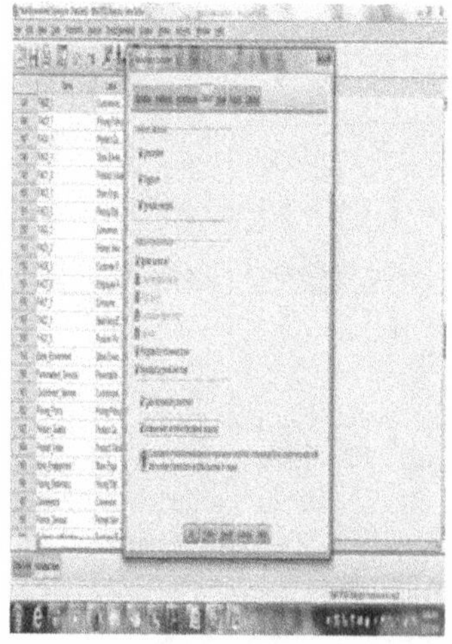

Figure - 5

Step - 3: From the Radical basis function (RBF) dialog, select the variables to include in the model. Transfer the dependent variable i.e. consumer preference into dependent variables box followed by select standardized in rescaling of scale dependent variables. Transfer the independent variables such as Store_Environment, Personalised_Services, Customised_Services, Pricing_Policy and Product_Quality in covariates box and select standardized in

Rescale of covariates box by drag-and-dropping the variables into their respective boxes or by using the SPSS Right Arrow Button. The result is shown below in figure -3.

Step - 4: Click the OK button to generate the output.

SPSS Output of the Neural Network – Radical Basis Function

Table 1 Case Processing Summary

		N	Percent
Sample	Training	359	69.6%
	Testing	157	30.4%
Valid		516	100.0%
Excluded		0	
Total		516	

Table 2 Network Information

Input Layer	Covariates	1	Store Environment
		2	Personalized Service
		3	Customized Service
		4	Pricing Policy
		5	Product Quality
	Number of Units		5
	Rescaling Method for Covariates		Standardized
Hidden Layer	Number of Units		6[a]
	Activation Function		Softmax
Output Layer	Dependent Variables	1	Consumer Preference
	Number of Units		1
	Rescaling Method for Scale Dependents		Standardized
	Activation Function		Identity
	Error Function		Sum of Squares

a. Determined by the testing data criterion: The "best" number of hidden units is the one that yields the smallest error in the testing data.

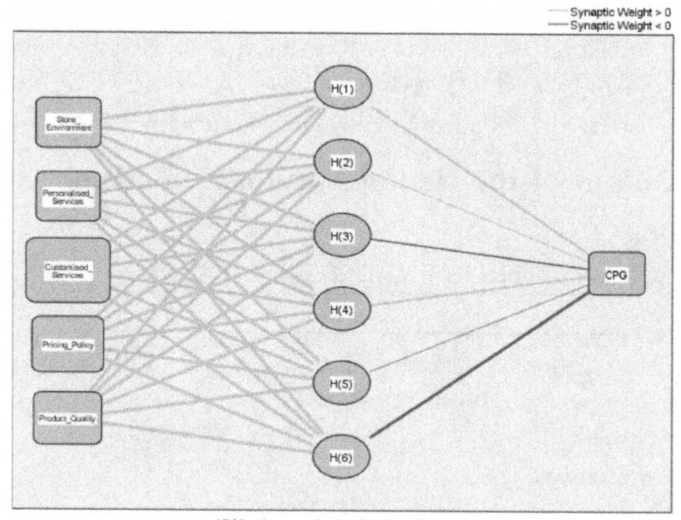

Hidden layer activation function: Softmax

Output layer activation function: Identity

Figure - 6

Table 3 Model Summary

Training	Sum of Squares Error	153.159
	Relative Error	.856
	Training Time	0:00:01.26
Testing	Sum of Squares Error	67.174[a]
	Relative Error	.953

Dependent Variable: Consumer Preference

a. The number of hidden units is determined by the testing data criterion: The "best" number of hidden units is the one that yields the smallest error in the testing data.

Table 4 Parameter Estimates

Predictor		Predicted						
		Hidden Layer[a]						Output Layer
		H(1)	H(2)	H(3)	H(4)	H(5)	H(6)	CPG
Input Layer	Store_ Environment	.797	.676	-.280	-.164	1.023	-1.515	
	Personalised_ Services	1.287	.475	-.089	-.010	.382	-1.879	
	Customised_ Services	1.290	.409	-.114	.374	-.258	-1.857	
	Pricing_Policy	1.024	.973	.337	-1.033	-.495	-.893	
	Product_ Quality	1.423	.412	.050	.202	-.657	-1.759	
Hidden Unit Width		.570	.589	.800	.786	.782	.612	
Hidden Layer	H(1)							.661
	H(2)							.262
	H(3)							-.464
	H(4)							.541
	H(5)							-.131
	H(6)							-.676

a. Displays the center vector for each hidden unit.

Table 5 Independent Variable Importance

	Importance	Normalized Importance
Store Environment	.174	66.1%
Personalized Service	.169	64.2%
Customized Service	.263	100.0%
Pricing Policy	.206	78.4%
Product Quality	.188	71.3%

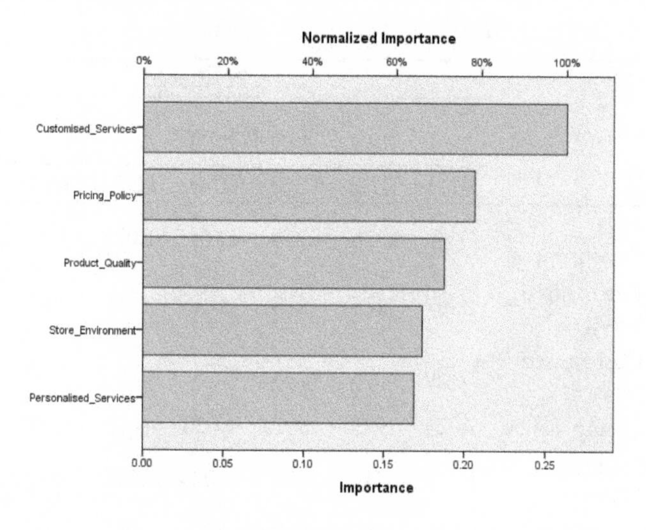

Figure - 7

Results and Discussion

Case Processing Summary

The case processing summary shows that 359 cases were assigned to the training sample, 157 to the testing sample. No cases were excluded from the analysis.

Network Information

The network information table displays information about the neural network and is useful for ensuring that the specifications are correct. Here, note in particular that:

- The number of units in the input layer is the number of covariates is Store Environment, Personalized Service, Customized Service, Pricing Policy, Product in modeling procedures.
- Likewise, a separate output unit is created for each category of Customer category, for a total of 5 units in the output layer.
- Covariates are rescaled using the standardized method.
- Automatic architecture selection has chosen 6 units in the hidden layer.
- All other network information is default for the procedure.

Hidden Layer Activation Function: Softmax

Output Layer Activation Function: Identity

Model Summary

The model summary displays information about the results of training, testing, and applying the final network to the holdout sample.

- Sum of squares error is displayed because that is always used for RBF networks. This is the error function that the network tries to minimize during training and testing.
- The percentage of incorrect predictions is taken from the classification table and will be discussed further in that topic.

Table -3 displays the corresponding information from the RBF network. There appears to be more error in the consumer preference of organised retail outlet in the training and holdout samples. The difference between the average overall relative errors of the training (0.856), and holdout (.953) samples, must be due to the small data set available, which naturally limits the possible degree of complexity of the model.

Parameter Estimates

Independent Variable Importance

Table -5 displays the independent variable importance i.e. consumer prefer organised retail outlet for customized services(100%), followed by pricing policy (78.4%), Product quality (71.3%), store environment (66.1%) and personalized services(64.2%). The normalize important chart gives pictorial representation of the table.

Normalized Importance Chart

The importance chart shows that the results are dominated by the customer services of organised retail outlet, followed by pricing policies, product quality, store environment and personalized services of the outlet.

Reference

IBM SPSS Neural Networks 21

Chapter - 12

Decision Trees

Learning Objectives

This chapter helps to understand the following

- Meaning of Decision Tree
- Advantage and disadvantage of Decision Tree
- Methods of Decision Tree
- Properties of CHAID Vs CART
- Procedures to Run Decision Tree Using SPSS
- Results and Discussions of CHAID Output Using SPSS
- Results and Discussions of Exhaustive CHAID Using SPSS
- Results and Discussions of CRT USING SPSS
- Results and Discussions of QUEST Using SPSS

Introduction

Decision Trees helps to identify groups, discover relationships between them and predict future events with highly visual classification and decision trees that enable to present categorical results in an intuitive manner, as well explain categorical analysis to non-technical audiences.

A decision tree is a classifier expressed as a recursive divider of the instance space. The decision tree consists of a node that figures a rooted tree, meaning it is intended for tree with a node called "root" that has no incoming edges. All other nodes have exactly one incoming edge. A node with outgoing boundaries is called an internal or test node. All other nodes are called leaves which is also known as terminal or decision nodes. In a decision tree, each internal node splits the instance space into two or more sub-spaces according to a certain discrete function of the input attributes values. In the simplest and most recurrent case, each test considers a single attribute, such that the concurrent space is partitioned according to the attribute's value. In the case of numeric attributes, the condition refers to a range.

Each leaf is assigned to one class representing the most appropriate target value. Alternatively, the leaf may hold a probability vector indicating the probability of the target attribute having a certain value. Instances are classified by navigating them from the root of the tree down to a leaf, according to the outcome of the tests along the path.

Internal nodes are represented as circles, whereas leaves are denoted as triangles. Note that this decision tree incorporates both nominal and numeric attributes. Given this classifier, the analyst can predict the response of a potential customer (by sorting it down the tree), and understand the behavioral characteristics of the entire potential customers population regarding direct mailing. Each node is labeled with the attribute it tests, and its branches are labeled with its corresponding values.

The IBM SPSS Decision Trees procedure creates a tree-based classification model. Decision Trees can be used as predictive models to predict the values of a dependent (target) variable based on values of independent (predictor) variables. This approach is often used as an alternative to methods such as Logistic Regression. Because the Decision Trees module is frequently used to correctly categorise cases into a target group, it may be applied in segmentation and profiling applications where the analysts wish to describe customers who are more likely to be more dissatisfied than others. It can also be used to describe cluster membership where the target field is the resultant cluster variable of an SPSS cluster analysis. It includes four tree-growing algorithms, with the ability to try different types and find the one that best fits the data. SPSS provides specialized tree-building techniques for classification within the IBM SPSS Statistics environment. The four tree-growing algorithms are

- Chi Square Automatic Interaction Detector (CHAID)
- Exhaustive CHAID
- Classification and Regression Tree (CRT)
- Quick Unbiased Efficient Statistical Tree (QUEST)

Advantages of decision tree

1. Self-explanatory and easy to follow.
2. Handles both nominal and numeric input attributes.

3. Higher representation of any discrete- value classifier.
4. Competent of handling datasets that may have errors.
5. Capable of handling datasets that may have missing values.
6. It is considered to be a nonparametric method i.e. which has no assumptions about the distribution space and Classifier structure.

Disadvantages of Decision tree

1. The target attributes must be with discrete values.
2. As decision trees use the "divide and conquer" method, they tend to perform well if a few highly relevant attributes exist, but less so if many complex interactions are present.
3. This is its over-sensitivity to the training set, to irrelevant attributes and to noise

Growing Methods of Decision Tree

Chi Square Automatic Interaction Detection (CHAID)

The CHAID Analysis (Chi Square Automatic Interaction Detection) is a form of analysis thatdetermines the best combination of variable to interpret the outcome i.e. Dependent variable. The CHAID algorithm is originally proposed by Kass in 1980; this method allows multiple splits of a node with three steps i.e. Merging, Splitting and Stopping. These trees grow repeatedly by three steps on each nodes starting from the root node.

This model is widely used in marketing for predicting the scenario by the available or collected data and interprets the same for varied research problems. The specialty of CHAID technique is it visualizes the relationship between the dependent and other related variable in tree image. It is especially useful for data expressing categorized values instead of continuous values. For this kind of data, some common statistical tools such as regression are not applicable and CHAID analysis is a perfect tool to discover the relationship between variables. As well it explores data quickly and efficiently, and builds segments and profiles with respect to the desired outcome

The algorithm only accepts nominal or ordinal categorical predictors. When predictors are continuous, they are transformed into ordinal predictors before using it.

Exhaustive CHAID

Exhaustive CHAID algorithm is a modification of CHAID Model, which examines all Possible Splits for each Predictor. This Model is developed by Biggs et al in 1991. Like CHAID Model, this one also follows the three steps - Merging, Splitting and Stopping. Like CHAID, Exhaustive CHAID only accepts nominal or ordinal Categorical Predictors, Continuous Predictor are first transformed into ordinal predictor before using them.

Classification and regression trees (C&RT)

A complete binary tree algorithm that partitions data and produces accurate homogeneous subsets. The Classification and Regression (C&R) Tree node generates a decision tree that allows you to predict or classify future observations. The method uses recursive partitioning to split the training records into segments by minimizing the impurity at each step, where a node in the tree is considered "pure" if 100% of cases in the node fall into a specific category of the target field. Target and input fields can be numeric ranges or categorical (nominal, ordinal, or flags); all splits are binary (only two subgroups).

Quick Unbiased Efficient Statistical Tree (QUEST)

A statistical algorithm that selects variables without bias and builds accurate binary trees quickly and efficiently. The QUEST node provides a binary classification method for building decision trees, designed to reduce the processing time required for large C&R Tree analyses while also reducing the tendency found in classification tree methods to favor inputs that allow more splits. Input fields can be numeric ranges (continuous), but the target field must be categorical. All splits are binary.

Properties of CHAID Vs CART

- CHAID uses multi-way splits by default (multi-way splits means that the current node is splitted into more than two nodes). This may or may not be desired (it can lead to better segments or easier interpretation). What it definitely does, though, is thin out the sample size in the nodes and thus lead to less deep trees. When used for segmentation purposes this can backfire soon as CHAID needs a large sample sizes to work well. CART does binary splits (each node is split into two daughter nodes) by default.
- CHAID is intended to work with categorical/discretized targets (XAID was for regression but perhaps they have been merged since then). CART can definitely do regression and classification.

- CHAID uses a pre-pruning idea. A node is only split if a significance criterion is fulfilled. This ties in with the above problem of needing large sample sizes as the Chi-Square test has only little power in small samples (which is effectively reduced even further by a Bonferroni correction for multiple testing). CART on the other hand grows a large tree and then post-prunes the tree back to a smaller version.
- Thus CHAID tries to prevent overfitting right from the start (only split is there is significant association), whereas CART may easily overfit unless the tree is pruned back. On the other hand this allows CART to perform better than CHAID in and out-of-sample (for a given tuning parameter combination).
- The most important difference in my opinion is that split variable and split point selection in CHAID is less strongly confounded as in CART. This is largely irrelevant when the trees are used for prediction but is an important issue when trees are used for interpretation: A tree that has those two parts of the algorithm highly confounded is said to be "biased in variable selection" (an unfortunate name). This means that split variable selection prefers variables with many possible splits (say metric predictors). CART is highly "biased" in that sense, CHAID not so much.
- With surrogate splits CART knows how to handle missing values (surrogate splits means that with missing values for predictor variables the algorithm uses predictor variables that are not as "good" as the primary split variable but mimic the splits produced by the primary splitter). CHAID has no such thing.

So it is depending upon the analyzers to choose the perfect method in decision tree, using of CHAID is simpler if the sample is of some size and the aspects of interpretation are more important. Also, if multi-way splits or smaller trees are desired CHAID is better. CART on the other hand is well working prediction methods so if prediction is aim, go for CART.

Decision Tree Using SPSS

CHAID, or Chi-squared Automatic Interaction Detection, is a classification method for building decision trees by using chi-square statistics to identify optimal splits. CHAID first examines the cross tabulations between each of the input fields and the outcome, and tests for significance using a chi-square independence test. If more than one of these relations is statistically significant, CHAID will select the input field that is the most significant (smallest p value). If an input has more than two categories, these are compared, and categories that show no differences in the outcome are collapsed together. This is done by successively joining the pair of categories showing the least significant difference. This category-merging process stops when all remaining categories differ at the specified testing level. For nominal input fields, any categories can be merged; for an ordinal set, only contiguous categories can be merged. Exhaustive CHAID is a modification of CHAID that does a more thorough job of examining all possible splits for each predictor but takes longer to compute.

Requirements. Target and input fields can be continuous or categorical; nodes can be split into two or more subgroups at each level. Any ordinal fields used in the model must have numeric storage (not string). If necessary, the Reclassify node can be used to convert them.

Strengths. Unlike the C&R Tree and QUEST nodes, CHAID can generate non-binary trees, meaning that some splits have more than two branches. It therefore tends to create a wider tree than the binary growing methods. CHAID works for all types of inputs, and it accepts both case weights and frequency variables.

Decision Tree Using SPSS

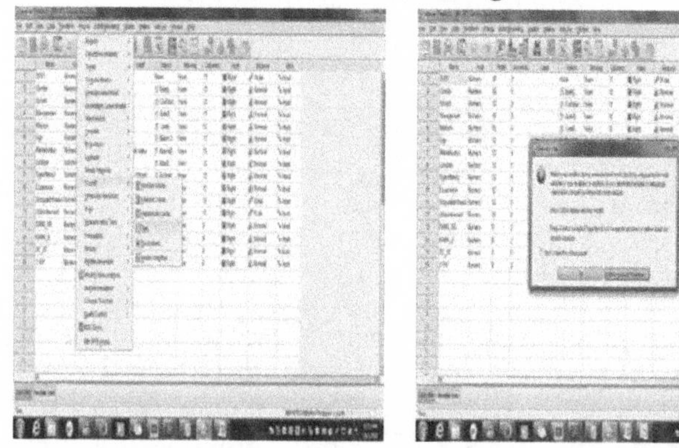

Figure -1 **Figure -2**

Step - 1: Click on Analyse > Classify > Tree as given in figure-1, Decision tree dialogue box will appear as given in Figure -2. Click ok, another dialogue box named decision tree will appear as given in Figure -3.

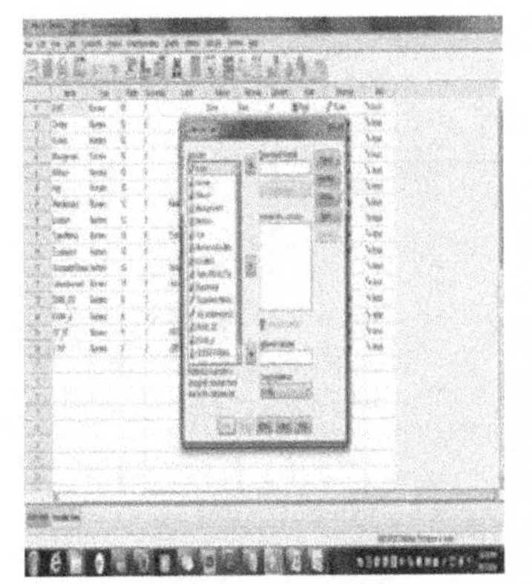

Figure – 3

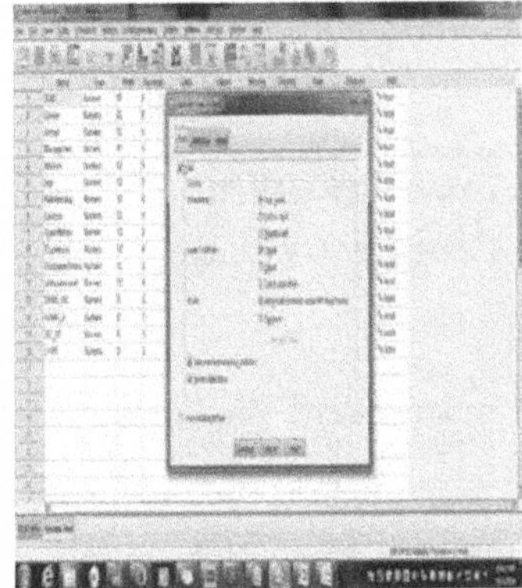

Figure – 4

Step - 2: In the decision tree dialogue box, From variable box transfer the required variable in Dependent variable and independent variable, if required transfer influence variable. Select CHAID in Growing method.

Step - 3: Click on Output tab, if needed change the required specifications. Likewise validation, criteria, save table. The output is generated by default option figures – 4 & 5 and click continue to return to decision tree dialogue box.

Step - 4: Click ok Tab in the Decision Tree dialogue box to generate output (figure -6).

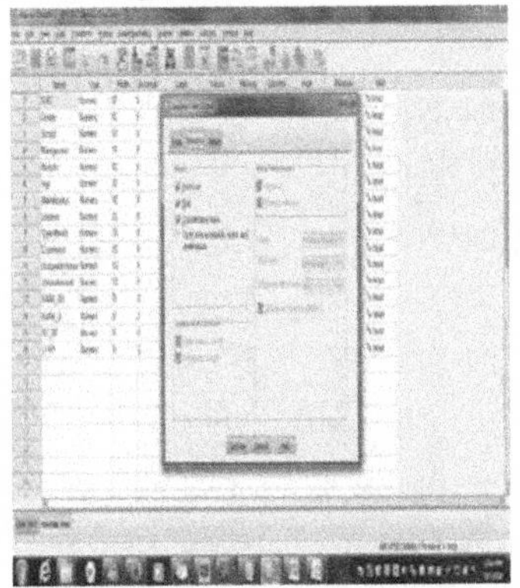

Figure – 5

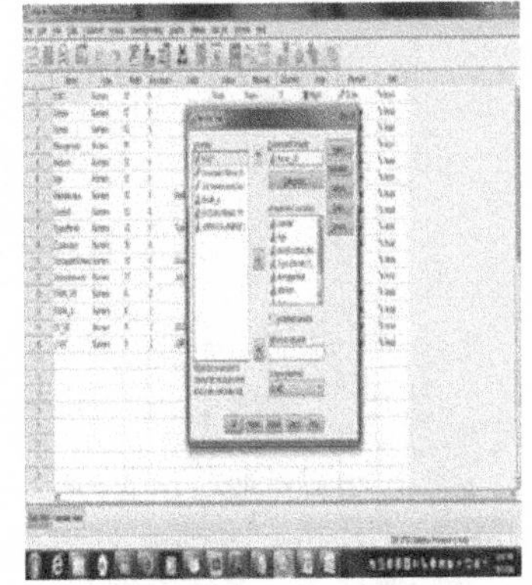

Figure – 6

CHAID Output Using SPSS

Table 1 Model Summary

Specifications	Growing Method	CHAID
	Dependent Variable	RANK_OS
	Independent Variables	Gender, Age, Marital status, Type of family, Management, Medium, School, Location, Experience
	Validation	None
	Maximum Tree Depth	3
	Minimum Cases in Parent Node	100
	Minimum Cases in Child Node	50
Results	Independent Variables Included	School, Management, Gender
	Number of Nodes	7
	Number of Terminal Nodes	4
	Depth	2

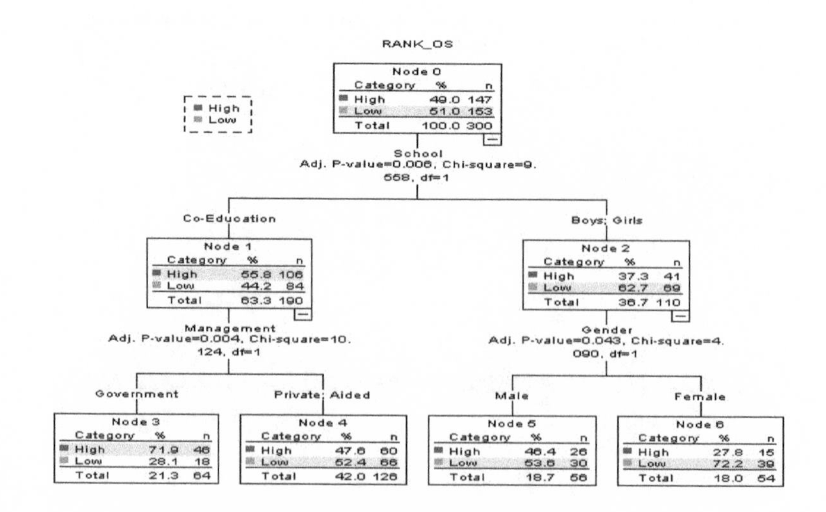

Figure - 7

Table 2 Risk

Estimate	Std. Error
.397	.028

Growing Method: CHAID
Dependent Variable: RANK_OS

Table 3 Classification

Observed	Predicted		
	High	Low	Percent Correct
High	46	101	31.3%
Low	18	135	88.2%
Overall Percentage	21.3%	78.7%	60.3%

Growing Method: CHAID
Dependent Variable: RANK_OS

Results and Discussion

The Model Summary Table provides the basic information about the CHAID Analysis. In this CHAID growing Method was used. The dependent variable of this example is Occupational Stress. The independent Variables are Gender, Age, Marital Status, Family Type, management type, Medium of the school, Location of the School, Experience of the teacher is taken. The Maximum Tree depth is 3, Minimum Cases in Parent node is 100. Minimum Cases in Child node is 50. In this example - school, Management and Gender is included for generating Results, Number of nodes is 7, Number of Terminal Nodes is 4 and the depth is 2.

The decision tree clearly depicts the results. 49% of the teachers feel, they are in High level of stress, 51% feels the stress level is lower in the occupation. Based the School Level classification, The teacher working in co-education schools have high level of stress in the occupation, teachers working in boys/ girls schools have low level of stress in occupation. The teacher responses are further classified with school management such as Government and Private schools - 71.9% of teachers working in government schools are highly stressed in the job, in private schools 52.4% of teachers are stressed in the job. Gender wise 46.4% of males are stressed and 72.2% are low stressed in the job.

The risk table shows that estimate is .397 with standard error of 0.28.

The classification Table predicted the correctness of prediction results; the result reveals that only 60.3% of the results are correctly predicted.

Exhaustive CHAID Using SPSS

The Steps are same as CHAID, to do Exhaustive CHAID, have to change the type in Growing Method i.e. Exhaustive CHAID, Everything is taken as default as in Decision Tree. Click ok to generate output

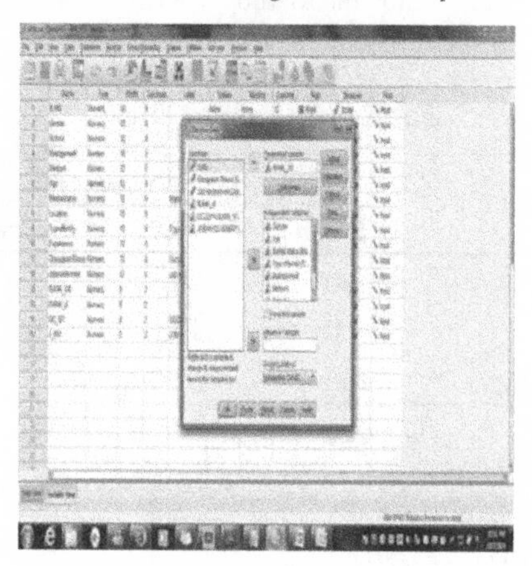

Figure - 8

Exhaustive CHAID Output Using SPSS

Table 4 Model Summary

Specifications	Growing Method	EXHAUSTIVE CHAID
	Dependent Variable	RANK_OS
	Independent Variables	Gender, Age, Marital status, Type of family, Management, Medium, School, Location, Experience
	Validation	None
	Maximum Tree Depth	3
	Minimum Cases in Parent Node	100
	Minimum Cases in Child Node	50
Results	Independent Variables Included	Medium, Management
	Number of Nodes	5
	Number of Terminal Nodes	3
	Depth	2

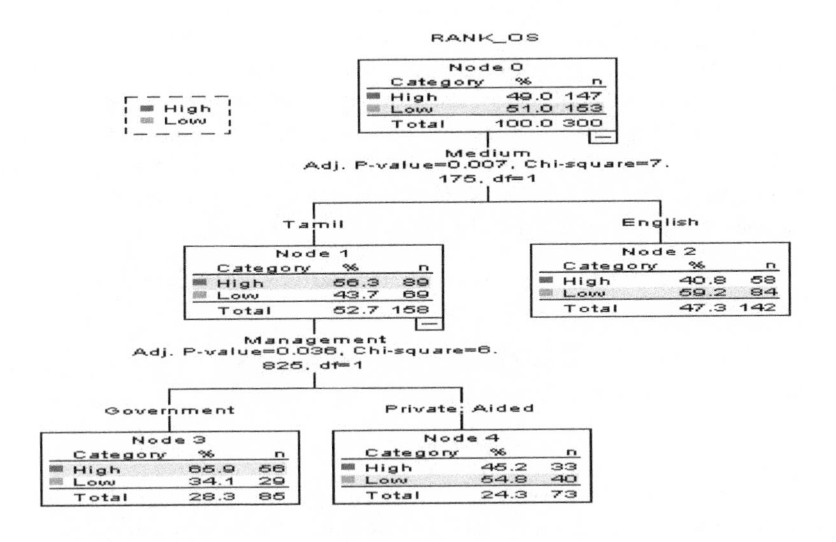

Figure - 9

Table 5 Risk

Estimate	Std. Error
.400	.028

Growing Method: EXHAUSTIVE CHAID
Dependent Variable: RANK_OS

Table 6 Classification

Observed	Predicted		
	High	Low	Percent Correct
High	56	91	38.1%
Low	29	124	81.0%
Overall Percentage	28.3%	71.7%	60.0%

Growing Method: EXHAUSTIVE CHAID
Dependent Variable: RANK_OS

Results & Discussion

The Model summary is like CHAID method, The Specifications are same here, but for generating output Medium and management is taken, Number of Nodes are 5, Number of terminal nodes are 3 and the depth is 2 (Table -4).

The Decision tree diagram (Figure -9) explains the Level of occupation stress among the teachers working in different management. As in CHAID example, The level of occupation stress is low i.e 51%, 49% are accounted for high level of occupational Stress. Based on the medium of schools - teachers working in Tamil medium are highly stressed i.e. 56.3% and people working in English medium are low stressed i.e. 59.2%. With respect to management i.e. Government or Private aided schools, Government school teachers are highly stressed compared to private aided schools.

The risk estimates are accounted for .40 with standard error of 0.28 (Table -5).

The classification Table (Table -6) reveals 60% of correctness of prediction result in Exhaustive CHAID method.

CRT USING SPSS

The Classification and Regression (C&R) Tree node is a tree-based classification and prediction method. Similar to C5.0, this method uses recursive partitioning to split the training records into segments with similar output field values. The C&R Tree node starts by examining the input fields to find the best split, measured by the reduction in an impurity index that results from the split. The split defines two subgroups, each of which is subsequently split into two more subgroups, and so on, until one of the stopping criteria is triggered. All splits are binary (only two subgroups).

Pruning

C&R Trees gives the option to first grow the tree and then prune based on a cost-complexity algorithm that adjusts the risk estimate based on the number of terminal nodes. This method, which enables the tree to grow large before pruning based on more complex criteria, may result in smaller trees with better cross-validation properties. Increasing the number of terminal nodes generally reduces the risk for the current (training) data, but the actual risk may be higher when the model is generalized to unseen data. In an extreme case, with a separate terminal node for each record in the training set the risk estimate would be 0%, since every record falls into its own node, but the risk of misclassification for unseen (testing) data would almost certainly be greater than 0. The cost-complexity measure attempts to compensate for this.

Requirements

To train a C&R Tree model, it is necessary to have one or more Input fields and exactly one Target field. Target and input fields can be continuous (numeric range) or categorical. Fields set to Both or None are ignored. Fields used in the model must have their types fully instantiated, and any ordinal (ordered set) fields used in the model must have numeric storage (not string). If necessary, the Reclassify node can be used to convert them.

Strengths

C&R Tree models are quite robust in the presence of problems such as missing data and large numbers of fields. They usually do not require long training times to estimate. In addition, C&R Tree models tend to be easier to understand than some other model types--the rules derived from the model have a very straightforward interpretation. Unlike C5.0, C&R Tree can accommodate continuous as well as categorical output fields.

The Steps are same as in Decision tree, to do CRT Method, change the type in Growing Method i.e. CART as given in Figure -10, Everything is taken as default as in Decision Tree. Click ok to generate output.

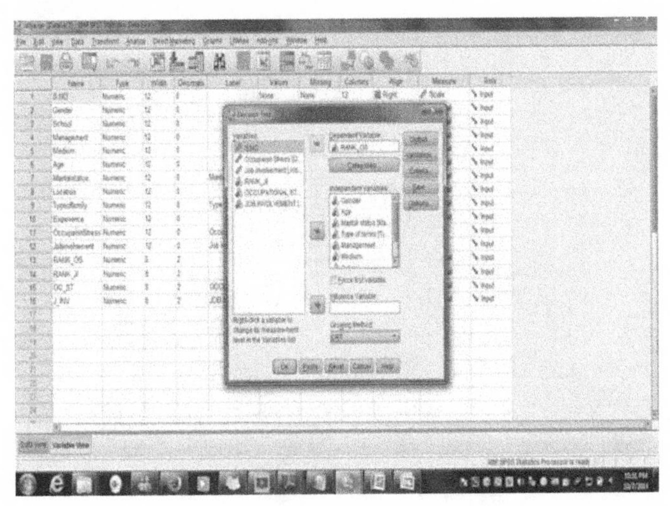

Figure - 10

CRT Output Using SPSS

Table 7 Model Summary

Specifications	Growing Method	CRT
	Dependent Variable	RANK_OS
	Independent Variables	Gender, Age, Marital status, Type of family, Management, Medium, School, Location, Experience
	Validation	None
	Maximum Tree Depth	5
	Minimum Cases in Parent Node	100
	Minimum Cases in Child Node	50
Results	Independent Variables Included	School, Management, Experience, Medium, Gender, Marital status, Age, Type of family, Location
	Number of Nodes	9
	Number of Terminal Nodes	5
	Depth	3

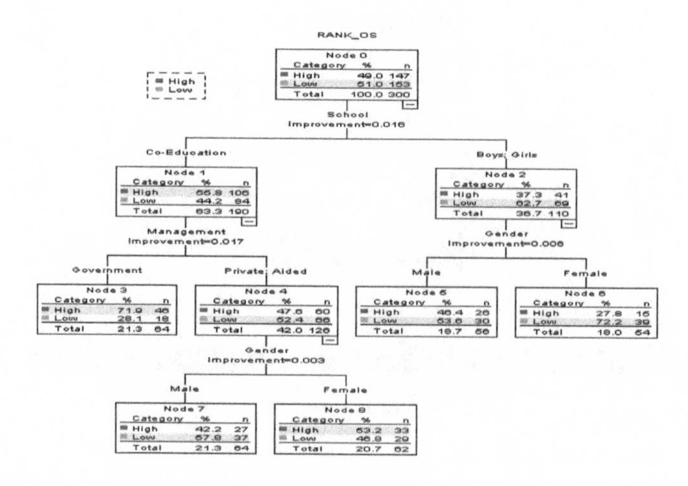

Figure - 11

Table 8 Risk

Estimate	Std. Error
.383	.028

Growing Method: CRT
Dependent Variable: RANK_OS

Table 9 Classification

Observed	Predicted		
	High	Low	Percent Correct
High	79	68	53.7%
Low	47	106	69.3%
Overall Percentage	42.0%	58.0%	61.7%

Growing Method: CRT
Dependent Variable: RANK_OS

Results and Discussion

Like CHAID and Exhaustive CHAID, the results are same here. But in model summary it is to be noted that all the independent variables are considered for Output generation. As like the above models, the decision tree, Risk estimates and Correctness of Prediction were given.

QUEST Using SPSS

QUEST—or Quick, Unbiased, Efficient Statistical Tree—is a binary classification method for building decision trees. A major motivation in its development was to reduce the processing time required for large C&R Tree analyses with either many variables or many cases. A second goal of QUEST was to reduce the tendency found in classification tree methods to favor inputs that allow more splits, that is, continuous (numeric range) input fields or those with many categories.

QUEST uses a sequence of rules, based on significance tests, to evaluate the input fields at a node. For selection purposes, as little as a single test may need to be performed on each input at a node. Unlike C&R Tree, all splits are not examined, and unlike C&R Tree and CHAID, category combinations are not tested when evaluating an input field for selection. This speeds the analysis. Splits are determined by running quadratic discriminant analysis using the selected input on groups formed by the target categories. This method again results in a speed improvement over exhaustive search (C&R Tree) to determine the optimal split.

Requirements

Input fields can be continuous (numeric ranges), but the target field must be categorical. All splits are binary. Weight fields cannot be used. Any ordinal (ordered set) fields used in the model must have numeric storage (not string). If necessary, the Reclassify node can be used to convert them.

Strengths

Like CHAID, but unlike C&R Tree, QUEST uses statistical tests to decide whether or not an input field is used. It also separates the issues of input selection and splitting, applying different criteria to each. This contrasts with CHAID, in which the statistical test result that determines variable selection also produces the split. Similarly, C&R Tree employs the impurity-change measure to both select the input field and to determine the split.

The Steps are same as in Decision tree. To do QUEST Method, change the type in Growing Method i.e. QUEST as given in Figure -12, Everything is taken as default as in Decision Tree. Click ok to generate output.

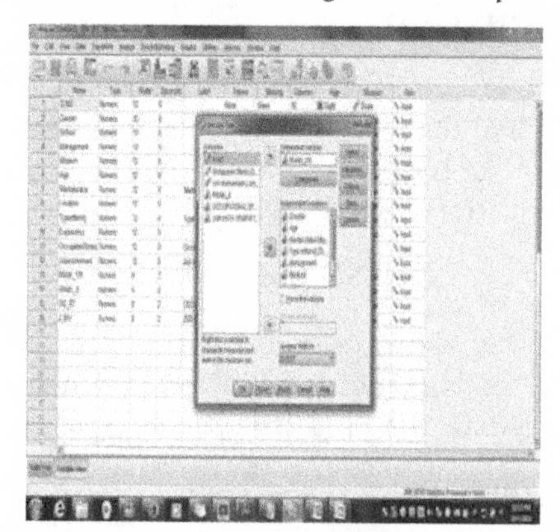

Figure -12

QUEST Output Using SPSS

Table 10 Model Summary

Specifications	Growing Method	QUEST
	Dependent Variable	RANK_OS
	Independent Variables	Gender, Age, Marital status, Type of family, Management, Medium, School, Location, Experience
	Validation	None
	Maximum Tree Depth	5
	Minimum Cases in Parent Node	100
	Minimum Cases in Child Node	50
Results	Independent Variables Included	No Independent Variable Included
	Number of Nodes	1
	Number of Terminal Nodes	1
	Depth	0

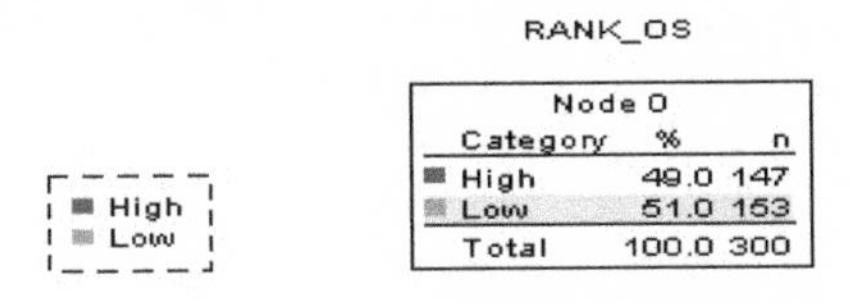

Figure - 13

Table 11 Risk

Estimate	Std. Error
.490	.029

Growing Method: QUEST
Dependent Variable: RANK_OS

Table 12 Classification

Observed	Predicted		
	High	**Low**	**Percent Correct**
High	0	147	0.0%
Low	0	153	100.0%
Overall Percentage	0.0%	100.0%	51.0%

Growing Method: QUEST
Dependent Variable: RANK_OS

SPSS OUTPUT FOR QUEST

As Given in CHAID, Exhaustive CHAID, CART, the same set of variables are given in QUEST model. In independent variable, no variables are taken, the number of Nodes and terminal nodes are 1 each. In Decision tree, only the level of stress classification is specified. The estimates of risk are .490 for .29 standard error. The classification results of correction of prediction have shown only the low stress level of respondents alone.

Chapter - 13

Path Analysis

Learning Objectives

This Chapter helps to understand the following

- Meaning of Path Analysis
- Historical Background of Path Analysis
- Discussion about Path Diagram Features and its Practices.

Introduction

Path analysis is an extension of multiple regressions. It goes beyond regression in which it allows for the analysis of more complicated models. Path analysis is a variant of multivariate regression analysis in which causal relationship between several variables are represented by a path or flow diagram and path coefficients. It provides estimates of the strength of relationship between two variables when all the other variables are held constant. Path analysis allows the researcher to examine the causal effects of several variables simultaneously. In this way, it is used to examine the degree of 'fit' between the model and the data.

In particular, it can examine situations in which there are several final dependent variables and those in which there are "chains" of influence, in that variable A influences variable B, which in turn affects variable C. Despite its previous name of "causal modelling," path analysis cannot be used to establish causality or even to determine whether a specific model is correct; it can only determine whether the data are consistent with the model. However, it is extremely powerful for examining complex models and for comparing different models to determine which one best fits the data.

Path analysis begins with the researcher developing a diagram with arrows connecting variables and depicting the causal flow, or the direction of cause and effect. The precursor to path analysis is a simpler version of causal modelling in which the only effects represented are direct causal effects. Path analysis has a substantial advantage over the simpler model in that both direct and indirect causal effects can be estimated. Path analysis can be used to test the fit between two or more causal models, which are hypothesized by the researcher to fit the data.

Since path analysis assesses the comparative strength of different effects on an outcome, the relationships between variables in the path model are expressed in terms of correlations and represent hypotheses proposed by the researcher. One of the advantages of using path analysis is that it forces researchers to explicitly specify how the variables relate to one another and thus encourage the development of clear and logical theories. Path analysis is also advantageous in that it allows researchers to break apart or decompose the various factors affecting an outcome into direct effects and indirect components.

Historical Background

Path Analysis is the statistical technique used to examine causal relationships between two or more variables. It is based upon a linear equation system and was first developed by Sewall Wright in the 1930s for use in phylogenetic studies. Path Analysis was adopted by the social sciences in the 1960s and has been used with increasing frequency in the ecological literature since the 1970s. In ecological studies, path analysis is used mainly in the attempt to understand comparative strengths of direct and indirect relationships among a set of variables. In this way, path analysis is unique from other linear equation models: In path analysis mediated pathways (those acting through a mediating variable, i.e., "Y," in the pathway X Y Z) can be examined. Pathways in path models represent hypotheses of researchers, and can never be statistically tested for directionality. Numerous articles deal with the use of path analysis in ecological studies.

Path analysis is a subset of Structural Equation Modeling (SEM), the multivariate procedure that, as defined by Ullman (1996), "allows examination of a set of relationships between one or more independent variables, either continuous or discrete, and one or more dependent variables, either continuous or discrete." SEM deals with measured and latent variables. A *measured variable* is a variable that can be observed directly and is measurable. Measured variables are also known as observed variables, indicators or manifest variables. A *latent variable* is a variable that cannot be observed directly and must be inferred from measured variables. Latent variables are implied by the covariances among two or more measured variables. They are also known as factors (i.e., factor analysis), constructs or unobserved variables. SEM is a combination of multiple regression and factor analysis. Path analysis deals only with measured variables.

Path analysis was developed as a method of decomposing correlations into different pieces for interpretation of effects (e.g., how does parental education influence children's income 40 years later?). Path analysis is closely related to multiple regressions; you might say that regression is a special case of path analysis. Some people call this stuff (path analysis and related techniques) "causal modeling." The reason for this name is that the techniques allow us to test theoretical propositions about cause and effect without manipulating variables. However, the "causal" in "causal modeling" refers to an assumption of the model rather than a property of the output or consequence of the technique. That is, people assume some variables are causally related, and test propositions about them using the techniques. If the propositions are supported, it does NOT prove that the causal assumptions are correct.

Path analysis models can be estimated using the following special features:

- Single or multiple group analysis
- Missing data
- Complex survey data
- Random slopes
- Linear and non-linear parameter constraints
- Indirect effects including specific paths
- Maximum likelihood estimation for all outcome types
- Bootstrap standard errors and confidence intervals
- Wald chi-square test of parameter equalities

The Path Diagram

Social scientific theories of causal relationships often specify a system of relationships in which some variables affect other variables and these in turn influences still other variables in the model. A single multiple regression model can only specify one response variable at a time. However, path analysis estimates as many regression equations as are needed to relate all the proposed theoretical relationships among the variables in the explanation. A path diagram represents the hypothesized causal model in path analysis.

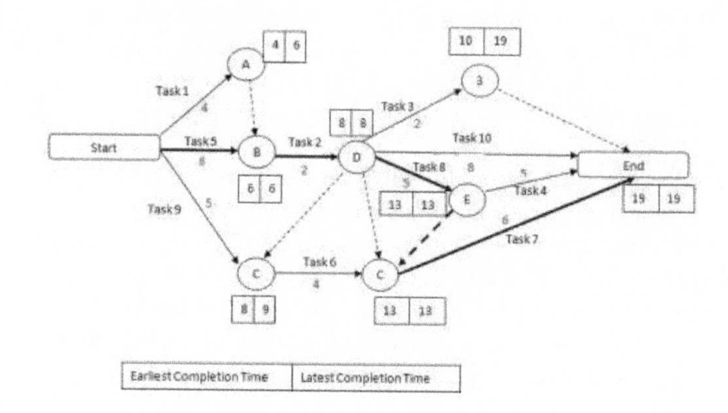

Exogenous and Endogenous Variables

Variables often play more than one role in path models and this is reflected in the analytic language used in path analysis. Exogenous variable is a variable whose variation is explained by factors outside the model and which also explains other variables within the model.

Endogenous variable is a variable whose variation is explained by one or more variables within the model.

Residual Error

Residuals or error terms are exogenous independent variables that are not directly measured and reflect unspecified causes of variability in the outcome or unexplained variance plus any error due to measurement. Residual error is assumed to have a normal distribution with a mean of zero and to be uncorrelated with other variables in the model.

Assumptions

- The assumptions for the type of path analysis we will be doing are as follows (some of these will be relaxed later):
- All relations are linear and additive. The causal assumptions (what causes what) are shown in the path diagram.
- The residuals (error terms) are uncorrelated with the variables in the model and with each other.
- The causal flow is one-way.
- The variables are measured on interval scales or better.
- The variables are measured without error (perfect reliability).

Following is the set of path analysis:

- Path analysis with continuous dependent variables
- Path analysis with categorical dependent variables
- Path analysis with categorical dependent variables using the Theta parameterization
- Path analysis with a combination of continuous and categorical dependent variables
- Path analysis with a combination of censored, categorical, and unordered categorical (nominal) dependent variables
- Path analysis with continuous dependent variables, bootstrapped standard errors, indirect effects, and confidence intervals
- Path analysis with a categorical dependent variable and a continuous mediating variable with missing data
- Moderated mediation with a plot of the indirect effect

Path analysis in practice

Bryman and Cramer give a clear example using four variables from a job survey: age, income, autonomy and job satisfaction. They propose that age has a **direct effect** on job satisfaction. However **indirect effects** of age on job satisfaction are also suggested; age affects income which in turn affects satisfaction, age affects autonomy which in turn affects satisfaction and age affects autonomy which affects income which affects satisfaction. Autonomy and income have direct effect on satisfaction.

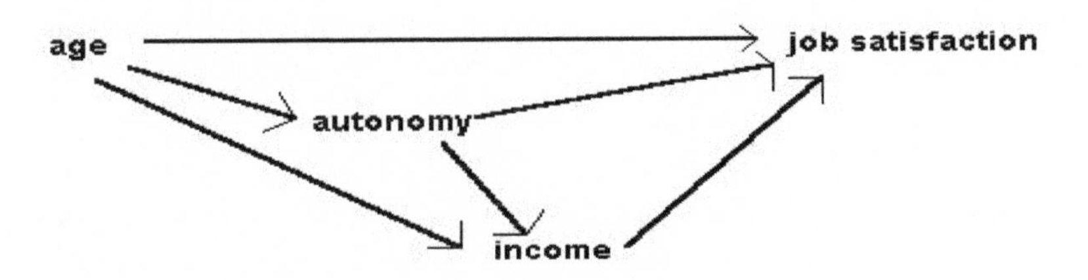

Input diagram of causal relationships in the job survey, after Bryman & Cramer (1990)

To move from this input diagram to the output diagram, we need to compute path coefficients. A path coefficient is a **standardized regression coefficient** (**beta weight**). We compute these by setting up structural equations, in this case:

$$\text{satisfaction} = b_{11}\text{age} + b_{12}\text{autonomy} + b_{13}\text{ income} + e_1$$

$$\text{income} = b_{21}\text{age} + b_{22}\text{autonomy} + e_2$$

$$\text{autonomy} = b_{31}\text{age} + e_3$$

We have used a different notation for the coefficients from Bryman and Cramer's, to make it clear that b_{11} in the first equation is different from b_{21} in the second. The terms e_1, e_2, and e_3 are the **error** or unexplained variance terms. To obtain the path coefficients we simply run three regression analyses, with satisfaction, income and autonomy being the dependent variable in turn and using the independent variables specified in the equations

Estimation and testing

Estimation of path models

The model is recursive since all the causal linkages flow in one direction and none of the variables represent both cause and effect the same time. This causal model is the only kind of model which can properly be called path analysis. In models where the hypothesized causality flows in a single direction, the estimation can be done relatively simply by using ordinary least squares regression or maximum likelihood estimation to solve the equations for each endogenous or outcome variable in the model.

Model Specification

Path analysis is particularly sensitive to model specification, the inclusion of irrelevant variables or the exclusion of important causal variable changes the value of path coefficients. The strength of direct and indirect effects on the outcome variables in the model are evaluated using the path coefficients. This method is most helpful in the testing of well specified theories about the relationships between variables and not for exploratory purposes.

Goodness of fit

Often researchers have more than one theory regarding which variables or what path to include in the path model. Competing theories can be evaluated by estimating separately path models and then assessing the goodness of fit statistics to determine which hypothesized model best fits the correlation matrix in the observed data. Alternatives theories can also be combined into a single path model and the researcher can assess which pathways are more significant by comparing the relative strength of different pathways within the same path model.

Strengths and limitations

Strengths of path analysis

The questions that are the subject of social inquiry often involve multiple causal influences. Strengths of the path analysis method is that it estimated a system of equations that specify all the possible causal linkages among a set of variables. In addition path analysis enables researchers to break down or decompose correlations among variables into causal and noncausal components; thus, path analysis helps researchers disentangle the complex interrelationships among variables and identify the most significant pathways involved in prediction and outcome.

Path analysis can also play a vital role in the theoretical or hypothesis testing stage of social research. Path analysis requires researcher to explicitly specify how they think the variables relate to one another within the path diagram. This method forces researchers to develop detailed and logical theoretical models to explain the outcome of interest. Thus, researchers using non experimental quantitative, or correlation data can test whether their hypotheses about the relationships between variables are plausible and supported by the data and represent underlying processes.

Limitations of path analysis

Path analysis is an extension of multiple regressions; it follows all the usual assumptions of regression. However, it is often difficult to meet the assumptions in social scientific research, particularly those of reliability and recursively or unidirectional causal flow. The path model has to assume that each variable is an exact manifestation of the theoretical concepts underlying them and reasonably free of measurement error.

Path analysis is a statistical tool used to evaluate whether the correlations between variables in a given data set reflect the causal hypotheses specified in the model. Since the models are based on correlations, path analysis cannot demonstrate causality or the direction of causal effects. However, as stated previously the path analytic method can indicate which of the path models best fits the pattern of correlations found in the data.

Chapter - 14

Structural Equation Modeling

Learning Objectives

This chapter helps to understand the following

- Meaning of Structural equation modeling
- History of structural equation modeling
- Need, Merits and Demerits of SEM Application
- Prerequisites of SEM
- Running SEM through AMOS Application
- Results and Discussion of SEM Output

Introduction

Structural Equation Modelling is one of the statistical technique used for establishing the relationship among the variables. SEM uses variance and covariances of the variables to establish the linear relationships. SEM is also known as LISREL Models (Linear Structural Relations), simultaneous equation model or multivariate regression model. Usually the relations are formulated by linear regression equations and this is also graphically represented by path diagrams. It deals with a system of regression equation. The geneticist Sewall Right, the economist Trygve Haavelmo and the cognitive scientist Herbert A.Simon, defined SEM as a statistical technique for testing and estimating causal relations using combination of statistical data and qualitative causal assumptions.

SEM stands for structural equation modelling. SEM is a notation for specifying structural equations, a way of thinking about them, and methods for estimating their parameters. Factor analysis, path analysis and regression all represent special cases of SEM.

Structural equation modeling (SEM)

- is a comprehensive statistical approach to testing hypotheses about relations among observed and latent variables (Hoyle, 1995).
- is a methodology for representing, estimating, and testing a theoretical network of (mostly) linear relations between variables (Rigdon, 1998).
- tests hypothesized patterns of directional and nondirectional relationships among a set of observed (measured) and unobserved (latent) variables (MacCallum & Austin, 2000).

Two goals in SEM are

- to understand the patterns of correlation/covariance among a set of variables and
- to explain as much of their variance as possible with the model specified (Kline, 1998).

Procedures for testing set of variances and covariance matrix fits a specified structure as follows:

1. Make an association between variables – probably with the use of path diagrams
2. Find out the implication of the variable on variance and covariances.
3. Check whether variances and covariances fit the model.
4. Also use various statistical testing, parameters estimates, standard errors for the numerical coefficient in the linear equations.
5. On the above tests – find whether the model is goodfit for the data prescribed.

A Short History of SEM

SEM can trace its history back more than 100 years.

At the beginning of 20th century Spearman laid the foundation for factor analysis and thereby for the measurement model in SEM (Spearman,1904). Spearman tried to trace the different dimensions of intelligence back to general intelligence factor. In the thirties Thurstone invented multi-factor analysis and factor rotation (more or less in opposition to Spearman), and thereby founded modern factor analysis, whereby e.g. intelligence , was thought of as being composed of several different intelligence dimensions (Thurstone and Thurstone, 1941; Thurstone 1947).

About 20 years after Spearman, Wright started the development of the so called path analysis (Wright,1918,1921). Based on box and arrow-diagram, he formulated the series of rules that connected correlations among the variables with parameters in the assumed data-generating model. Most of his work was on models with only manifest variables, but a few also include models with latent variables.

Wright was a biometrician and it is amazing that his work was more or less unknown to scientists outside this area, until taken up by social researchers in the sixties (Blalock 1961,1971;Duncan,1975).

In economics a parallel development took place in what was to be known as econometrics. However, this development was unaffected by Wright's ideas, and was characterized by absence of latent variables- atleast in the sense of the word used in this book. However, in the fifties econometricians became aware of Wright's work and some of them found to their surprise that he had pioneered estimation of supply and demand functions and in several respects was far ahead of econometricians of his time (Goldberger , 1972).

In the early seventies path analysis and factor analysis were combined to form the general SEM of today. Foremost in its development was Joreskog, who created the well known LISREL (Linear Structural Relations) program for analyzing such models.

However, LISREL is not alone on the scene. Among other similar computer programs mention can be made of EQS (EQuationS) (Bentler,1985) and RAM (Reticular Action Model) (McArdle and McDonald, 1984) included in the SYSTAT package of statistics programs under the name of RAMONA (Reticular Action Model or Near Approximation), and of course AMOS (Arbuckel, 1989).

Need for SEM

Generally social sciences researches seems to be complex, because of its multiple outcomes. All these outcomes can't be fit in a single model. As a solution of this problem SEM arise which allows the representation of complex theory in a single model. The primary need for SEM arises, where a researcher needs to study the relationship among latent constructs indicated by multiple measures. The most prominent feature in SEM is it has the capability to deal with latent variables.

For example, if a researcher wants to know the relationship between isolated life and suicide – First he must conceptually define isolated life. There are several forms of isolation. There may be physical isolation and psychological isolation. So he has to form several hypothetical constructs to establish the relationship. SEM allows the researcher to represent these hypothetical constructs explicitly and to distinguish the measurement of construct from key relationships among the constructs.

Features of SEM

Byrne (2001) compared SEM against other multivariate techniques and listed four unique features of SEM

- SEM takes a confirmatory approach to data analysis by specifying the relationships among variables a priori. By comparison, other multivariate techniques are descriptive by nature (e.g. exploratory factor analysis) so that hypothesis testing is rather difficult to do.
- SEM provides explicit estimates of error variance parameters. Other multivariate techniques are not capable of either assessing or correcting for measurement error. For example, a regression analysis ignores the potential error in all the independent (explanatory) variables included in a model and this raises the possibility of incorrect conclusions due to misleading regression estimates.
- SEM procedures incorporate both unobserved (i.e. latent) and observed variables. Other multivariate techniques are based on observed measurements only.
- SEM is capable of modeling multivariate relations, and estimating direct and indirect effects of variable under study.

Applications of Structural Equation Modelling

SEM is used extensively in the fields of

1. Counselling Psychology research: confirming the factor structure of a psychological assessment instrument. Analysing potential mediator and moderator effect.
2. Market research: relative importance of service and product quality to overall satisfaction
3. Conservation Biology: evaluating factors that determine reproductive success in plants.
4. However, SEM is relatively unfamiliar among the researchers and statiscians.

Advantages of SEM

- SEM has the capability to deal with latent variables (i.e variables that cannot be measured directly)
- SEM is a flexible statistic technique which could deal with the system of regression equations.

- More specifically the model implies a specific structure of the covariance which can be compared to the co-variances in the sample.
- It transfers substantive theory into workable model.

Limitations of SEM

- SEM relatively requires a large sample size (i.e N is 200 0r more.)
- SEM techniques only look at first order (linear) relationships between variables.
- Since SEM is a confirmatory technique, researcher must know the number of parameters needed to estimate including covariances, path coefficients and variances to begin the analyses.

Similarities between Traditional Statistical Methods and SEM

- Traditional model and SEM are based on linear statistical models.
- Traditional methods follow normality in distribution and SEM also follows multivariate normality.
- Test of causality is not offered in both methods.

Differences Between Traditional and SEM Methods

SEM differs from traditional approaches in various areas.

S. no	Traditional Methods	SEM Methods
1	Specify a default model	SEM offers no default model. It requires a formal specification
2	Analyses only measured variables	It is a multivariate technique incorporating observed and unobserved variables
3	These methods assumes measurement occurs without error	SEM explicitly specifies errors and also allows researcher to recognize the imperfect nature of their measures.
4	Provides straight forward significance tests to determine group differences, relationship between variables or the amount of variance explained	Provides no straightforward siginificance tests to determine model fit. The best strategy for evaluating model fit is to examine multiples tests such as chi-square, Comparative Fit Index (CFI), Bentler-Bonett Nonnormed Fit Index (NNFI), Root Mean Square of Approximation (RMSEA).

S. no	Traditional Methods	SEM Methods
5	Multicollinearity cannot occur because unobserved variables represent distinct latent constructs.	Resolves problems of multi collinearity. Multiple measures are required to describe a latent constructs.
6	No graphical language is there to present complex relationships.	SEM uses graphical language to provide a convenient and powerful way to present complex relationships. Model specification involves formulating statements about a set of variables. A diagram, a pictorial representation of a model is transformed into a set of equations. The set of equations is solved simultaneously to test model fit and estimate parameters.

Prerequisites of SEM

Covariance & Correlation

Covariance is the measure of relationship between two variables. It is designed to show the degree of co-movement between two variables. It is calculated by averaging the set of variables's deviation from its average value. (i.e total of actual – expected / no of observations-1)

- If covariance is positive, the variables tend to move in same direction.
- If covariance is negative, the variables move in opposite direction.
- If covariance is zero, no relationship.

Correlation is a concept related to covariance. It also gives a degree to which two variables are associated. It always ranges from -1 to +1.

- -1 indicates a perfectly inverse relationship.
- +1 indicates a perfectly positive relationship
- 0 indicates no relationship one way or another.

The uniform scale is always -1 to +1. If correlation value moves closer to 1, it indicates more positive relationship. This is in contrast to covariance where a value between two variables is large but indicates little actual relationship. Correlation matrices and covariance matrices describe the pair wise relationships between a set of variables.

Exploratory Factor Analysis (EFA): EFA is a tool for assessing the factors that underlie a set of variables. It is a technique within factor analysis whose goal is to identify the relationship between measured variable. It is commonly used by researchers to assess which items should be grouped together to form a scale. It introduces the distinction between the observed variables (e.g., questions on a test) and latent variables (the underlying factor that we are usually interested in). This is extended within the SEM context when performing confirmatory factor analysis.

Multiple Linear Regression: This is a statistical technique that uses several explanatory variables to predict the outcome of the response variable. The objective of Multiple linear regression is to model the relationship between the explanatory and response variable. The predictor variables are weighted in order to form a composite variable that aims to maximise prediction of the outcome variable. Regression coefficients are used to indicate the expected increase in the outcome variable for an increase of one on the predictor variable holding all other predictor variables constant. Standardised regression coefficients are frequently used to indicate the relative importance of predictors. Structural equation modelling has all these elements. Sometimes we are interested in assessing the relative importance of different latent predictor variables in outcome variable.

Causal inference: Causal inference is strongest in controlled experiments involving random allocation of subjects to conditions. Longitudinal data and other quasi-experimental designs can also provide evidence for causal claims, although the threats to such an influence are typically stronger than in experiments. Finally, correlational cross-sectional designs which represent the majority of applications of SEM designs make provide the weakest form of evidence of causality. It is for this reason that we need to be cautious in the interpretation of directional arrows in structural equation modelling.

SEM Model

Math Review

To get started with SEM – we should also understand about Covariance and Basic Matrix operations.

Covariance is a statistic describing the degree of linear relationship between two variables in joint units. The standardized form of covariance is the correlation, which can be defined as the covariance divided by the standard deviations of both variables.

Matrix operations: A matrix is a rectangular table of numbers, usually represented by a symbol (like a single letter). Much like scalar algebra, matrix algebra can be used to carry out operations on matrices, like addition and multiplication. The order of a matrix is defined in terms of the number of rows and columns a matrix has (m x n). Rows are typically indexed by the letter i, and columns by j, and cells by their position ij.

Other Matrix operations are Matrix addition , Matrix subtraction, Matrix multiplication, Matrix transposition and Matrix inversion.

A Covariance Matrix is just a matrix where any element is the covariance of the variables indexing that row and column.

SEM is fundamentally about expressing relationships between variables as simple as possible. SEM takes as input a covariance matrix (or matrices) and reproduces that matrix with a series of structural equations defined by number of parameters.

Moderation in SEM is a situation where, there exists three or more variables, and change in one of those variable affects the relationship between the other two.

Mediation in SEM refers to the situation that include three or more variables, but there is a causal process between three variables

Symbols represented in SEM

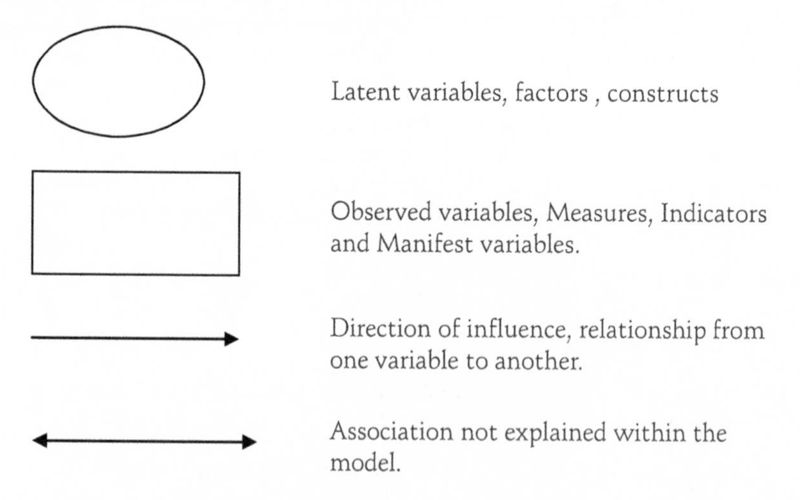

Types of Variables in SEM

Types of variables that can be included in structural equation models.

- Exogeneous variable: variables which are not influenced by the other variable is known as exogeneous variable. This is also known as independent variable.

- Endogeneous variable: variables that are influenced by the other variable is known as endogeneous variable, which is also termed as dependent variable.
- Manifest variables: variables that are directly measured and observed are called manifest variables.
- Latent variables: variables that are not directly measured are termed as latent variables.

The Model thus consists of two parts:

- ***The structural model***: describing the causal connections among the latent variables. Mapping of these connections is the main purpose of analysis.
- ***The Measurement model*** describing the connections between the latent variables and their manifest indicators.

Steps in performing SEM Analysis

1. Model specification
2. Model identification
3. Model estimation
4. Model evaluation
5. Model modification
6. Interpretation

Model specification

To build a sound model the researchers should have sound knowledge on theory and concepts of SEM. Two main components of the model are first the structural model and measurement model. The structural model shows causal dependencies between exogeneous and endogeneous variable. Path diagrams are set as an example for structural model. Second the measurement model shows relationships between latent and manifest variables. Exploratory and confirmatory factor analysis are examples of measurement model. For easy communications the models are described in graphical form. The variables with an arrow pointing is called endogeneous variable. Variables with no arrow pointing are exogeneous variables. Unexplained variables are represented by curved arrows. Because of flexible nature of SEM variety of models can be conceived.

Models cannot have large numbers of unknown parameters to be estimated. Since SEM is a confirmatory technique one must clearly mention the number of parameters estimated, variances and covariances to be calculated, number of latent and manifest variables to be included in this model.

Model identification

It determines whether model estimates are unique. The identification problems shall be related to the either structural or measurement portion of the model. Check whether each item is connected only to its respective constructs and error terms are not correlated. It should have structural connections among all the constructs.

A model is said to be,

- overidentified - if it contains fewer parameters to be estimated than the number of variances and covariances.
- Just identified - if it contains same number of parameters estimated and variances and covariances.
- Under identified - if it contains more parameters to be estimated than the number of variances and covariances.

In any event the results of model is uninterpretable, such models requires respecification.

Data should have passed through reliability test and validity tests like any other techniques of statistics. The sample size required basically depends on the model characteristics such as model size, scores of the measured variables etc. Sample size should be determined well priori. As a general rule minimum sample size required should be atleast 200. It shall also be 5 to 20 times the number of parameters to be estimated. Sample size shall be calculated using two methods: i) as a function of the ratio of indicator variables to the latent variables and ii) function of minimum effect, power and significance.

Model estimation

A specific structural equation model has both fixed and free parameters for evaluation. Variety of methods are available for estimating the model fit. Some of the examples are Maximum Likelihood (ML), Generalised Least Squares (GLS), Weighted Least Squares (WLS), Ordinary Least Squares (OLS), Arbitary Distruibution Free (ADF) etc. Researchers shall select any of these methods based on the data conditions such as sample size and data distribution. Generally maximum likelihood method is used when data is distributed normally. ADF shall be used where assumptions are minimal and sample size is large.

The procedures of SEM shall be stopped at estimation stage itself, if the data provided is problematic or out of range. For Ex: Correlation greater than 1, sample size is too small, variables are highly correlated etc., Methods for solving multicollinearity problems established for multiple regression shall also be applied in SEM.

Model evaluation

Once the model estimation is complete the next step is to test the goodness of fit and decide whether to accept or reject the hypothesized model. For model evaluation various statistical techniques shall be used based on the data.

Statistical tool	When /where to apply	Interpretatation/ table value
Chi-square test	The sample size is relatively small	Df -13 , 21.21
Normed Fit Index (NFI)	Observed variables are uncorrelated	.95
Comparitive Fit Index (CFI)	Observed variables are uncorrelated	.95
Goodness of Fit Index (GFI)	Test the extent to which specified model of interest reproduces the sample covariance matrix.	.95
Standardised Root Mean Square Residual (SRMR)	Test the extent to which specified model of interest reproduces the sample covariance matrix.	.032
Root Mean Square Error of Approximation (RMSEA)	Test the extent to which specified model of interest reproduces the sample covariance matrix.	.056

The researcher has to appropriately select the best statistical test to suit his data model. Usually the combination of the tools shall be used to test. It is prudent for the researchers to examine individual parameter estimates as well as their estimated standard errors. If model fits the data well and the estimated solution is deemed proper, individual parameter estimates can be interpreted and examined for statistical significance. The test for individual parameters are Z value, t value. At 95% level the value shall be 1.96.

Model modification

When the model is rejected based on the goodness of fit, the researcher shall be ready to modify the model to suit the goodness test. Many program provides modification indices to make minor modifications. At each step the parameter is freed that produces largest improvement in fit, and this is continued till an adequate fit is reached. The modifications shall be generally applied only if there is a theoretical justification for the same. Otherwise if multiple parameters are changed simultaneously the order of change may matter which will result in different final models when the same initial model is modified by different analysis.

It is always precautionary for a researcher to always have multiple alternative models priori instead of post adhoc modifications in a single model. Such alternatives models should be set prior to model evaluation. Multiple models can be generated for same set of variables.

Interpretation

SEM allows the testing for causal hypothesis, but a well fit SEM model cannot prove causal relations without satisfying necessary conditions for causal inference. The researchers are adviced not to make unwarranted causal claims, since many other untested models may also produce the same or better level of fit.

Most commonly used Software programmes for SEM

LISREL (Linear Structural Relationships)

This is the pioneer and most frequently used SEM software. It has three components PRELIS, SIMPLIS and LISREL. The functions of these components is to check distributional assumptions such as univariate and multivariate normality, calculating summary statistics etc.,

PRELIS - used as a stand-alone program in conjunction with other programmes.

SIMPLIS or LISREL - for the estimation of SEM models.

EQS

Developed by one of the leading authorities on the subject, Dr.Peter M.Bentler. It provides simple method for conducting the full range of structural equations models including multiple regression, multivariate regression, confirmatory factor analysis and path analysis. Various data exploration plots such as scatter plot, histogram and matrix plot are readily available in EQS. Data screening and model estimation are performed in one run.

AMOS

Amos (Analysis of Momentum Structure) is add-on module for SPSS. It is designed primarily for SEM, Path analysis and covariance structural modeling. It has two components Amos graphics and Amos basic. Amos Graphics permits the specification of models by diagrams and Amos basic allows specification of models by equations.

Mplus

Mplus is a special purpose software package that estimates statistical models for observed and unobserved variables. Mplus module includes base program and 3 add-on module. This program can analyse the single level models. Add on modules in Mplus can analyse multilevel models and models with latent variables.

Quick Glance to draw the model Structural Equation Model with AMOS

1. Select the required variables using the icon List of variables in the dataset, and drag it to the drawing screen in Amos graphics, in this example – Product Quality, Service Quality, Price Quality, Satisfaction, and Customer Loyalty is taken for drawing the model.
   - Independent variable – Product Quality, Service Quality, Price Quality
   - Dependent variable – Customer satisfaction and Customer Loyalty
2. Draw covariance line using between Product Quality, Service Quality, Price Quality.
3. Draw regression path using from product quality to customer satisfaction, service quality to customer satisfaction, price quality to customer satisfaction.
4. Draw regression path using between Customer satisfaction and customer loyalty.
5. Place error term, using error icon in the observed variable i.e. customer satisfaction and Customer loyalty. (Check figure -1)

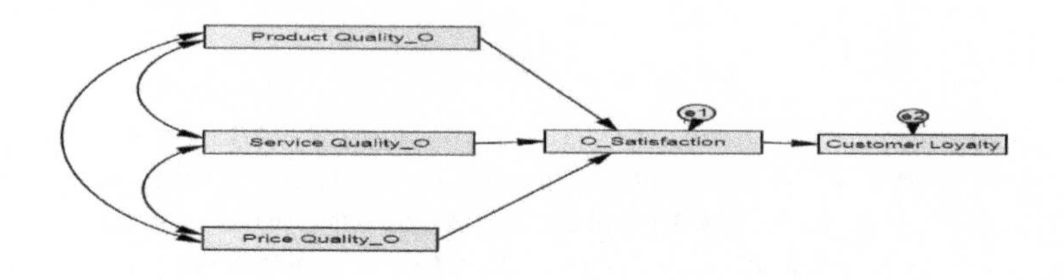

Figure - 1

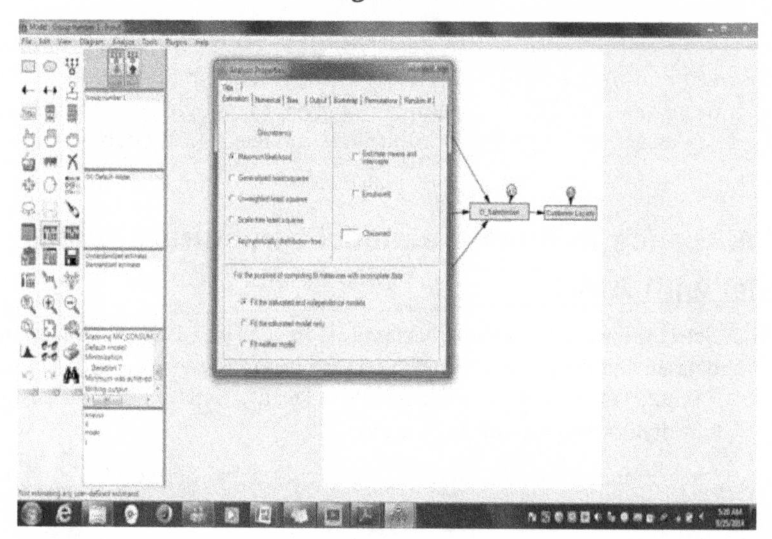

Figure - 2

6. Click on analysis properties from the menu or drawing toolbar , a dialogue box will appear as given in Figure -2, in that click output tab, select standard estimates, squared multiple correlation, modified indices, indirect, direct & total effect, factor scores weights, and correlation estimates, as given in Figure -3 and close the dialogue box.

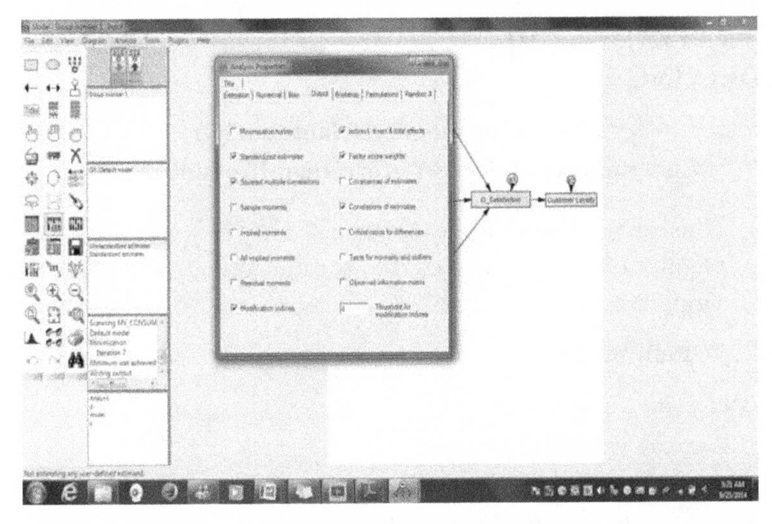

Figure - 3

7. Save the diagram using popup menu or by clicking .
8. Calculate the estimates by clicking on calculate estimates icon.
9. View the output by clicking on text output.
10. To view the output on graphics, click on view graphic output icon, as well click on standard estimates in the Amos Graphic diagram screen.

The Results and SEM output are given below.

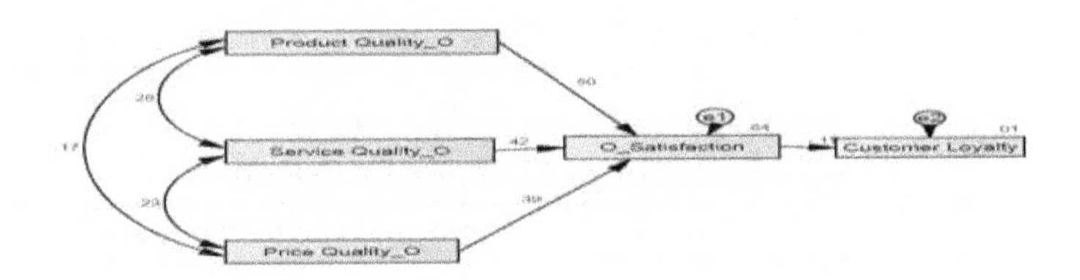

Figure 4 Structural Equation Modeling

AMOS OUTPUT FOR STRUCTURAL EQUATION MODELLING

Notes for Model (Default model)

Computation of degrees of freedom (Default model)

Number of distinct sample moments:	15
Number of distinct parameters to be estimated:	12
Degrees of freedom (15 - 12):	3

Result (Default model)

Minimum was achieved

Chi-square = 6.753

Degrees of freedom = 3

Probability level = .080

Scalar Estimates (Group number 1 - Default model)

Maximum Likelihood Estimates

Regression Weights: (Group number 1 - Default model)

			Estimate	S.E.	C.R.	P	Label
O_ Satisfaction	<---	SQO	.282	.012	22.692	***	par_4
O_ Satisfaction	<---	PQO	.282	.010	27.165	***	par_5
O_ Satisfaction	<---	PQO_A	.227	.011	21.360	***	par_7
TCB	<---	O_Satisfaction	-.143	.056	-2.552	.011	par_6

Standardized Regression Weights: (Group number 1 - Default model)

			Estimate
O_Satisfaction	<---	SQO	.423
O_Satisfaction	<---	PQO	.501
O_Satisfaction	<---	PQO_A	.389
TCB	<---	O_Satisfaction	-.112

Covariances: (Group number 1 - Default model)

			Estimate	S.E.	C.R.	P	Label
PQO	<-->	SQO	.192	.032	6.062	***	par_1
SQO	<-->	PQO_A	.152	.030	5.028	***	par_2
PQO	<-->	PQO_A	.137	.035	3.882	***	par_3

Correlations: (Group number 1 - Default model)

			Estimate
PQO	<-->	SQO	.277
SQO	<-->	PQO_A	.227
PQO	<-->	PQO_A	.174

Variances: (Group number 1 - Default model)

	Estimate	S.E.	C.R.	P	Label
PQO	.820	.051	16.047	***	par_8
SQO	.585	.036	16.047	***	par_9
PQO_A	.760	.047	16.047	***	par_10
e1	.041	.003	16.047	***	par_11
e2	.419	.026	16.047	***	par_12

Squared Multiple Correlations: (Group number 1 - Default model)

	Estimate
O_Satisfaction	.841
TCB	.012

Model Fit Summary

CMIN

Model	NPAR	CMIN	DF	P	CMIN/DF
Default model	12	6.753	3	.080	2.251
Saturated model	15	.000	0		
Independence model	5	1034.783	10	.000	103.478

RMR, GFI

Model	RMR	GFI	AGFI	PGFI
Default model	.015	.995	.974	.199
Saturated model	.000	1.000		
Independence model	.145	.639	.458	.426

Baseline Comparisons

Model	NFI Delta1	RFI rho1	IFI Delta2	TLI rho2	CFI
Default model	.993	.978	.996	.988	.996
Saturated model	1.000		1.000		1.000
Independence model	.000	.000	.000	.000	.000

Parsimony-Adjusted Measures

Model	PRATIO	PNFI	PCFI
Default model	.300	.298	.299
Saturated model	.000	.000	.000
Independence model	1.000	.000	.000

NCP

Model	NCP	LO 90	HI 90
Default model	3.753	.000	15.420
Saturated model	.000	.000	.000
Independence model	1024.783	922.867	1134.081

FMIN

Model	FMIN	F0	LO 90	HI 90
Default model	.013	.007	.000	.030
Saturated model	.000	.000	.000	.000
Independence model	2.009	1.990	1.792	2.202

RMSEA

Model	RMSEA	LO 90	HI 90	PCLOSE
Default model	.049	.000	.100	.429
Independence model	.446	.423	.469	.000

AIC

Model	AIC	BCC	BIC	CAIC
Default model	30.753	31.036	81.706	93.706
Saturated model	30.000	30.354	93.692	108.692
Independence model	1044.783	1044.900	1066.013	1071.013

ECVI

Model	ECVI	LO 90	HI 90	MECVI
Default model	.060	.052	.082	.060
Saturated model	.058	.058	.058	.059
Independence model	2.029	1.831	2.241	2.029

HOELTER

Model	HOELTER .05	HOELTER .01
Default model	596	866
Independence model	10	12

RESULTS AND DISCUSSION

The proposed or hypothesized model is assessed by producing estimates of the unknown parameters. These findings are then compared with the relationships (the correlation or covariance matrices) existence in the actual or observed data. According to Stevens, the model assessment is classified into two categories," those that measures the overall fit of the model and those that are concerned with individual model parameter". There are 24 fits to assess the model, as of Hair Anderson, Tatham and Black, Structural equation modeling has no single statistical test that best described the strength of the model's prediction. Researchers have proposed various classification schemas to organize the fit indexes. The most cited organization system appears to be the three classification scheme i.e. absolute, relative and parsimonious.

Absolute fit Measures

This indicates how well the proposed interrelationships between the variable match the interrelationships between the actual or observed interrelationships. The most common four absolute fit measures assess the general features such as chi square, goodness of fit index(GFI), the root mean square residual (RMSR) and the root mean square error of approximation (RMSEA).

The criterion are chi square - P>.05, GFI >.90, RMSR <.05, RMSEA <.10. In this example Chi square p value is 0.080 (which is >.05),GFI is .995 which is above the threshold limit, RMSR is .015 (which is <.50), RMSEA is .049 ,(which is <.10). It is concluded that all the absolute fit indices conditions are fulfilled.

Relative fit Measures

This is also called as comparisons with baseline measures or incremental fit measure. These are measures of fit relative to the independence model, which assumes that there are no relationship in the data (thus a poor fit) and the saturated model, which assumes a perfect fit. This indicate the relative position on this continuum between worst fit to perfect fit, with values greater than .90 suggesting acceptable fit between the model and the data. The relative fit measures are CFI >.95, NFI >.90, IFI >.90, RFI >.90. In this example NFI is .993, RFI is .978, IFI is .988 and CFI is .996, all the values are above the threshold limit which is indicating an acceptable fit.

Parsimonious Fit Measures

Parsimonious fit measures are sometimes called adjusted fit measures. These fit statistics are similar to the adjusted R2 in multiple regression analysis, the parsimonious fit statistics penalize larger models with more estimated parameters. These measures can be used to compare models with differing number of parameters to determine the impact of adding additional parameters to the model. Adjusted goodness of fit and Parsimonious goodness of fit is common parsimonious fit measures. Ideally, values greater than .90 indicates an acceptable model; however typically parsimony based measures have lower acceptable value (e.g., .50 or greater is deemed acceptable). in our example PNFI is .298 and PCFI is .299 which is lower the threshold limit.

The result of this study supports the theory that product quality, price quality and service quality influence customer satisfaction which has effects on customer loyalty. All the measured variables correlated with their respective factors at a reasonably strong level. This study illustrated that even though the overall fit of a model to the data appears acceptable, some paths were not supported by the data. This acceptable fit was obtained simply because measured paths had extremely high coefficients i.e product quality (.50, price quality (.39) and service quality (.49). To assess the accuracy of the prediction in the structural equations, the researcher calculated the proportion of variance accounted for R^2 is 84%. In this weak effect size was reported for Customer Loyalty i.e. 0.12 and high effect size was found for customer satisfaction.

Chapter - 15

Canonical Correlation

Learning Objectives

This chapter helps to understand the following

- Meaning of Canonical Correlation
- Need of Canonical Correlation
- Assumptions of Canonical Correlation

Introduction

In Canonical correlation (multiple correlation), one has two or more X variables and two or more Y variables. Canonical correlation analysis is a multivariate statistical model that facilitates the study of interrelationships among sets of multiple dependent variables and multiple independent variables (X & Y) whereas multiple regression predicts a single dependent variable from a set of multiple independent variables, canonical correlation simultaneously predicts multiple dependent variables from multiple independent variables. Canonical correlation places the fewest restrictions on the types of data on which it operates. Because the other techniques impose more rigid restrictions, it is generally believed that the information obtained from them is of higher quality and may be presented in a more interpretable manner. For this reason, many researchers view canonical correlation as a last-ditch effort, to be used when all other higher-level techniques have been exhausted. But in situations with multiple dependent and independent variables, canonical correlation is the most appropriate and powerful multivariate technique. It has gained acceptance in many fields and represents a useful tool for multivariate analysis, particularly as interest has spread to considering multiple dependent variables.

Definition

The canonical correlation technique examines several linear combinations of X variables and the same number of linear combination of Y variables in such a way that these linear combination best express the correlation between the two sets.

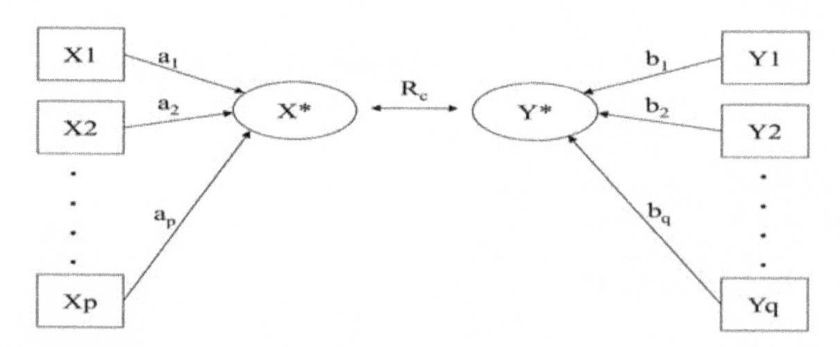

Objectives

The appropriate data for canonical correlation analysis are two sets of variables. We assume that each set can be given some theoretical meaning, at least to the extent that one set could be defined as the independent variables and the other as the dependent variables. Once this distinction has been made, canonical correlation can address a wide range of objectives. These objects may be any or all of the following:

1. Determining whether two sets of variables (measurements made on the same objects) are independent of one another or, conversely, determining the magnitude of the relationships that may exist between the two sets.

2. Deriving a set of weights for each set of dependent and independent variables so that the linear combinations of each set are maximally correlated. Additional linear functions that maximize the remaining correlation are independent of the preceding set(s) of linear combinations.

3. Explaining the nature of whatever relationships exist between the sets of dependent and independent variables, generally by measuring the relative contribution of each variable to the canonical functions (relationships) that are extracted.

The inherent flexibility of canonical correlation in terms of the number and types of variables handled, both dependent and independent, makes it a logical candidate for many of the more complex problems addressed with multivariate techniques.

Need of Canonical Correlation

As exploratory tool to see if two sets of variables are related If significant overall shared variance, several layers of analysis can explore variables & variates involved in a modeling type approach where you have a theoretical reason for considering the variables as sets, and that one predicts another See if one set of two or more variables relate longitudinally across two time points Variables at t1 are predictors; same variables at t2 are DVs See layers of analysis for tentative causal evidence.

As the most general form of multivariate analysis, canonical correlation analysis shares basic implementation issues common to all multivariate techniques. Some principal features that are discussed in the text (particularly multiple regression, discriminant analysis, and factor analysis) are also relevant to canonical correlation analysis.

These include

1. Appropriate Sample Size,
2. Variables And Their Conceptual Linkage,
3. Absence Of Missing Data And Outliers.

Sample Size

Issues related to the impact of sample size (both small and large) and the necessity for a sufficient number of observations per variable are frequently encountered with canonical correlation. Researchers are tempted to include many variables in both the independent and dependent variable set, not realizing the implications for sample size. Sample sizes that are very small will not represent the correlations well, thus obscuring meaningful relationships. Very large samples have a tendency to result in statistical significance in all instances, even where practical significance is not indicated. The appropriate sample size is related to the reliability of the variables. Different disciplines have different expectations regarding reliability but for social science and business researchers reliability is generally expected to be .7 or higher, and they are encouraged to maintain at least 10 observations per independent variable to avoid "overfitting" the data.

Variables and Their Conceptual Linkage

Canonical correlation analysis is the most liberal form of multivariate analysis in that both metric and nonmetric variables can be included in the latent variables. The classification of variates as dependent or independent is of little importance for the statistical estimation of the canonical functions, however, because the method calculates weights for both variates to maximize the correlation and places no particular emphasis on either variate. Yet because the technique produces variates to maximize the correlation between them, a variable in either set relates to all other variables in both sets. The result is that the addition or deletion of a single variable may affect the entire solution, particularly the other variate. The composition of each variate, either independent or dependent, thus becomes critical. Researchers must have conceptually linked the sets of variables before applying canonical correlation analysis. This makes the specification of dependent versus independent variates essential to establishing a strong conceptual foundation for the variables.

Missing Data and Outliers

Canonical correlation analysis is sensitive to changes in the data set. Therefore, different procedures for handling missing data can create substantial changes in canonical solutions. Similarly, outliers can also substantially impact canonical analysis results. Missing data can be replaced by estimating values or removing cases with missing data. Outliers can be detected by univariate, bivariate, and multivariate diagnostic methods.

Assumptions of Canonical Correlation

The assumption of linearity affects two aspects of canonical correlation results. First, the correlation coefficient between any two variables is based on a linear relationship. If the relationship is nonlinear, then one or both variables should be transformed, if possible. Second, the canonical correlation is the linear relationship between the variates. If the variates relate in a nonlinear manner, the relationship will not be captured by canonical correlation. Thus, while canonical correlation analysis is the most generalized multivariate method, it is still constrained to identifying linear relationships. Canonical correlation analysis can accommodate any metric variable without the strict assumption of normality. Normality is desirable because it standardizes a distribution to allow for a higher correlation among the variables. But in the strictest sense, canonical correlation analysis can accommodate even non normal variables if the distributional form (e.g., highly skewed) does not decrease the correlation with other variables. This allows for transformed nonmetric data (in the form of dummy variables) to be used as well. However, multivariate normality is required for the statistical inference test of the significance of each canonical function. Because tests for multivariate normality are not readily available, the prevailing guideline is to ensure that each variable has univariate normality. Thus, although normality is not strictly required, it is highly recommended that all variables be evaluated for normality and transformed if necessary.

Homoscedasticity, to the extent that it decreases the correlation between variables, should also be remedied. Finally, multicollinearity among either variable set will confound the ability of the technique to isolate the impact of any single variable, making interpretation less reliable.

Relationships of Canonical Correlation Analysis to Other Multivariate Techniques

Canonical correlation analysis is the most generalized member of the family of multivariate statistical techniques. It is directly related to several dependence methods, such as multiple regression analysis, which can predict the value of a single metric dependent variable from a linear function of a set of independent variables. Similar to regression, the goal of canonical correlation is to quantify the strength of the relationship, in this case between the two sets of variables (independent and dependent). Whereas multiple regression predicts a single dependent variable from a set of multiple independent variables, canonical correlation simultaneously predicts multiple dependent variables from multiple independent variables. Canonical correlation analysis also resembles discriminant analysis in its ability to determine independent dimensions (similar to discriminant functions) for each variable set. Discriminant analysis is a method of estimating the relationship between a single nonmetric dependent variable and a set of metric independent variables. In sum, canonical correlation analysis is more general than multiple regression and discriminant analysis because it can handle multiple dependent variables that can be metric or nonmetric. Canonical correlation analysis is also closely related to principal components analysis, which is included in factor analysis. The primary purpose of which is to define the underlying structure among the variables in the analysis. Canonical correlation analysis corresponds to principal components analysis and factor analysis in the creation of the optimum structure or dimensionality of each variable set that maximizes the relationship between independent and dependent variable sets. Whereas principal components analysis and factor analysis attempt to explain the linear relationship among a set of observed variables and an unknown number of factors/variates, canonical correlation analysis focuses more on the linear relationship between two variates. As such, it is similar in purpose to PLS, a variant of structural equation modeling, which is discussed in the text as well. Canonical correlation places the fewest restrictions on the types of data on which it operates and can be used for both metric and nonmetric data. Because the other techniques impose more rigid restrictions, it is generally believed that the information obtained from them is more robust statistically and may be presented in a more interpretable manner. But in situations with multiple dependent and independent variables, canonical correlation is the most appropriate and powerful multivariate technique. It has gained acceptance in many fields and represents a useful tool for multivariate analysis, particularly with the expanding interest in considering relationships between multiple dependent variables.

Deriving Canonical Functions

The derivation of successive canonical variates is similar to the procedure used with unrotated factor analysis. The first canonical function that is extracted accounts for the maximum amount of variance in the set of variables. The second function is then computed so that it accounts for as much as possible of the variance not accounted for by the first function, and so forth, until all functions have been extracted. Therefore, successive functions are derived from residual or leftover variance from earlier functions. Canonical correlation analysis follows a procedure similar to factor analysis but focuses on accounting for the maximum amount of the relationship between the two sets of variables, rather than within a single set (e.g., a single factor). The result is that the first pair of canonical variates is derived so as to have the highest inter correlation possible between the two variates. The second pair of canonical variates is then derived so that it exhibits the maximum relationship between the two sets of variables not accounted for by the first pair of variates. In short, successive canonical functions estimate pairs of canonical variates based on residual variance from the previous canonical functions, and their respective canonical correlations (which reflect the interrelationships between the variates) become smaller as each additional function is extracted. That is, the first pair of canonical variates exhibits the highest intercorrelation, the next pair the second-highest correlation, and so forth. Because canonical correlation analysis is closely linked with principal components analysis, rotation of canonical variates can be considered as an aid to increase the interpretability of canonical results. Rotation does not change the sums of the squared canonical correlation coefficients but it will lead to a simpler structure. As noted, successive pairs of canonical variates are based on residual variance. Therefore, each pair of variates is orthogonal and independent of all other variates derived from the same set of data. The selection of rotation methods is therefore limited to those that are orthogonal and Varimax rotation is the most obvious choice given the close relationship between canonical correlation analysis and principal components analysis. Rotation is only possible when there are at least two canonical functions. It is available in some computer programs, including SPSS and Statistics. However, some researchers do not recommend rotation for canonical correlation analysis for two reasons. First, rotation can reduce the optimality of the canonical correlations when each pair of canonical variates is derived to maximize their correlation. Second, rotation introduces correlations among succeeding canonical variates. Therefore, even though rotation may increase the interpretability of the canonical results, this gain may be offset by the increased complexity due to interrelationships among the pairs of canonical variates. Researchers therefore need to be careful when using rotation for canonical correlation analysis.

CPSIA information can be obtained
at www.ICGtesting.com
Printed in the USA
LVHW041304231119
638277LV00003B/463/P